Recent Approaches in Enhanced Oil Recovery - New Perspectives and Future Outlook

Edited by Manar Elsayed Abdel-Raouf, Rasha S. Kamal, Mohamed Keshawy and Reham El Shazly

Published in London, United Kingdom

Recent Approaches in Enhanced Oil Recovery - New Perspectives and Future Outlook
http://dx.doi.org/10.5772/intechopen.1005985
Edited by Manar Elsayed Abdel-Raouf, Rasha S. Kamal, Mohamed Keshawy and Reham El Shazly

Contributors
Abdelaziz El-hoshoudy, Abdulraheim M.A. Hasan, Adel Salem, Arturo Palacio-Pérez, Bongiwe Ladyfair Hem, Esbeydi Villicaña-García, Fatemeh Hosseinpour, Federico Vagnon, Glenda Taddia, Juan Carlos García-Mayorga, Juan Ernesto Juri, Khalaf Gad Salem, Manar E. Abdel-Raouf, Manar El-Sayed Abdel-Raouf, Martina Gizzi, Mohamed Keshawy, Neeraj Tandan, Nouratan Singh, Poonam Rani, Rasha S. Kamal, Reham El-Shazly, Reham I. El-Shazly, Sayed Gomaa, Silvia Raquel García-Benítez, Stefano Lo Russo, Vladimir Escobar-Barrios

First published in London, United Kingdom, 2025 by IntechOpen
IntechOpen is the global imprint of INTECHOPEN LIMITED, registered in England and Wales, registration number: 11086078, 167-169 Great Portland Street, London, W1W 5PF, United Kingdom

For EU product safety concerns: IN TECH d.o.o., Prolaz Marije Krucifikse Kozulić 3, 51000 Rijeka, Croatia, info@intechopen.com or visit our website at intechopen.com.

British Library Cataloguing-in-Publication Data
A catalogue record for this book is available from the British Library

Recent Approaches in Enhanced Oil Recovery - New Perspectives and Future Outlook
Edited by Manar Elsayed Abdel-Raouf, Rasha S. Kamal, Mohamed Keshawy and Reham El Shazly
p. cm.
Print ISBN 978-1-83634-539-8
Online ISBN 978-1-83634-538-1
eBook (PDF) ISBN 978-1-83634-540-4

If disposing of this product, please recycle the paper responsibly.

Meet the editors

Professor Manar El-Sayed Abdel-Raouf is a pioneering Professor of Petroleum Chemistry at the Egyptian Petroleum Research Institute. She joined the program in 1992 as a Research Assistant and earned her Ph.D. in 2000. Since 2011, she has been a Professor and has supervised several M. Sc. and Ph.D. theses. She has published more than 75 research papers, case studies, and review articles in reputable journals, supervised 17 M. Sc. and Ph.D. theses, and attended 32 international conferences. She has granted four scientifically sound research projects, registered three patents, and is a major contributor to ESEM. She has participated as a senior member in some international joint projects. She is an active member of national and international scientific societies. She has participated in several international conferences with seminars and poster presentations. She has three patents under consideration. Her primary research interests focus on environmental issues, including sludge processing, wastewater remediation, and the modification of natural polymers into sustainable products for green packaging and energy storage materials.

Professor Rasha Samir Mohamed Kamal is a distinguished professor in petroleum chemistry at the Egyptian Petroleum Research Institute, where she has served since 1999. She has supervised numerous M. Sc. and Ph.D. theses across leading Egyptian universities. She has authored over 35 peer-reviewed research papers, review articles, and book publications in reputable international journals and publishing houses. She holds one registered patent in the field of lube oil additives based on thiourea derivatives. She has actively participated in several national and international scientific projects, presenting her work at major conferences through oral presentations and posters. Professor Kamal serves on the editorial boards of several international journals and is a recognized reviewer for over 25 high-impact scientific publications. She has been honored multiple times for her academic and community contributions, including recognitions from Qassim University and the Egyptian Petroleum Research Institute. She is also a grand award judge at the Regeneron International Science and Engineering Fair (ISEF). Her primary research interests focus on the development of multifunctional petroleum additives, sustainable lubricants, rheological modifiers, and environmentally friendly polymeric materials for various applications, including wastewater treatment and oil spill remediation. She is an active member of national and international scientific societies, including OWSD, and plays a leading role in promoting women's participation in science and innovation.

Dr. Mohamed Keshawy has been a Full-Time Professor and Associate Researcher at the Egyptian Petroleum Research Institute since 2023. He specializes in advanced polymer applications for petroleum engineering and environmental sustainability, with a focus on polymer synthesis and design for use in the petroleum field. Nanocellulose Production: Extraction from agricultural waste for eco-friendly materials. Pollutant Remediation: Separation of hazardous organic/inorganic contaminants from water. Development of copolymer and natural polymer sorbents for selective pollutant removal. He has supervised several M. Sc. and Ph.D. theses. He has published more than 30 research papers, review articles, and book chapters in reputed journals. He has secured three scientific research projects and registered two patents. He is an active member of national and international scientific societies. He has presented at several international conferences. During his postdoctoral and Ph.D. research, he developed foundational expertise in polymer functionalization, nanomaterial synthesis, and environmental remediation. His work emphasizes practical solutions to challenges in the petroleum industry and water purification.

Dr. Reham Ibrahim Ibrahim El Shazly is a lecturer in petroleum chemistry at the Egyptian Petroleum Research Institute, where she has served since 2016. She has authored over nine peer-reviewed research papers and book publications in reputable international journals and publishing houses. She has presented her work at major conferences through oral presentations and posters. Dr. El Shazly serves on the editorial boards of several international journals. She is also a grand award judge at the International Science and Engineering Fair (ISEF). Her primary research interests focus on the development of multifunctional petroleum additives, sustainable lubricants, rheological modifiers, and environmentally friendly polymeric materials for various applications, including wastewater treatment and oil spill remediation. She is an active member of national and international scientific societies, including OWSD, and plays a leading role in promoting women's participation in science and innovation.

Contents

Preface

The global energy sector stands at a pivotal crossroads, where the demand for fossil fuels persists even as the world accelerates its transition toward sustainability. In this complex landscape, Enhanced Oil Recovery (EOR) emerges as a critical discipline, not only for maximizing hydrocarbon extraction but also for aligning the oil and gas industry with environmental stewardship and circular economy principles. This edited volume, *Recent Approaches in Enhanced Oil Recovery – New Perspectives and Future Outlook*, is a timely contribution that explores the intersection of traditional EOR techniques and innovative, eco-conscious methodologies. By blending historical perspectives, latest research, and forward-looking insights, this book aims to guide researchers, industry professionals, and policymakers toward sustainable practices in oil and gas recovery.

The need for sustainable EOR

The petroleum industry has long been a cornerstone of global energy supply, yet its environmental footprint has sparked urgent calls for reform. Traditional EOR methods, although effective, often involve high carbon emissions, excessive water usage, and the use of chemical additives that pose ecological risks. This book addresses these challenges by advocating for the integration of green materials and smart technologies into EOR processes. The introductory chapter lays the foundation by framing EOR within the circular economy, a system that minimizes waste, optimizes resources, and mitigates environmental impact. It underscores how sustainable EOR can coexist with economic viability, offering a blueprint for the industry's future.

A journey through innovation

The subsequent chapters take the reader on a comprehensive journey through the evolution of EOR, with a particular emphasis on green chemistry and advanced materials. Chapter 2 provides a historical overview of chemical agents used in EOR, tracing their development from conventional surfactants and polymers to today's biodegradable and low-toxicity alternatives. This retrospective analysis is complemented by a forward-looking discussion on emerging green materials that promise to revolutionize the field.

Technological innovation is a recurring theme throughout the book. Chapter 3 examines how sustainability and technological advancements are transforming EOR, with a focus on data analytics and the integration of renewable energy. The chapter highlights case studies where solar thermal energy and AI-driven optimization have reduced the carbon footprint of EOR operations, demonstrating that environmental responsibility and operational efficiency are not mutually exclusive.

The role of smart technologies

One of the most exciting developments in EOR is the application of artificial intelligence and nanotechnology, which are thoroughly examined in Chapters 4 and 5. AI is transforming reservoir management by enabling real-time monitoring, predictive analytics, and cost-effective decision-making. Meanwhile, nanotechnology introduces functionalized nanoparticles and smart fluids that enhance oil mobility while reducing the need for harmful chemicals. These chapters not only elucidate the science behind these technologies but also discuss their practical implementation and scalability.

Chapter 6 delves into nanoparticle-assisted polymer flooding, a groundbreaking approach that overcomes the limitations of traditional polymer flooding. By improving rheological properties and interfacial tension, polymeric nanofluids significantly boost recovery rates while minimizing environmental impact. The chapter also addresses the economic and logistical challenges of deploying these advanced materials, providing a balanced view of their potential.

Practical solutions for heavy oil and hybrid techniques

Heavy oil reserves represent a vast yet challenging resource, and Chapter 7 offers innovative solutions through viscosity-reducing additives. These materials, including imidazoline derivatives, enable more efficient transportation and processing of heavy oils, reducing energy consumption and operational costs. The chapter includes a bibliometric analysis of recent research, underscoring the growing interest in sustainable additives.

Chapter 8 presents a novel approach to polymer injection through hybrid strategies, such as Burst-Water-Alternating-Polymer (Burst-WAP) and modular plant standardization. By combining hysteresis principles with flexible injection cycles, this method maximizes recovery while optimizing costs. The chapter draws on field applications to demonstrate its success, offering actionable insights for industry adoption.

A vision for the future

The concluding chapter of the book reflects on the broader implications of sustainable EOR, including policy and regulatory support and stakeholder engagement. The integration of green materials and innovative technologies is not merely a technical endeavor but a cultural shift that requires collaboration across academia, industry, and government. By embracing these innovations, the petroleum sector can reduce its environmental impact while maintaining its economic relevance in a rapidly changing energy landscape.

Acknowledgments

This volume is the culmination of contributions from leading experts in the field, whose dedication to advancing sustainable EOR is deeply appreciated. We also extend our gratitude to the reviewers and editorial teams whose meticulous efforts ensured the quality and coherence of this work. Finally, we thank the readers for engaging with this material, whether you are a researcher, engineer, or policymaker, your role in driving the industry toward sustainability is invaluable.

We hope that this book will serve as both a reference and an inspiration, encouraging further innovation and collaboration in the pursuit of greener, more efficient, enhanced oil recovery.

Manar El-Sayed Abdel-Raouf and Rasha S. Kamal
Professor of Petroleum Science,
Egyptian Petroleum Research Institute,
Cairo, Egypt

Mohamed Keshawy
Associate Professor of Polymer Science,
Egyptian Petroleum Research Institute,
Cairo, Egypt

Reham El-Shazly
Lecturer of Applied Chemistry,
Egyptian Petroleum Research Institute,
Cairo, Egypt

Section 1

Foundations and Sustainability in EOR

Chapter 1

Introductory Chapter: Towards Greener Processes for Environmental Concerns and Circular Economy in Relevant to EOR

Manar El-Sayed Abdel-Raouf, Mohamed Keshawy, Reham El-Shazly and Rasha S. Kamal

1. Introduction

The oil produced from mature and maturing oil fields by conventional primary and secondary production processes is not sufficient to meet the leap of the energy demand and oil-based industries. It was estimated that world's average recovery factor is only about 30–50% [1–3]. Since the oil is entrapped inside the porous rocky structures and is hard to evacuate, emphasis was made on tertiary recovery of oils. Methods for crude oil recovery are illustrated in **Figure 1**. It is observed that while primary and secondary methods encompass only few techniques, EOR methods deal with the extraction of residual oil in the well after primary and secondary recoveries by several approaches, which greatly improves the recovery factor and maximizes the amount of recovered oil [4, 5]. The cost of EOR method(s) majorly

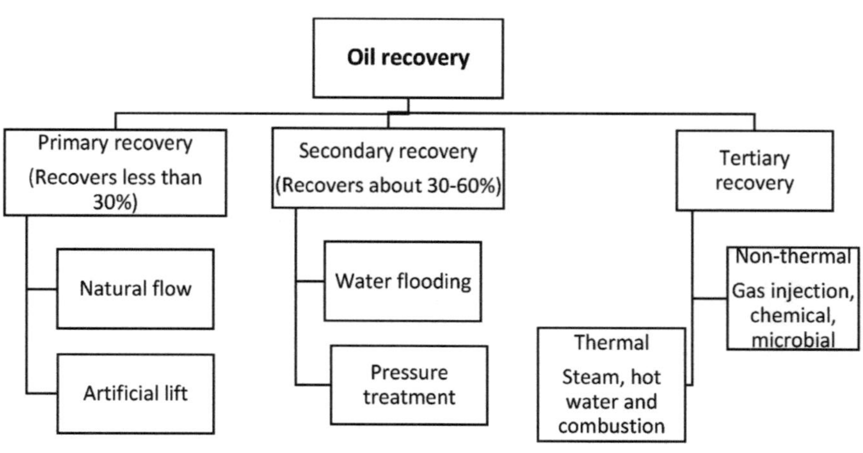

Figure 1.
Different methods for oil recovery.

depends on the available facilities and the amount of the oil left in the oil field. However, the selected methods become economical when compared to the cost of new field development [6, 7]. Meanwhile, the economic viability of the enhanced oil recovery (EOR) relies on the total cost of production, and it is accomplished with augmentation of oil prices [8]. The EOR methods have been roughly categorized as thermal and nonthermal. Many methods have been recommended by various researchers with different techniques to activate the oil left in the well. Several parameters are to be considered for proper selection of the recovery method with lower cost and higher efficiency. These include the nature of crude oil, its rheology, surface tension, and mobility ratio [9].

However, oil companies enhance the practices of EOR to extend their lifespan and boost output. Thus, the global market of EOR is increased tremendously with

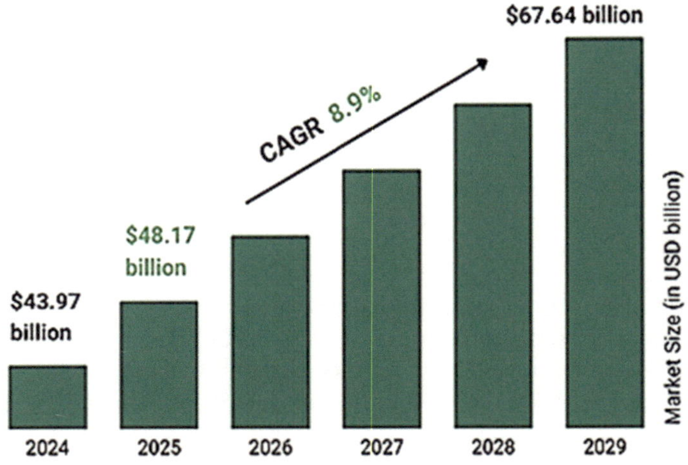

Figure 2.
Enhanced oil recovery global market report, 2025 [https://www.thebusinessresearchcompany.com/report/enhanced-oil-recovery-global-market-report].

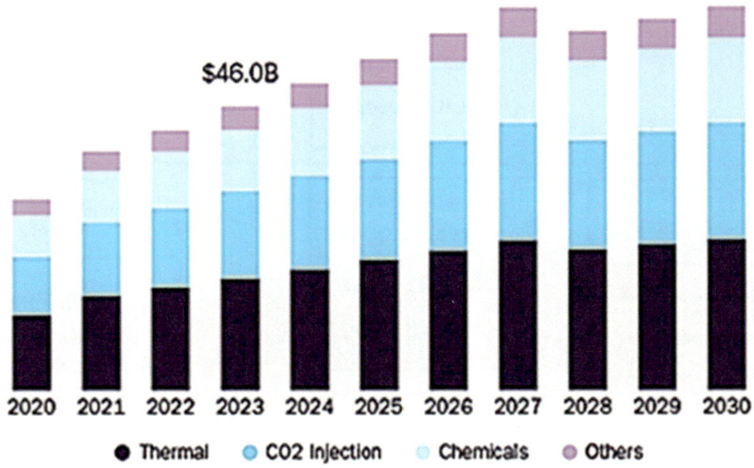

Figure 3.
Global Growing in Different EOR practices [https://www.grandviewresearch.com/industry-analysis/enhanced-oil-recovery-eor-market].

increasing compound annual growth rate (CAGR) of about 8.9%, **Figure 2**. It has been established that green methodologies such as thermal methods are highly encouraged [10]. The most (green) protocols are given in **Figure 3**.

2. EOR and circular economy

In regard to economic points of views, a quick comparison between linear and circular economy is given in **Figure 4**. Linear economy is usually described as "take, make, dispose." It involves extracting raw materials from their original source, manufacturing different products, consuming these products, and finally discarding them as waste at the end of their lifecycle [11]. This approach results in substantial resource depletion and environmental degradation with elevated production costs, as the waste generated is not systematically reintegrated into the production process [12–14].

On the other hand, the circular economy is linked to recycling of different raw materials to enter the production wheel. Therefore, it can be concluded that thermal and gas injection oil recoveries are among the approaches that assist the circular economy. On the other hand, integrating smart technologies such as artificial intelligence (AI), nanotechnology, and sustainable practices enhances EOR and reduces the overall operational costs [15]. AI is a revolutionary protocol that reduces the cost of EOR operations by optimizing reservoir conditions and improving real-time monitoring. In addition, artificial intelligence tools such as fuzzy logic can be used to categorize oil wells based on some geophysical variables relating to crude oil and well conditions such as pressure, production, saturation, viscosity, API gravity, compressibility, and volume factor. Thus, the historical data of the well are now not required to determine the case and recovery status of the well [16]. Moreover, AI is effectively applied to design innovative and economic surfactants harnessed in EOR by creation of an automated workflow through combination of patent search, analysis, structure generation and simulation, structure

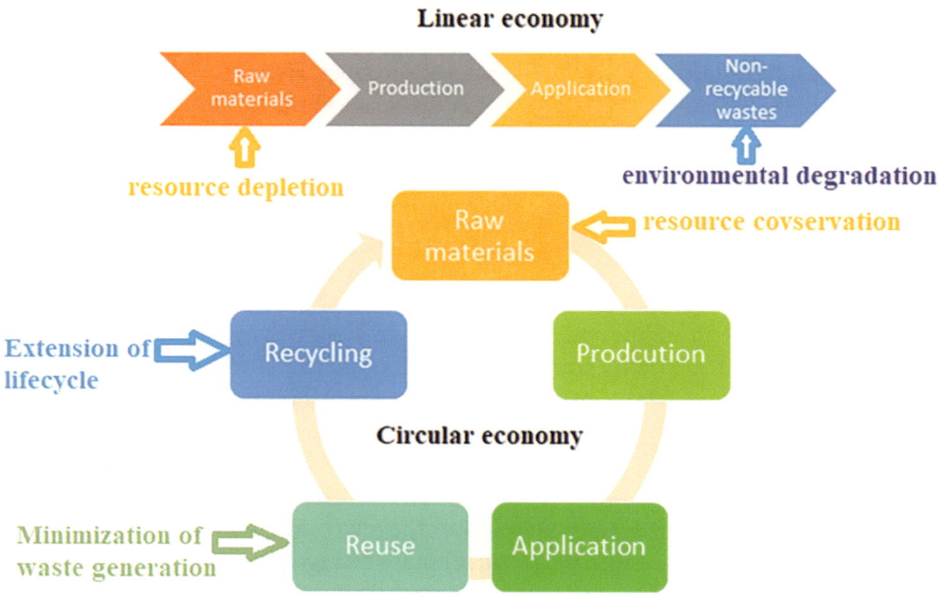

Figure 4.
Linear vs. circular economy.

analysis, and confirmation. This can be conducted through developing and adapting the power of computational chemistry software. AI can also introduce an assortment of different automated structure creations and molecule parameterization through the optimization of structures collected from various patents. These chemical formulae can be created within an hour. Also, The AI can give precisely the work conditions required for their application [17]. At last, AI can provide smart sensor designs that can accurately inform the field conditions for better production.

On the other hand, nanotechnology boosts oil recovery through the addition of functionalized cost-effective nanoparticles and smart fluids, which enhance the physical features of the entrapped oil and improve its mobility and thus reduce chemical consumption [18]. In addition, recent experimental investigations have declared that the synergistic effects of nanoparticles (NPs) and polymer can enhance flooding performance and increase oil recovery compared to conventional polymer flooding due to better enhancement of rheological properties, IFT reduction, and altering oil wettability. From an economic point of view, nanoparticle-assisted polymer flooding achieves sustainability and promotes circular economy [19, 20].

3. EOR and green economy

From a practical point of view, the two terms (circular economy and green economy) are interconnected. In oil and gas industry, establishing circular economy approaches promotes several benefits that can discourse environmental and economic challenges. Environmentally, the circular economy subsidizes to noteworthy waste minimization and resource preservation [21–23]. The reduction in waste generation and decreasing the amount of wastes sent to landfills can be achieved by executing benign strategies such as reusing and recycling by-products from different oil-gas operations [24]. These practices regulate the need for new raw materials, minimize the environmental footprint of the industrial operation, and help eliminate the adverse effects associated with resource extraction and waste disposal [25]. In addition, integrating circular economy principles in oil and gas industry can lead to significant perfections in resource organization and operational competence. For instance, approaches such as waste-to-energy structures and enhanced oil recovery systems contribute to a more circular tactic by reusing and recycling by-products and leftover materials [26]. Additionally, implementing circular economy strategies can help the industry comply with increasingly stringent environmental regulations and meet stakeholder expectations for sustainability [27–29]. Indeed, the development in enhanced oil recovery is subjected to rapid improvement in the increase of sustainability efforts and technological innovation. This enhancement and EOR optimization are related to data analytics and nanotechnology. In addition, renewable energy sources, such as solar-thermal, are being combined to EOR operations, completing worldwide decarbonization policies. The strategies and legal regulations are also mandatory for sustainable EOR, including water treatment and emission reductions. While innovation is unending, EOR could also link conventional fossil fuel production and clean energy sources. Furthermore, sustainable platforms such as CO_2-EOR, and biodegradable chemicals, are being adopted to lower the environmental impact of EOR, especially in terms of carbon emissions and climate changes. In spite of the continued efforts in this field, some challenges remain—such as the elevated cost of processes and changeable oil prices—the future of EOR holds promise through continuous technological innovation and a growing emphasis on sustainability and greenness.

4. Conclusions

Oil exploration and production are among the most important operations in oil industry that directly affect the economy of nations and the environment of the planet. This industry is subjected to continuous improvement aiming to go parallel with economic and environmental considerations. This book introduces recent approaches associated with EOR for production maximization, resource conservation, and environmental protection.

Conflict of interest

The authors declare no conflict of interest.

Author details

Manar El-Sayed Abdel-Raouf*, Mohamed Keshawy, Reham El-Shazly and Rasha S. Kamal
Egyptian Petroleum Research Institute, Nasr City, Cairo, Egypt

*Address all correspondence to: drmanar770@yahoo.com; drmanar770@epri.sci.eg

References

[1] Aashima K, Shaivya S, Srishti B, Mudita S, Jagadeeshwar K. Overview and methods in Enhanced Oil Recovery. Journal of Physics: Conference Series. IOP Publishing; 2021;**2070**:012061. DOI: 10.1088/1742-6596/2070/1/012061

[2] Ayirala SCC, Yousef AAA. A state-of-the-art review to develop injection-water-chemistry requirement guidelines for IOR/EOR projects. SPE Production & Operations. 2015;**2015**(30):26-42. DOI: 10.2118/169048-PA

[3] Shafiai SH, Gohari A. Conventional and electrical EOR review: The development trend of ultrasonic application in EOR. Journal of Petroleum Exploration and Production Technologies. 2020;**10**:2923-2945. DOI: 10.1007/s13202-020-00929-x

[4] Marcin K. Improving the efficiency of oil recovery in Research and Development. Energies. 2022;**15**:4488. DOI: 10.3390/en15124488

[5] Sławomir F, Andrzej U, Marek S, Marcin M. A new strategy for pre-selecting gas wells for the water shut-off treatment based on geological integrated data. Energies. 2021;**14**:7316. DOI: 10.3390/en14217316

[6] Lieder M, Rashid A. Towards circular economy implementation: A comprehensive review in context of manufacturing industry. Journal of Cleaner Production. 2016;**115**:36-51. DOI: 10.1016/j.jclepro.2015.12.042

[7] Mouboua PD, Atobatele FA, Akintayo OT. Cross-Cultural Competence in Global HRD: Strategies for Developing an Inclusive and Diverse Workforce. International Journal of Science and Research Archive. 2024;**12**(01):103-113

[8] Lüdeke-Freund F, Gold S, Bocken NM. A review and typology of circular economy business model patterns. Journal of Industrial Ecology. 2018;**23**(1):36-61

[9] Guo K, Li H, Yu Z. In-situ heavy and extra-heavy oil recovery: A review. Fuel. 2016;**185**:886-902. DOI: 10.1016/j.fuel.2016.08.047

[10] Geraci M, Jehangeer Ali S, Romolt C, Rossmann R. The environmental risks and oversight of enhanced oil recovery in the United States. 2017;**63**

[11] Mansoor Z, Madjid J. Qualitative Analysis of Enhanced Oil Recovery: Impacts on Air, Surface Water and Groundwater. International Conference on Chemical, Environment & Biological Sciences (CEBS-2014) Sept. 17-18, Kuala Lumpur (Malaysia). 2024

[12] Abdul-Azeez O, Ihechere AO, Idemudia C. SMEs as catalysts for economic development: Navigating challenges and seizing opportunities in emerging markets. GSC Advanced Research and Reviews. 2024;**19**(3):325-335

[13] Abdul-Azeez OY, Nwabekee US, Agu EE, Ijomah TI. Sustainability in product life cycle management: A review of best practices and innovations. International Journal of Applied Research in Social Sciences. Sep 2024;**6**(9):2018-2043

[14] Abdul-Azeez T, Nwabekee I, Agu US, Ijomah EE. Strategic approaches to sustainability in multinational corporations: A comprehensive review. International Journal of Frontline Research in Science and Technology. 2024;**3**(2):038-054

[15] Dwivedi YK, Hughes L, Ismagilova E, Aarts G, Coombs C,

Crick T, et al. Artificial intelligence (AI): Multidisciplinary perspectives on emerging challenges, opportunities, and agenda for research, practice and policy. International Journal of Information Management. 2021;**57**:101994. DOI: 10.1016/j.ijinfomgt.2019.08.002

[16] Natale S, Ballatore A. Imagining the thinking machine: Technological myths and the rise of artificial intelligence. Convergence. 2020;**26**(1):3-18. DOI: 10.1177/1354856517715164

[17] Qian X, Yoon BJ, Arróyave R, Qian X, Dougherty ER. Knowledge-driven learning, optimization, and experimental design under uncertainty for materials discovery. Patterns. 2023;**4**:11. DOI: 10.1016/j.patter.2023.100863

[18] Tsekhmistrenko OS, Bityutskyy VS, Tsekhmistrenko SI, Kharchishin VM, Melnichenko OM, Rozputnyy OI. Nanotechnologies and environment: A review of pros and cons. Ukrainian Journal of Ecology. 2020;**10**(3):162-172. DOI: 10.15421/2020_149

[19] Foroozesh J, Kumar S. Nanoparticles behaviors in porous media: Application to enhanced oil recovery. Journal of Molecular Liquids. 2020;**316**:113876

[20] Saleh TA. Nanomaterials: Classification, properties, and environmental toxicities. Environmental Technology and Innovation. 2020;**20**:101067. DOI: 10.1016/j.eti.2020.101067

[21] Nikolaou IE, Jones N, Stefanakis A. Circular Economy and Sustainability: the Past, the Present and the Future Directions. Circular Economy and Sustainability. 2021;**1**:1-20. DOI: 10.1007/s43615-021-00030-3

[22] Saavedra YM, Iritani DR, Pavan AL, Ometto AR. Theoretical contribution of industrial ecology to circular economy. Journal of Cleaner Production. 2018;**170**:1514-1522

[23] Hanumante NC, Shastri Y, Hoadley A. Assessment of circular economy for global sustainability using an integrated model. Resources, Conservation and Recycling. 2019;**151**:104-460

[24] Korhonen J, Nuur C, Feldmann A, Birkie SE. Circular economy as an essentially contested concept. Journal of Cleaner Production. 2018a;**175**:544-552

[25] Prieto-Sandoval V, Jaca C, Ormazabal M. Towards a consensus on the circular economy. Journal of Cleaner Production. 2018;**179**:605-615

[26] Ellen MacArthur Foundation. Circularity indicators. 2016. Available from: https://www.ellenmacarthurfoundation.org/programmes/insight/circularity-indicators [Accessed: March 03, 2021]

[27] Ghosh S, Rajan J. The business case for SDGs: An analysis of inclusive business models in emerging economies. International Journal of Sustainable Development & World Ecology. 2019;**26**(4):344-353

[28] Bai C, Sarkis J, Yin F, Dou Y. Sustainable supply chain flexibility and its relationship to circular economy-target performance. International Journal of Production Research. 2019;**58**:1-18

[29] Bag S, Yadav G, Dhamija P, Kataria KK. Key resources for industry 4.0 adoption and its effect on sustainable production and circular economy: An empirical study. Journal of Cleaner Production. 2021;**81**:125233

Chapter 2

Green Materials Used in Enhanced Oil Recovery: A Historical Background and a Future Outlook

Rasha S. Kamal, Reham I. El-Shazly, Mohamed Keshawy, Abdulraheim M.A. Hasan and Manar E. Abdel-Raouf

Abstract

Petroleum crude oil is still the major source of fossil energy. In addition, it is a predominant spring of infinite raw materials. However, increased oil consumption due to enhanced industrial development necessitates establishing innovative protocols that ensure rational production and fair distribution with environmental protection. The extraction of oils left in reservoirs is a preliminary process of high economic value. This process is generally known as enhanced oil recovery. Recovering crude oil from aged oil fields is of utmost economic importance. Among all the methods that are being applied in this regard, the addition of chemical agents was the most facile. This chapter introduces a comprehensive review of all these methods, focusing on the green chemical agents as a competitive, ecofriendly protocol.

Keywords: biopolymers, biosurfactants, polymer nanoparticles, economic impacts, environmental impacts

1. Introduction

Enhanced oil recovery (EOR) is a term used in the petroleum industry to describe advanced techniques for extracting additional oil from a reservoir after the primary (natural pressure) and secondary recovery methods (water or gas injection) have been exhausted [1]. These techniques are employed to increase the amount of oil that can be recovered from an oil field, often extending its productive life [2]. EOR is crucial for maximizing oil recovery, especially in mature or challenging reservoirs, and plays a key role in the energy industry [3]. While primary recovery relies on natural reservoir pressure and secondary recovery utilizes water or gas injection to maintain that pressure, these methods typically leave a significant portion of oil up to 60–80% trapped in the reservoir (**Figure 1**). EOR methods aim to access this remaining oil, significantly boosting recovery rates and extending the productive life of oil fields [5]. In addition to increasing oil recovery, EOR plays a critical role in addressing energy challenges by optimizing production from mature and unconventional reservoirs [4, 6]. Its application is particularly valuable in a world seeking to balance growing energy demands with sustainable resource management.

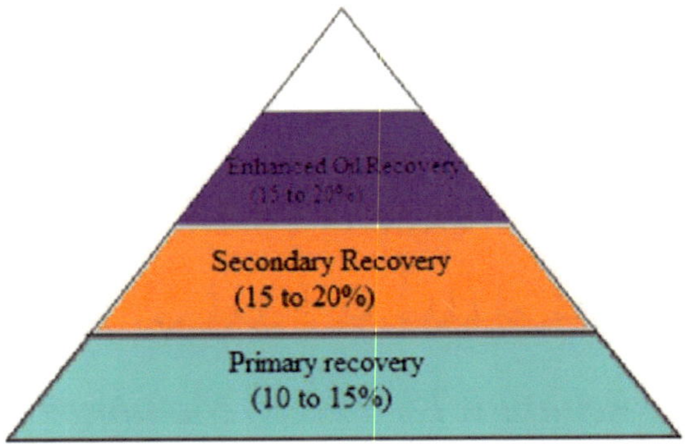

Figure 1.
Percentage of oil recovery by the various processes [4].

Figure 2.
The history of enhanced oil recovery (EOR).

The history of enhanced oil recovery (EOR) dates back over a century and reflects the evolution of technology and innovation in petroleum extraction [7]. Given below is a historical overview of EOR development, (**Figure 2**).

2. Historical overview of EOR

2.1 Early beginnings (1800s: Early 1900s)

- *First experiments with EOR*: The earliest efforts to enhance oil recovery involved basic water flooding, where water was injected into reservoirs to maintain pressure and displace oil [8].

- *Steam injection concepts*: By the early 1900s, researchers explored heating reservoirs to improve heavy oil recovery, laying the foundation for thermal EOR methods [9].

2.2 Thermal EOR development (1920s: 1950s)

- *Steam flooding*: In the 1920s, the first documented steam injection experiments were conducted in California oil fields [10].
- *In-situ combustion*: The 1940s saw the development of in-situ combustion techniques, where a portion of the oil in the reservoir is burned to generate heat and improve flow [11].
- *Commercial use*: By the 1950s, thermal EOR became commercially viable, particularly in heavy oil reservoirs like those in California [12].

2.3 Chemical EOR emergence (1950s: 1970s)

- *Introduction of surfactants and polymers*: In the 1950s and 1960s, scientists began using chemicals like surfactants and polymers to reduce surface tension and improve oil displacement [13].
- *Alkaline flooding*: Research into alkaline flooding, which creates soap-like substances within reservoirs, emerged in the 1960s [14].
- *Increased adoption*: Chemical EOR gained traction in the 1970s as oil prices rose, making these methods economically attractive [15].

2.4 Gas injection advances (1970s: 1980s)

- *CO_2 injection*: The first large-scale CO_2 injection project was implemented in 1972 in the Permian Basin of West Texas, marking the beginning of a widely used EOR method [16].
- *Miscible gas techniques*: Gases like CO_2 and nitrogen were used to create miscible conditions, allowing for more efficient oil recovery [17].
- *Environmental considerations*: CO_2 injection also gained attention for its potential to sequester carbon while enhancing oil production [16].

2.5 Modern innovations (1990s – Present)

- *Integration of advanced technologies*: Improved reservoir modeling, horizontal drilling, and real-time monitoring enhanced the efficiency of EOR methods [16].
- *Microbial EOR (MEOR)*: In the 1990s, biotechnological approaches using microbes to improve recovery were explored [18].
- *Nanotechnology and hybrid EOR*: Recent years have seen the introduction of nanoparticles and hybrid EOR methods to address complex reservoirs [19].
- *Environmental and economic focus*: The industry has increasingly focused on reducing the environmental impact of EOR, including integrating renewable energy sources for thermal methods and maximizing CO_2 sequestration [16].

2.6 Early awareness and challenges (mid-twentieth century)

- In the mid-1900s, EOR primarily relied on conventional chemicals and processes, such as synthetic polymers and surfactants, which were effective but raised environmental concerns due to toxicity, persistence in ecosystems, and groundwater contamination risks [20].

- The environmental movement of the 1960s and 1970s highlighted the need for greener practices in industrial operations, including oil recovery [5, 21].

3. Progress of biological resources and environmental concerns

3.1 Introduction of biopolymers and biosurfactants (1970s: 1980s)

- Researchers began exploring biodegradable alternatives like *biopolymers* (e.g., xanthan gum, guar gum, and Arabic gum) and *biosurfactants* produced by microorganisms [22]. These materials were environmentally friendly and performed comparably to synthetic options in some EOR processes [23].

- Early successes in pilot projects demonstrated the potential of biobased materials for reducing environmental harm [24].

3.2 Advances in biotechnology and green chemistry (1990s: 2000s)

- Advances in microbiology and biotechnology enabled the large-scale production of microbial-derived materials such as biosurfactants and bioengineered polymers [25].

- Simultaneously, the concept of *carbon capture and utilization* (CCU) gained momentum, integrating captured CO_2 in gas injection EOR projects to both enhance oil recovery and sequester greenhouse gases [26].

3.3 Emergence of nanotechnology and renewable resources (2000s: 2010s)

- *Nanoparticles* synthesized using green methods (e.g., plant-based processes) began being explored for their ability to enhance recovery by altering reservoir wettability and reducing oil–water interfacial tension [27].

- Algae-derived chemicals and other biobased solutions were developed as renewable, sustainable sources for EOR agents [28].

3.4 Modern era and sustainability focus (2010s to present)

- Growing global focus on sustainability, climate change, and the environmental impact of the oil industry accelerated research into green EOR materials.

- Governments, environmental agencies, and companies emphasized the development of materials that reduce toxicity, enhance biodegradability, and leverage renewable sources.

- The integration of renewable energy in thermal EOR methods, such as using solar energy for steam generation, emerged as a green enhancement to traditional techniques [16, 29, 30].

3.5 Current role in the industry

EOR has become a critical component of modern oil production, especially in mature fields and unconventional reservoirs. It is driven by the need to maximize recovery efficiency and address global energy demands while balancing environmental concerns [15]. The history of EOR reflects the interplay of technological innovation, economic drivers, and environmental considerations in the petroleum industry.

Today, the use of green materials in EOR is increasingly recognized as a way to balance efficient oil recovery with environmental stewardship, addressing both energy demands and sustainability goals [31–34].

4. Types of EOR methods

EOR can be broadly categorized into three main groups – based on the technique employed – as displayed in **Figure 3**.

4.1 Thermal EOR

Thermal enhanced oil recovery (thermal EOR) involves introducing heat into an oil reservoir to reduce the viscosity of heavy oil and facilitate its flow toward production

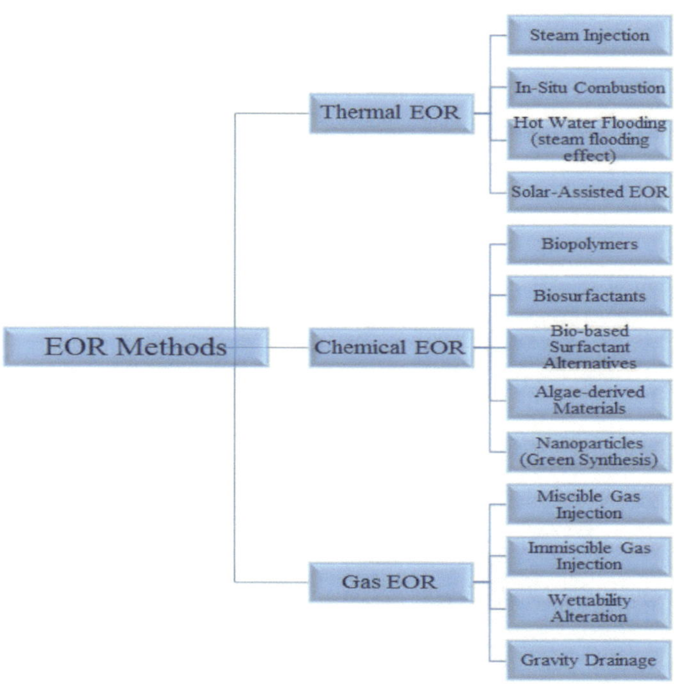

Figure 3.
Types of EOR methods.

wells [12]. This technique is particularly effective in heavy and extra-heavy oil fields, where oil is too thick to flow under natural conditions [10, 35]. An illustration of the overall process is given in **Figure 4**.

4.1.1 Mechanism of thermal EOR

- *Heat introduction*: Heat is introduced into the reservoir through methods like steam injection or in-situ combustion [10, 36].
- *Viscosity reduction*: Heat reduces the viscosity of the oil, making it more fluid and easier to mobilize [7, 37].
- *Pressure and expansion effects*: In some cases, steam also generates pressure that pushes oil toward production wells [37, 38].
- *Reservoir changes*: Heat alters the reservoir's wettability and reduces oil–water interfacial tension, further aiding oil flow [39, 40].

4.1.2 Types of thermal EOR

1. *Steam injection*
    - *Cyclic steam stimulation (CSS)*: Also known as "huff and puff," steam is injected, left to soak, and then oil is produced [41].
    - *Steam flooding*: Continuous injection of steam to displace oil [12].
2. *In situ combustion*
    - A portion of the reservoir oil is ignited to generate heat, which reduces viscosity and mobilizes the remaining oil [42, 43].

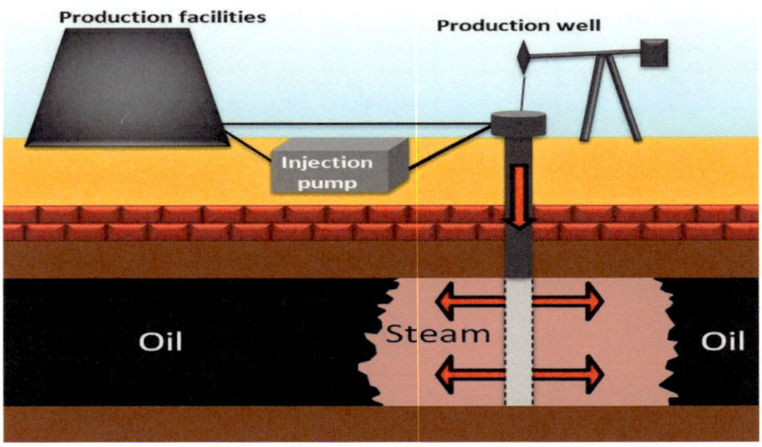

Figure 4
Thermal enhanced oil recovery (Modified from reference [35]).

3. *Hot water flooding (steam flooding effect)*

    - Injection of hot water instead of steam to achieve similar effects, though less commonly used [44, 45].

4. *Solar-assisted EOR*

    - Uses concentrated solar power to generate steam for injection, reducing reliance on fossil fuels [12, 46].

The overall advantages and disadvantages of thermal EOR are collected in **Table 1.**

4.1.3 Economic impacts of thermal EOR

Thermal enhanced oil recovery (EOR) has a notable economic impact, with high initial investments required for steam generation facilities and supporting infrastructure (**Figure 5**). Despite these costs, the method remains cost-competitive per barrel, particularly in heavy oil fields. However, its profitability is highly sensitive to oil price fluctuations, as lower prices can render it economically unfeasible. On the positive side, thermal EOR fosters significant job creation across engineering, operations, and maintenance sectors, contributing to local economic growth and providing employment opportunities in specialized fields.

- *High initial investment (40%)*: Capital costs for steam generation and infrastructure dominate initial expenditures. Indeed, steam generation facilities and infrastructure comprise substantial capital costs [54].
- *Cost per barrel (25%)*: This factor reflects competitiveness and efficiency in production, where the cost per barrel of oil produced through thermal EOR is competitive, especially in heavy oil fields [55].
- *Oil price sensitivity (20%)*: Fluctuations in oil prices significantly affect profitability but are secondary to fixed costs. Profits depend on oil prices; lower prices can make thermal EOR economically unfeasible [56].
- *Job creation (15%)*: While crucial for socio-economic impact, its weight in overall economic assessment is lower compared to the other factors. Thermal EOR projects generate employment in engineering, operations, and maintenance [57].

Advantages of thermal EOR	Disadvantages of thermal EOR
High efficiency: Particularly effective in recovering heavy oil and bitumen [47].	*Energy intensive*: Requires significant energy input for steam generation or combustion [50].
Enhanced recovery rates: Can recover up to 60–70% of the original oil in place in some cases [48].	*Water demand*: Consumes large amounts of water, which can be a challenge in arid regions [51].
Mature field application: Extends the productive life of mature and heavy oil reservoirs [12].	*Reservoir suitability*: Less effective in light oil reservoirs or highly fractured formations [52].
Economic viability: Steam generation is cost-effective compared to other methods, especially in heavy oil fields [49].	*Operational complexity*: Managing heat distribution and reservoir conditions is technically challenging [53].

Table 1.
Advantages and disadvantages of thermal EOR.

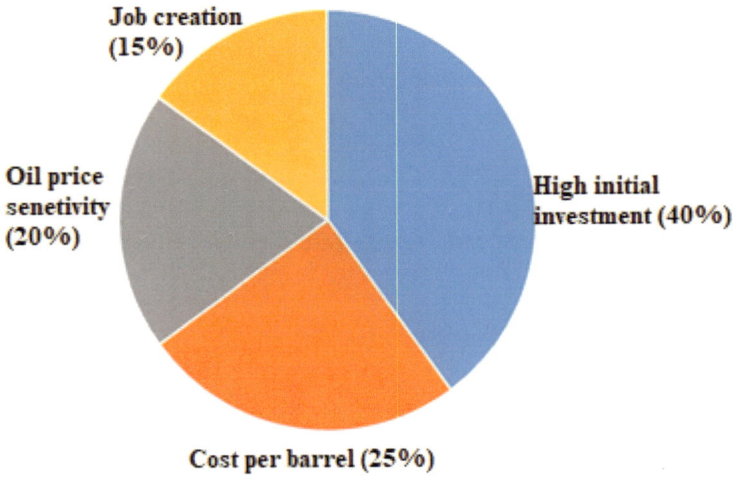

Figure 5
Economic impacts of thermal EOR.

4.1.4 Environmental impact of thermal EOR

Thermal EOR can induce both positive and negative environmental aspects. On the positive side, it reduces gas flaring by using produced gas for steam generation and can integrate CO_2 injection to enhance recovery while sequestering carbon [58, 59]. Negatively, steam generation contributes to greenhouse gas emissions, uses significant water resources, and risks groundwater contamination, alongside land disturbance from surface operations [60, 61].

4.1.5 Sustainability efforts

Sustainability efforts in thermal EOR focus on reducing its environmental footprint while maintaining efficiency. Solar-assisted thermal EOR uses renewable solar energy for steam generation, significantly lowering greenhouse gas emissions compared to traditional fossil fuels [62]. Additionally, advancements in water management include recycling and treating produced water for reuse, minimizing freshwater consumption, and reducing risks to local water resources. Efforts also include optimizing CO_2 injection, which not only enhances oil recovery but also aids in carbon sequestration. These initiatives aim to balance energy needs with environmental stewardship [63]. Innovations such as *solar-assisted thermal EOR* and recycling-produced water aim to mitigate environmental concerns and improve the sustainability of these methods [64]. Although thermal EOR is a powerful tool for maximizing oil recovery from heavy oil reservoirs, it requires careful management of its economic and environmental trade-offs.

4.1.6 Green materials in thermal enhanced oil recovery (thermal EOR)

Thermal enhanced oil recovery (thermal EOR) techniques, particularly steam injection, are commonly used to recover heavy oil and bitumen by increasing the temperature within the reservoir, which lowers the viscosity of the oil, making it easier to extract. However, the traditional thermal methods can be energy-intensive, potentially contributing to environmental pollution, including high carbon emissions. This has led to increased interest in *green materials* in thermal EOR to reduce environmental impacts, increase energy efficiency, and promote sustainability [12]. Green materials in thermal EOR aim to enhance oil recovery processes while reducing the reliance on fossil fuels and minimizing harmful environmental effects. These materials are typically sourced from renewable, biodegradable, and non-toxic resources [65, 66]. Types of green materials used in thermal EOR and their importance are tabulated in **Table 2**. In addition, the benefits and challenges of using green materials in thermal EOR are gathered in **Figure 6** [76].

4.2 Chemical EOR

Chemical EOR involves injecting chemicals into a reservoir to improve oil displacement efficiency. These chemicals alter the physical properties of the oil and reservoir rock, enabling more oil to flow toward production wells. (Using chemicals to alter the properties of the oil or the reservoir, improving oil displacement [77, 78]).

4.2.1 Common methods of chemical EOR

- *Polymer flooding*: Polymers increase the viscosity of water injected into the reservoir, improving sweep efficiency [79].

- *Surfactant-polymer flooding*: Surfactants reduce the surface tension between oil and water, making it easier for oil to flow [80].

- *Alkaline flooding*: Alkaline chemicals react with crude oil to produce soap-like substances that improve recovery [81].

- *Combination processes*: Techniques often combine surfactants, polymers, and alkaline agents to enhance recovery efficiency further [82].

However, the rewards and drawbacks of chemical EOR are given in **Table 3**. In addition, the economic [88, 89] and environmental impacts [90, 91] of chemical EOR are summarized in **Figure 7**. Chemical EOR is a powerful technique for enhancing oil recovery, especially in reservoirs where traditional methods have failed. While its high costs and environmental challenges are notable, ongoing innovation in sustainable chemical alternatives and optimized processes continues to address these limitations [92].

In recent years, there has been growing interest in using *green materials* for chemical enhanced oil recovery (EOR) processes [77]. These materials are derived from renewable sources and are designed to minimize environmental impact, reduce toxicity, and enhance biodegradability [93]. The use of green materials in EOR aligns with the global emphasis on sustainability and environmental responsibility while still addressing the need to improve oil recovery in mature fields and challenging reservoirs [93].

4.2.2 Types of green materials used in chemical EOR

- *Biopolymers*: They are polymers derived from biological sources, such as plants, microorganisms, or algae [94, 95]. They include gums from different sources such

Green materials	Description	Examples	Mechanism	Advantages
Biopolymers for viscosity control	Biopolymers are naturally occurring polymers that can be used to improve the efficiency of thermal EOR processes by controlling the viscosity of the injected fluids [67].	Guar gum (from guar plant) and xanthan gum (from bacteria).	Thickening the injected fluid, making it more effective in displacing oil. Modifying the rheology of water or steam injected into the reservoir, improving its mobility and distribution, thus enhancing the overall process efficiency [24, 68].	Biodegradable and non-toxic.
Surfactants from renewable resources	Surfactants are surface-active materials that can reduce the interfacial tension between oil and water, which can help reduce the viscosity of heavy oils, making them easier to extract when combined with thermal techniques.	Alkyl polyglycosides (from plant sugars) Fatty acid methyl esters (derived from plant oils).	Lowering the surface tension between oil and water improves the efficiency of oil displacement in thermal EOR processes [69, 70].	Renewable and biodegradable. Non-toxic and safer for the environment. Surfactants derived from renewable resources can replace synthetic surfactants that might be harmful to the ecosystem [24].
Green solvents	Ecofriendly solvents that can replace traditional solvents that are being used in thermal EOR.	Limonene (derived from citrus fruits) and terpines.	They can be injected alongside steam to help reduce the viscosity of heavy oils, improving the flow and recovery rate during the thermal EOR process [71].	Biodegradable, highly effective, and sustainable [72, 73].
Nanomaterials	They can be synthesized through environmentally friendly processes using renewable resources [31, 74, 75].	Silica nanoparticles derived from plant-based sources. Magnetic nanoparticles synthesized using biobased methods.	They can be used in conjunction with steam injection to enhance heat transfer, improve oil mobility, or modify reservoir wettability, enhancing the overall oil recovery process [19].	Small size and large surface area allow nanoparticles to interact more effectively with the oil and the reservoir rock. Can help in improving oil flow and increasing thermal recovery efficiency. Can be designed to interact with steam or heat to reduce energy requirements for the thermal EOR process.

Table 2.
Green materials used in thermal EOR.

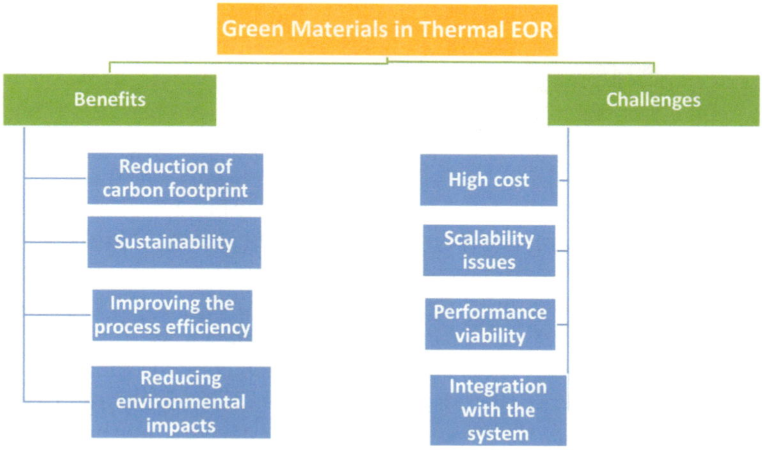

Figure 6.
Benefits and challenges of using green materials in thermal EOR.

Rewards of chemical EOR	Drawbacks of chemical EOR
Increased recovery efficiency: Enhances the recovery of residual oil trapped by capillary forces [83]. *Applicable to diverse reservoirs*: Effective in reservoirs with high heterogeneity and residual oil saturation [13]. *Customizable formulations*: Chemical mixtures can be tailored to reservoir conditions [84].	*High costs*: Chemicals, especially surfactants and polymers, can be expensive to produce and transport [85]. *Reservoir compatibility issues*: Not all reservoirs are suitable for chemical EOR; the injected chemicals may react negatively with the reservoir rock or brine [86]. *Potential plugging*: Polymers can clog the reservoir's pore spaces, reducing permeability [87].

Table 3.
Rewards and drawbacks of chemical EOR.

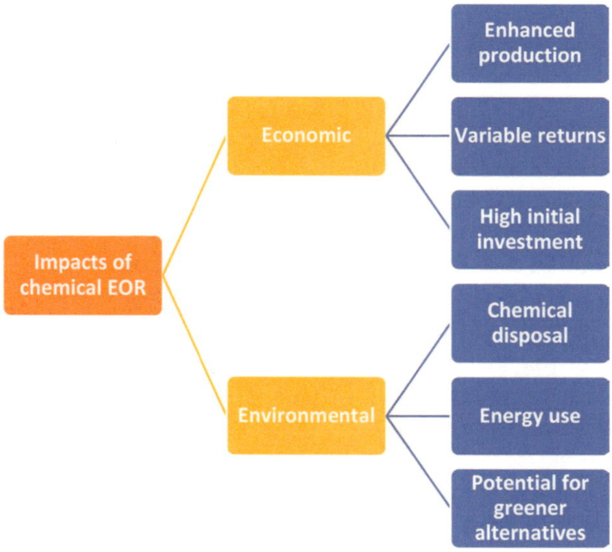

Figure 7.
Economic and environmental impacts of chemical EOR.

as xanthan gum (produced by bacteria), guar gum (from guar beans), and carboxymethyl cellulose (from plant cellulose). As natural polymers, they are benign to the environment. Most of these polymers are excellent viscosifiers and help improve oil displacement [96]. In chemical EOR, they are used to increase the viscosity of water, enhancing the efficiency of water flooding in reservoirs [79].

- *Biosurfactants*: These are surface-active compounds produced by microorganisms, capable of lowering the interfacial tension between oil and water, making oil more mobile and less viscous [97]. They involve many materials that are usually of microorganism origin such as *Rhamnolipids* (produced by certain bacteria), *Sophorolipids* (produced by yeast), and *Surfactin* (produced by Bacillus subtilis). They fulfill the criteria of natural materials, being biodegradable and environmentally friendly. They can reduce toxicity compared to synthetic surfactants. In addition, they are highly effective at reducing the surface tension between oil and water, improving oil displacement [98]. Their major role in chemical EOR is to improve the recovery of trapped oil by reducing surface tension and facilitating better oil flow [98].

- *Biobased surfactant alternatives*: These categories are simply surfactants derived from renewable sources like plants or animal fats such as *Fatty acid methyl esters* (from vegetable oils) and *Alkyl polyglycosides* (from plant sugars). They can replace petrochemical-based sources [99]. They are environmentally friendly and biodegradable [99]. They can be more effective in certain types of reservoirs, reducing environmental concerns [100]. As surface-active compounds, they act by reducing the interfacial tension between the oil and water, making it easier to displace oil from reservoir rocks [101].

- *Algae-derived materials*: These materials are polymeric compounds produced from different algal species. They are mainly polysaccharides such as alginate and lipids. They combine the ecofriendly nature with a high ability to reduce the viscosity of the crude oil, facilitating its extraction [102–104].

- *Nanoparticles (Green synthesis)*: These tiny particulates of 1–100 nm can be synthesized via green methods from plant extracts or from other renewable sources **Figure 8** [105]. They can be engineered to alter the wettability of reservoir rocks, improving oil mobility [106]. They can be used in both water flooding and gas injection techniques to improve oil recovery efficiency, considering the environmental impact of traditional methods [107]. Their green nature and their low toxicity make them very useful in controlling the environmental pollution, especially when they are discarded in the oil wells [31, 75]. However, applying this type of material is judged by some challenges such as the overall cost of the operation, performance variability according to the operation conditions, and the proper infrastructures required for scaling production of these materials [108, 109]. Indeed, the use of *green materials* in *chemical EOR* represents a promising step toward making oil recovery processes more sustainable and environmentally responsible. By replacing traditional, toxic chemicals with biodegradable, renewable alternatives, the oil industry can significantly reduce its environmental footprint while continuing to improve recovery rates [31]. The growing interest in green materials reflects a broader shift toward sustainability in the oil and gas sector, aligning energy production with ecological preservation and societal demands for cleaner energy practices [110].

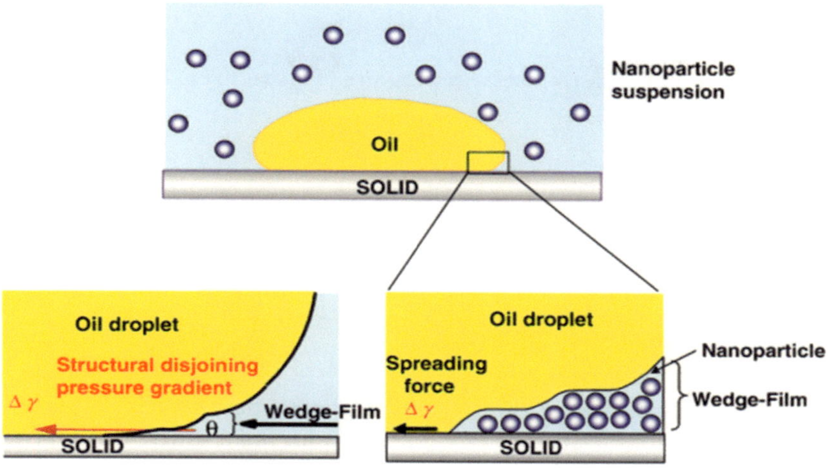

Figure 8.
Nanoparticles (green synthesis) in EOR.

5. Gas EOR

Gas EOR involves injecting some gases, such as carbon dioxide (CO_2), nitrogen (N_2), or natural gas, into an oil reservoir to enhance oil recovery. These gases either dissolve in the oil to reduce its viscosity or increase reservoir pressure to mobilize trapped oil [15, 111].

5.1 Common methods of gas EOR

5.1.1 CO_2 injection

When carbon dioxide mixes with oil, oil viscosity is reduced and its flowability is enhanced (**Figure 9**) [111].

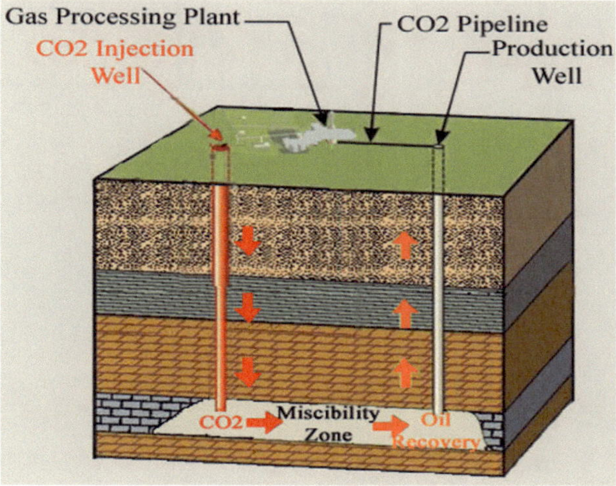

Figure 9.
CO_2 injection in gas enhanced oil recovery [109].

5.1.2 Nitrogen injection

Displaces oil and maintains reservoir pressure (**Figure 10**) [112].

5.1.3 Miscible gas injection

Gases mix with oil to form a single phase, enhancing flow [112].

5.2 Mechanism of gas EOR

1. *Miscible gas injection*

    - *Mechanism*: Gases like CO_2 are injected until they become miscible (blend seamlessly) with the oil at reservoir conditions. This reduces the interfacial tension between oil and gas, allowing the oil to flow more freely [113].

    - *Effect*: Enhances the displacement efficiency of oil from the reservoir [114].

2. *Immiscible gas injection*

    - *Mechanism*: When miscibility cannot be achieved due to pressure or temperature limitations, gases act as a pressurizing agent, forcing oil toward production wells [79].

    - *Effect*: Improves reservoir pressure and oil recovery, though less efficiently than miscible techniques [115].

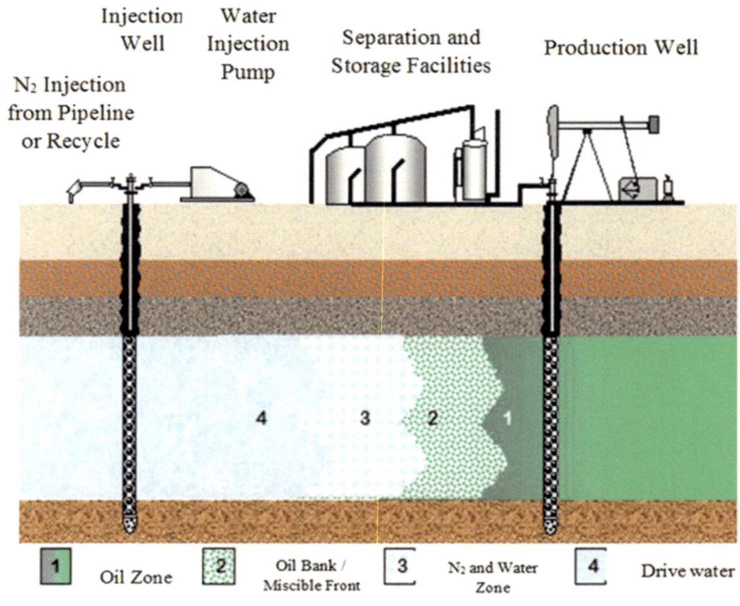

Figure 10.
N_2 injection in gas enhanced oil recovery [110].

3. *Wettability alteration*

   - In some cases, gas injection alters the rock's wettability, making it easier for oil to detach and flow [116].

4. *Gravity drainage*

   - Gases injected into the reservoir can flow to the top, aiding in oil movement through gravitational forces [117].

At last, cons and pros of gas EOR are given in **Table 4**.

5.3 Economic impact of gas EOR

1. *Positive economic outcomes*

   - Prolongs the productive life of reservoirs, maximizing the economic return on investment in existing fields [88].

   - Particularly viable in mature fields where primary and secondary methods are no longer effective [123].

2. *Cost challenges*

   - High costs for gas sourcing, compression, and injection infrastructure [124].

   - Economic feasibility is tied to oil prices; higher prices make gas EOR more attractive [125].

5.4 Environmental impact of gas EOR

1. *CO_2 utilization and storage*:

   - Gas EOR offers a dual benefit of enhanced recovery and carbon sequestration, mitigating greenhouse gas emissions [126].

   - However, there is a risk of leakage from improperly sealed reservoirs [127].

Returns of gas EOR	Shortcomings of gas EOR
High efficiency in light oil reservoirs: Particularly effective for light and moderately heavy oils [118]. *CO_2 sequestration potential*: CO_2 injection not only enhances oil recovery but also serves as a method for carbon capture and storage (CCS) [117]. *Reservoir pressure maintenance*: Gas injection helps sustain reservoir pressure, improving long-term productivity [119, 120].	*Limited miscibility conditions*: Achieving miscibility requires specific pressure and temperature conditions, which may not be present in all reservoirs [121]. *Gas availability and infrastructure*: Reliable sources of CO_2 or other gases and the infrastructure for injection and recycling are critical [112]. *Gas breakthrough*: Injected gas may bypass the oil, reaching production wells prematurely and reducing recovery efficiency [122].

Table 4.
Gas EOR: advantages and disadvantages.

2. *Energy consumption*:

   - Compressing and injecting gases require significant energy, contributing to the carbon footprint [128].

   - Recycling gases can reduce energy demands but adds operational complexity [129].

3. *Potential groundwater contamination*:

   - Improper handling of injected gases can risk contaminating groundwater reserves [130].

Recently, emerging technologies such as microbial EOR and nanotechnology have expanded the scope of these methods.

5.5 Green materials in gas EOR

Why green materials are needed:

1. *Reducing environmental risks*: Traditional gas injection processes involve potential greenhouse gas emissions, contamination risks, and resource-intensive operations. Green materials can mitigate these impacts [131].

2. *Improving sustainability*: Biobased or biodegradable materials can complement gas EOR, enhancing oil recovery with lower environmental footprints [132].

3. *Integration with renewable energy*: Renewable energy-powered CO_2 capture and compression technologies can make gas EOR more sustainable [133].

Examples of green innovations:

- *Biologically enhanced gas EOR*: Introducing microbes that interact with gases to improve recovery efficiency [134].

- *Nanoparticles in gas injection*: Nanoparticles that stabilize miscibility and enhance oil mobility [107].

- *Carbon-neutral gas sources*: Sourcing CO_2 from renewable energy-driven industrial processes [135].

Gas enhanced oil recovery (EOR) is an exceptionally effective method, especially for reservoirs containing light oils or exhibiting diminished pressures. Especially for reservoirs with light oils or depleted pressures, gas EOR is a very effective technology [30, 136]. Although its ability to sequester CO_2 is in line with global climate goals, there are environmental and economic obstacles. EOR is not only efficient but also sustainable for future energy needs because of these considerations, which highlight the significance of implementing green materials and methods [137, 138].

In enhanced oil recovery (EOR), green materials refer to ecologically acceptable and sustainable substitutes for traditional chemicals and procedures. While preserving or enhancing recovery efficiency, these materials seek to lessen their impact on the environment [31]. Here are a few instances:

1. *Biopolymers*: Natural, biodegradable polymers generated from microorganisms, such as xanthan gum and scleroglucan, are utilized for polymer flooding to improve oil mobility [22].

2. *Biosurfactants*: Made by microorganisms, these surfactants are safer for the environment than synthetic ones and increase oil displacement by lowering interfacial tension [25, 139].

3. *Nanoparticles*: By changing reservoir wettability and lowering oil viscosity, green-synthesized nanoparticles (such as silica derived from plants or biosynthesis) enhance oil recovery [107].

4. *Algae-derived chemicals*: Fatty acids and lipids, which can function as solvents or surfactants in EOR, are biobased chemicals that algae offer as a renewable resource [140].

5. *CO_2 from renewable sources*: Oil recovery and carbon sequestration are combined when captured CO_2 from bioenergy operations is used in gas injection for miscible flooding. These resources assist more environmentally friendly methods in the oil and gas sector by being in line with sustainability objectives [133].

5.6 Innovative and combined approaches

Innovative and combined approaches are being developed in addition to the conventional EOR methods:

- Microbial EOR (MEOR): By dissolving heavy oil molecules or lowering interfacial tension, microorganisms or their byproducts might enhance oil recovery [25].

- Nanotechnology in EOR: To improve recovery, nanoparticles modify fluid dynamics at the molecular level [141].

- Hybrid EOR methods: These methods combine gas, chemical, or thermal approaches to maximize recovery in intricate reservoirs [142].

6. Applications of EOR

The petroleum industry makes extensive use of enhanced oil recovery (EOR) techniques to maximize oil production from reservoirs that are challenging to exploit with traditional methods. EOR is used in a wide range of geological, industrial, and economic settings. Applications of enhanced oil recovery (EOR) (**Figure 11**).

EOR, which aims to maximize the effective use of existing reservoirs and address falling output, is a crucial part of contemporary petroleum engineering [143].

Numerous geological, environmental, and economic contexts are covered by the applications of enhanced oil recovery, or EOR. EOR methods are essential for optimizing

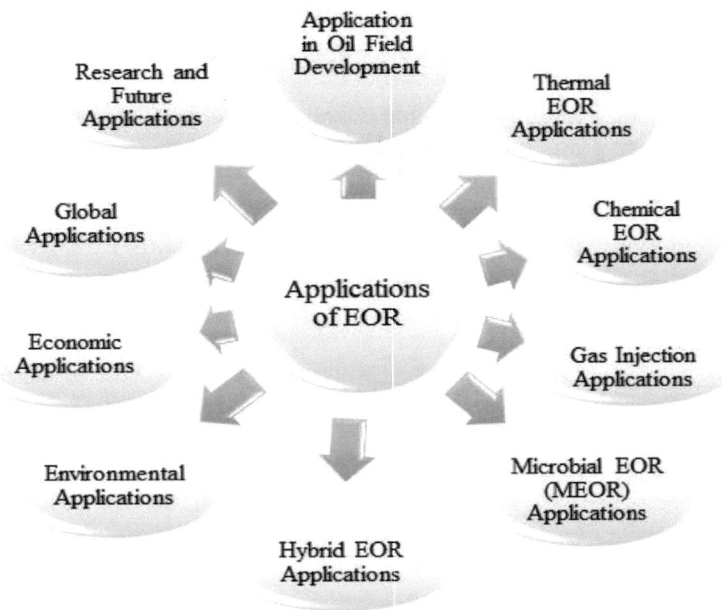

Figure 11
Applications of EOR.

hydrocarbon recovery, meeting energy needs, and facilitating sustainable oil production. The next step in making EOR procedures more effective and ecologically conscious is the incorporation of green materials and renewable energy sources [144].

7. An overview of the future of green materials in enhanced oil recovery (EOR)

Given the increased focus on minimizing the environmental impact of industrial operations, especially oil recovery, enhanced oil recovery's (EOR) future is strongly linked to sustainability. Green materials minimize negative environmental effects while increasing recovery efficiencies, providing an inventive way ahead. An outlook for the future of green material use in EOR is provided below, covering developments, obstacles, and the possibility of incorporating these materials into standard procedures. Incorporating environmentally friendly materials into enhanced oil recovery (EOR) is not only a technological breakthrough but also a crucial step in the oil and gas sector's transition to sustainability. With further development, these materials could make EOR a greener, cleaner process that strikes a balance between the need to produce energy and protect the environment [145]. Moreover, the most important green materials used in EOR and their specific role(s) are tabulated in **Table 5**.

However, the main challenges of utilizing natural materials in EOR are:

- Scalability: Natural nanomaterials sometimes require complex extraction.
- pH/salinity sensitivity: Performance varies in harsh reservoir conditions.
- Cost: Processing may offset sustainability benefits.

Material	Source	General roles	Key EOR mechanism	Advantages
Cellulose nanocrystals (CNCs)	Wood pulp, cotton, agricultural waste.	Interfacial tension reduction: Modifies oil–water interfaces. Emulsion stabilization: Forms Pickering emulsions for mobility control. Thermal stability: Withstands reservoir temperatures (up to 120°C).	Emulsion stabilization	Renewable, low-cost
Lignin nanoparticles	Paper/pulp industry waste, plant biomass.	Foam stabilizer: Enhances CO_2/N_2 foam durability for gas flooding. Antioxidant properties: Prevents crude oil oxidation. Heavy oil viscosity reduction: Disrupts asphaltene aggregates.	Foam/CO_2 flooding	Waste valorization
Chitosan nanofibers	Crab/shrimp shells, fungal biomass.	Wettability alteration: Biopolymer flooding: Improves sweep efficiency via viscosity enhancement. pH-responsive: Expands in low-pH reservoirs for pore blockage.	Wettability alteration	Biodegradability
Silica nanoparticles (Bio-SiO_2)	Rice husk, sugarcane bagasse.	Nanofiltration and thermal conductivity.	Permeability modification	High abundance
Carbon quantum dots (CQDs)	Biochar, coconut shells.	Tracers and catalyst carriers.	Reservoir monitoring	Non-toxic, optical properties

Table 5.
The most important green materials used in EOR.

8. Sustainable development goals (SDGs) and green materials in enhanced oil recovery (EOR)

A number of the sustainable development goals (SDGs) of the UN are strongly related to the incorporation of green materials into enhanced oil recovery (EOR), **Figure 12**. These objectives seek to advance social well-being, environmental stewardship, and sustainability on a worldwide scale. By encouraging sustainability and environmental stewardship in the energy industry, the use of green materials into enhanced oil recovery (EOR) is in line with the sustainable development goals (SDGs) of the UN. By increasing energy efficiency, prolonging reservoir life, and integrating renewable energy sources like solar power into extraction processes, green EOR materials such as biosurfactants, biopolymers, and nanoparticles based on renewable resources support SDG 7. By encouraging innovation and sustainable industrial practices through environmentally friendly technologies, these developments also support SDG 9. Green materials help achieve SDG 12 by reducing environmental impact and guaranteeing responsible production by substituting biodegradable chemicals for hazardous ones [24]. Furthermore, gas EOR that makes use of carbon capture, utilization, and storage (CCUS) greatly lowers greenhouse gas emissions, which is consistent with SDG 13 and international climate action programs. By lowering the ecological dangers of conventional chemical pollutants, the implementation of green EOR techniques also safeguards terrestrial habitats (SDG 15) and aquatic

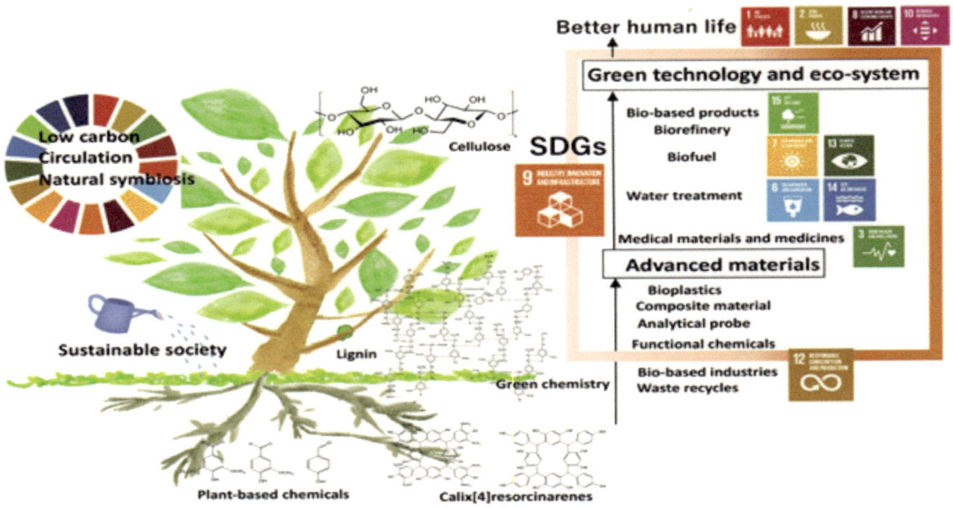

Figure 12.
EOR in relevance to UN-SDGs [146].

ecosystems (SDG 14). Additionally, the creation of green materials encourages collaborations across sectors, uniting governments, businesses, and academic institutions to support sustainable energy solutions (SDG 17). By optimizing resource utilization, these advances protect ecosystems, lessen the environmental impact of oil extraction, and enhance global energy security. In addition to supporting the larger SDG framework, green EOR materials collectively represent a revolutionary step toward tackling climate issues, creating resilience in the oil and gas sector, and striking a balance between energy demands and environmental sustainability [147].

9. Governmental policies and industry regulations for green EOR

Several governmental regulations have been established to ensure the achievement of the best EOR practices. They include—but not limited to—regulatory drivers, industrial standardization for the materials used in the process, bans and phasing out potentially harmful materials, and funding giant research for innovative EOR projects. Some of these regulations and their impacts on EOR are given in **Table 6**.

Policy type	Brief impacts on EOR	Example
Carbon pricing and taxes	Enhancing the economic viability of CO_2-EOR	Alberta's *CCUS Hub*
Biobased mandates	Forces the implementation of plant-derived nanomaterials and chemicals	Brazil's *RenovaBio* program
Emission caps	Adoption of cleaner, low-emission EOR fluids	Norway's CLIMIT program
API/ISO guidelines	Standards like *ISO 14001* push for lifecycle assessments (LCA) of EOR chemicals (e.g., banning toxic solvents)	U.S. DOE's NETL funds

Table 6.
The impact of some governmental policies and industry regulations on EOR.

However, the most impactful challenges associated with these *regulations* are

- *Policy fragmentation*: Inconsistent regulations across regions (e.g., U.S. state-by-state rules).

- *High compliance costs*: Small operators struggle with establishing green tech.

10. Case studies for different green EOR techniques

At last, some case studies for EOR mandates are introduced in **Table 7** to illustrate their role in enhancing the credibility and economic viability of EOR [148].

Technique	Mechanism	Case study	Performance data	Policy driver
CO_2-EOR	CO_2 injection with sequestration	*Petra Nova, Texas (USA)*	• *Storage*: 1.6 M tonnes CO_2/yr. • *Oil gain*: 15% recovery boost • *Cost*: $60/barrel	45Q Tax Credit ($85/ tonne CO_2)
Microbial EOR	In-situ biosurfactant production	*ONGC, Ankleshwar (India)*	• *Incremental oil*: 8–10% • *Cost*: $2.5/barrel • *Carbon footprint*: 0.05 kg CO_2/barrel	India's biofuel policy subsidies
Solar-SAGD	Solar-thermal steam generation	*GlassPoint, Oman*	• *Emissions*: 80% lower than gas-fired SAGD • *Steam cost*: 3.5/ MBTU vs. 3.5/ MBTU vs. 8.5 • *Limitation*: Land-intensive	Oman's renewable energy targets
Lignin nanoparticles	Foam stabilization for facile gas flood	*EcoEOR, North Sea (Norway)*	• *Foam half-life*: 3x longer than synthetic • *Cost*: 30% lower than PEG surfactants • *Scalability*: 500 bbl/ day pilot	EU REACH restrictions on synthetics

Table 7.
Some case studies on EOR regulations and their economic impacts.

11. Conclusions

In conclusion, the use of green materials in enhanced oil recovery (EOR) represents a promising and sustainable approach to improving oil extraction efficiency while minimizing environmental impact. As discussed throughout this chapter, natural polymers, biobased nanomaterials, and biodegradable surfactants have demonstrated significant potential to replace conventional chemicals that often pose ecological and health risks. These environmentally friendly alternatives not only enhance oil recovery performance but also align with global efforts toward sustainable development and cleaner energy production. Advances in biotechnology and nanotechnology have further improved the functionality and resilience of green materials under harsh reservoir conditions, making them increasingly viable for field applications. However, large-scale implementation still faces challenges, including the need for comprehensive field studies, economic feasibility assessments, and a deeper understanding of material-reservoir interactions. Despite these obstacles, the shift toward green EOR technologies reflects a growing awareness of the need to reduce the carbon footprint of the oil and gas industry. As environmental regulations tighten and societal demand for eco-conscious solutions increases, green materials are expected to play a pivotal role in shaping the future of EOR practices. Their integration could mark a key step toward achieving more sustainable and responsible resource recovery in the years to come.

Author details

Rasha S. Kamal, Reham I. El-Shazly, Mohamed Keshawy, Abdulraheim M.A. Hasan and Manar E. Abdel-Raouf*
Department of Petroleum Applications, Egyptian Petroleum Research Institute, Nasr City, Cairo, Egypt

*Address all correspondence to: drmanar770@yahoo.com

References

[1] Thomas S. Enhanced Oil Recovery-an overview. Oil and Gas Science and Technology – Revue d'IFP Energies. 2008;**63**:9-19. DOI: 10.2516/ogst:2007060

[2] Alvarado V, Manrique E. Enhanced oil recovery: An update review. Energies. 2010;**3**:1529-1575. DOI: 10.3390/EN3091529

[3] Etc RTJ, Lake LW, Johns R, Rossen B, Lake L. Fundamentals of Enhanced Oil Recovery. Richardson, Texas, USA: Penn Well Publishing Co; Society of Petroleum Engineers; 2014. DOI: 10.2118/9781613993286

[4] Behera US, Sangwai JS. Interaction of nanoparticles with reservoir fluids and rocks for enhanced oil recovery. In: Green Energy Technology. Cham: Springer; 2020. pp. 299-328. DOI: 10.1007/978-3-030-33774-2_13

[5] Ogbu A, Ozowe W, Ikevuje A. Data science's pivotal role in enhancing oil recovery methods while minimizing environmental footprints: An insightful review. Computer Science & IT Research Journal. 2024;**5**:1621-1633. DOI: 10.51594/CSITRJ.V5I7.1348

[6] Khan A, Saxena S, Baloni S, Sharma M, Kodavaty J. Overview and methods in enhanced oil recovery. Journal of Physics: Conference Series. 2021;**2070**. DOI: 10.1088/1742-6596/2070/1/012061

[7] Xue L, Liu P, Zhang Y. Development and research status of heavy oil enhanced oil recovery. Geofluids. 2022;**1**(4):1-13. DOI: 10.1155/2022/5015045

[8] Speight JG. Introduction to enhanced recovery methods for heavy oil and Tar Sands. In: Introduction to Enhanced Recovery Methods for Heavy Oil and Tar Sands. Gulf Professional Publishing; 2016. pp. 1-563. DOI: 10.1016/C2014-0-01296-8

[9] Dong X, Liu H, Chen Z, Wu K, Lu N, Zhang Q. Enhanced oil recovery techniques for heavy oil and oilsands reservoirs after steam injection. Applied Energy. 2019;**239**:1190-1211. DOI: 10.1016/J.APENERGY.2019.01.244

[10] Hascakir B. Introduction to thermal enhanced oil recovery (EOR) special issue. Journal of Petroleum Science and Engineering. 2017;**154**:438-441. DOI: 10.1016/J.PETROL.2017.05.026

[11] Mahinpey N, Ambalae A, Asghari K. In situ combustion in Enhanced Oil Recovery (EOR): A review. Chemical Engineering Communications. 2007;**194**:995-1021. DOI: 10.1080/00986440701242808

[12] Mokheimer EMA, Hamdy M, Abubakar Z, Shakeel MR, Habib MA, Mahmoud M. A comprehensive review of thermal enhanced oil recovery: Techniques evaluation. Journal of Energy Resources Technology, Transactions of the ASME. 2019;**141**(3):030801. DOI: 10.1115/1.4041096/377484

[13] Gbadamosi AO, Junin R, Manan MA, Agi A, Yusuff AS. An overview of chemical Enhanced Oil Recovery: Recent advances and prospects. International Nano Letters. 2019;**9**:171-202. DOI: 10.1007/S40089-019-0272-8

[14] Pei H, Zhang G, Ge J, Tang M, Zheng Y. Comparative effectiveness of alkaline flooding and alkaline–surfactant flooding for improved heavy-oil recovery. Energy & Fuels. 2012;**26**:2911-2919. DOI: 10.1021/EF300206U

[15] Karimov D, Toktarbay Z. Enhanced oil recovery: Techniques, strategies, and advances. ES Materials and Manufacturing. 2024;**23**:1005. DOI: 10.30919/ESMM1005

[16] Ozowe W, Daramola GO, Ekemezie IO. Innovative approaches in Enhanced Oil Recovery: A focus on gas injection synergies with other EOR methods. Magna Scientia Advanced Research and Reviews. 2024;**11**:311-324. DOI: 10.30574/MSARR.2024.11.1.0095

[17] Wang L, Tian Y, Yu X, Wang C, Yao B, Wang S, et al. Advances in improved/Enhanced Oil Recovery technologies for tight and shale reservoirs. Fuel. 2017;**210**:425-445. DOI: 10.1016/J.FUEL.2017.08.095

[18] She H, Kong D, Li Y, Hu Z, Guo H. Recent advance of microbial enhanced oil recovery (MEOR) in China. Geofluids. 2019;**2019**:1871392. DOI: 10.1155/2019/1871392

[19] Hosny R, Zahran A, Abotaleb A, Ramzi M, Mubarak MF, Zayed MA, et al. Nanotechnology impact on chemical-enhanced oil recovery: A review and bibliometric analysis of recent developments. ACS Omega. 2023;**8**:46325-46345. DOI: 10.1021/ACSOMEGA.3C06206/ASSET/IMAGES/LARGE/AO3C06206_0024.JPEG

[20] E. and M. National Academies of sciences, environmental engineering for the 21st century: Addressing grand challenges. Environmental Engineering in the 21st Century Faces Grand Challenges. 2018:1-121. DOI: 10.17226/25121

[21] Ozowe W, Daramola GO, Ekemezie IO. Petroleum engineering innovations: Evaluating the impact of advanced gas injection techniques on reservoir management. Magna Scientia Advanced Research and Reviews. 2024;**11**:299-310. DOI: 10.30574/MSARR.2024.11.1.0094

[22] Abou-Alfitooh SAM, El-Hoshoudy AN. Eco-friendly modified biopolymers for enhancing oil production: A review. Journal of Polymers and the Environment. 2023;**32**:2457-2483. DOI: 10.1007/S10924-023-03132-1

[23] Tackie-Otoo BN, Ayoub Mohammed MA, Yekeen N, Negash BM. Alternative chemical agents for alkalis, surfactants and polymers for Enhanced Oil Recovery: Research trend and prospects. Journal of Petroleum Science and Engineering. 2020;**187**:106828. DOI: 10.1016/J.PETROL.2019.106828

[24] Shaikhah D, Loise V, Angelico R, Porto M, Calandra P, Abe AA, et al. New trends in biosurfactants: From renewable origin to green enhanced oil recovery applications. Molecules. 2024;**29**:301. DOI: 10.3390/MOLECULES29020301

[25] Sen R. Biotechnology in petroleum recovery: The microbial EOR. Progress in Energy and Combustion Science. 2008;**34**:714-724. DOI: 10.1016/J.PECS.2008.05.001

[26] Nwabueze QA, Leggett S. Advancements in the application of CO2 capture and utilization technologies—A comprehensive review. Fuel. 2024;**5**:508-532. DOI: 10.3390/FUELS5030028

[27] Sagala F, Hethnawi A, Kajjumba GW, Nassar NN. Nanoparticles as potential agents for enhanced oil recovery. Lecture Notes in Nanoscale Science and Technology. Nanoparticles: An Emerging Technology for Oil Production Process Application. 2021:41-95. DOI: 10.1007/978-3-319-12051-5_2

[28] Shasiya PS, Simi Pushpan K, Nair AB. Biopolymers as Engineering

Materials. Singapore: Springer; 2022. DOI: 10.1007/978-981-16-6603-2_23-1

[29] Contreras-Mateus M, Hethnawi A, Mheibesh Y, Montoya T, Hashlamoun K, Bakir M, et al. Applications of nanoparticles in energy and the environment: Enhanced oil upgrading and recovery and cleaning up energy effluents. ACS Symposium Series. 2022;**1412**:169-267. DOI: 10.1021/BK-2022-1412.CH005

[30] Malozyomov BV, Martyushev NV, Kukartsev VV, Tynchenko VS, Bukhtoyarov VV, Wu X, et al. Overview of methods for enhanced oil recovery from conventional and unconventional reservoirs. Energies. 2023;**16**:4907. DOI: 10.3390/EN16134907

[31] Massarweh O, Abushaikha AS. Towards environmentally sustainable oil recovery: The role of sustainable materials. Energy Reports. 2024;**12**:95-119. DOI: 10.1016/J.EGYR.2024.06.013

[32] Mariyate J, Bera A. Paradigm shift towards the sustainability in upstream oil industry for enhanced recovery - a state-of-art review. Journal of Cleaner Production. 2023;**386**:135784. DOI: 10.1016/J.JCLEPRO.2022.135784

[33] Islam R, Khan MM, Chhetri AB. Green Petroleum: How Oil and Gas Can Be Environmentally Sustainable. John Wiley and Sons; 2012

[34] Ranjbar S, Sarlak N, Rashidi A. Development of a green and sustainable Nanofluid based on carbon dots with sensing capability for enhanced oil recovery. Energy & Fuels. 2022;**36**:12126-12134. DOI: 10.1021/ACS.ENERGYFUELS.2C02065/SUPPL_FILE/EF2C02065_SI_001.PDF

[35] Naghizadeh A, Abbaspour M. Formation damage in thermal-Enhanced Oil Recovery processes. In: Therm. Methods Enhanc. Oil Recover. Ser. Gulf Professional Publishing; 2023. pp. 385-404. DOI: 10.1016/B978-0-12-821933-1.00006-9

[36] Medina OE, Olmos C, Lopera SH, Cortés FB, Franco CA. Nanotechnology applied to thermal enhanced oil recovery processes: A review. Energies. 2019;**12**:4671. DOI: 10.3390/EN12244671

[37] Zhao F, Liu Y, Lu N, Xu T, Zhu G, Wang K. A review on upgrading and viscosity reduction of heavy oil and bitumen by underground catalytic cracking. Energy Reports. 2021;**7**:4249-4272. DOI: 10.1016/J.EGYR.2021.06.094

[38] Huang S, Cao M, Cheng L. Experimental study on the mechanism of Enhanced Oil Recovery by multi-thermal fluid in offshore heavy oil. International Journal of Heat and Mass Transfer. 2018;**122**:1074-1084. DOI: 10.1016/J.IJHEATMASSTRANSFER.2018.02.049

[39] Khoramian R, Issakhov M, Pourafshary P, Gabdullin M, Sharipova A. Surface modification of nanoparticles for enhanced applicability of nanofluids in harsh reservoir conditions: A comprehensive review for improved oil recovery. Advances in Colloid and Interface Science. 2024;**333**:103296. DOI: 10.1016/J.CIS.2024.103296

[40] Punase A, Zou A, Elputranto R. How do thermal recovery methods affect wettability alteration? Journal of Petroleum Science and Engineering. 2014;**2014**:538021. DOI: 10.1155/2014/538021

[41] Razavifar M, Almasiyan P, Rezaei A. EOR potentials in the poststeam-injected heavy oil reservoirs. In: Therm. Methods. Gulf Professional Publishing; 2023. pp. 355-384. DOI: 10.1016/B978-0-12-821933-1.00005-7

[42] Minakov AV, Meshkova VD, Guzey DV, Pryazhnikov MI. Recent advances in the study of In situ combustion for enhanced oil recovery. Energies. 2023;**16**:4266. DOI: 10.3390/EN16114266

[43] Antolinez JD, Miri R, Nouri A. In situ combustion: A comprehensive review of the current state of knowledge. Energies. 2023;**16**:6306. DOI: 10.3390/EN16176306

[44] Askarova A, Turakhanov A, Markovic S, Popov E, Maksakov K, Usachev G, et al. Thermal Enhanced Oil Recovery in deep heavy oil carbonates: Experimental and numerical study on a hot water injection performance. Journal of Petroleum Science and Engineering. 2020;**194**:107456. DOI: 10.1016/J.PETROL.2020.107456

[45] Zhao F, Wang K, Li G, Zhu G, Liu L, Jiang Y. A review of high-temperature foam for improving steam flooding effect: Mechanism and application of foam. Energy Technology. 2022;**10**:2100988. DOI: 10.1002/ENTE.202100988

[46] Temizel C, Aydin H, Hosgor FB, Yegin C, Kabir CS. Green energy sources reduce carbon footprint of oil and gas industry processes: A review. Journal of Energy and Power Technology. 2023;**5**:1-25. DOI: 10.21926/JEPT.2301004

[47] Ghoodjani SH, Kharrat R, Vossoughi M, Bolouri. A review on thermal enhanced heavy oil recovery from fractured carbonate reservoirs. Journal of Petroleum & Environmental Biotechnology. 2011;**2**:109. Available from: https://www.google.com/search?q=A+review+on+thermal+enhanced+heavy+oil+recovery+from+fractured+carbonate+reservoirs.&sca_esv=15988f96c695c280&source=hp&ei=wpCOZ5D5M9W6i-gP8qGweQ&iflsig=AL9hbdgAAAAZ46e0se3kzLyPZVsxBFbgxOxeUoRtSQN&ved=0ahUKEwiQ4oSD8oSLAxVV [Accessed: January 20, 2025]

[48] Muggeridge A, Cockin A, Webb K, Frampton H, Collins I, Moulds T, et al. Recovery rates, Enhanced Oil Recovery and technological limits. Philosophical Transactions of the Royal Society A - Mathematical Physical and Engineering Sciences. 2014;**372**. DOI: 10.1098/RSTA.2012.0320

[49] Liu Y, Liu X, Hou J, Li HA, Liu Y, Chen Z. Technical and economic feasibility of a novel heavy oil recovery method: Geothermal energy assisted heavy oil recovery. Energy. 2019;**181**:853-867. DOI: 10.1016/J.ENERGY.2019.05.207

[50] Kovscek AR. Emerging challenges and potential futures for thermally Enhanced Oil Recovery. Journal of Petroleum Science and Engineering. 2012;**98-99**:130-143. DOI: 10.1016/J.PETROL.2012.08.004

[51] Sedighi M, Aminian M, Mohammadi M, Koohi AD. Overview of thermal concepts in Enhanced Oil Recovery. In: Therm. Methods. Gulf Professional Publishing; 2023. pp. 1-45. DOI: 10.1016/B978-0-12-821933-1.00012-4

[52] Mogensen K, Masalmeh S. A review of EOR techniques for carbonate reservoirs in challenging geological settings. Journal of Petroleum Science and Engineering. 2020;**195**:107889. DOI: 10.1016/J.PETROL.2020.107889

[53] Kairgeldina LK, Sarsenbekuly B. Alternative methods of thermal oil recovery: A review. Kazakhstan Journal for the Oil and Gas Industry. 2024;**6**(1):50-63. DOI: 10.54859/KJOGI108692-82064

[54] Abuov Y, Serik G, Lee W. Techno-economic assessment and life cycle

assessment of CO2-EOR. Environmental Science & Technology. 2022;**56**:8571-8580. DOI: 10.1021/ACS.EST.1C06834

[55] Sharma T, Chaturvedi KR, Ganat T, Ali I. Viability of EOR when oil prices are low. In: Adv. Chem. Enhanc. Oil Recover. Burlington, ON, Canada: Apple Academic Press; 2025. Available from: https://www.routledge.com/Advancements-in-Chemical-Enhanced-Oil-Recovery/Sharma-Chaturvedi-Ganat-Ali/p/book/9781774915080 [Accessed: January 21, 2025]

[56] Espinola O, Nunez F, Malibran PT. Proposal EOR fiscal incentives to increase the recovery factor in Mexican fields implementing enhanced oil recovery methods through sensitivity analysis with different contracts. In: Soc. Pet. Eng. - SPE Trinidad Tobago Sect. Energy Resour. Conf. 2021, TTCE 2021. OnePetro; 2021. DOI: 10.2118/200916-MS

[57] Bera A, Vij RK, Shah S. Impact of newly implemented enhanced oil and gas recovery screening policy on current oil production and future energy supply in India. Journal of Petroleum Science and Engineering. 2021;**207**:109196. DOI: 10.1016/J.PETROL.2021.109196

[58] Shahab-Deljoo M, Medi B, Kazi MK, Jafari M. A techno-economic review of gas flaring in Iran and its human and environmental impacts. Process Safety and Environment Protection. 2023;**173**:642-665. DOI: 10.1016/J.PSEP.2023.03.051

[59] Garcia-Garcia G, Fernandez MC, Armstrong K, Woolass S, Styring P. Analytical review of life-cycle environmental impacts of carbon capture and utilization technologies. ChemSusChem. 2021;**14**:995-1015. DOI: 10.1002/CSSC.202002126

[60] Lee J, Babadagli T. Mitigating greenhouse gas intensity through new generation techniques during heavy oil recovery. Journal of Cleaner Production. 2021;**286**:124980. DOI: 10.1016/J.JCLEPRO.2020.124980

[61] Yuan B, Wood DA. A holistic review of geosystem damage during unconventional oil, gas and geothermal energy recovery. Fuel. 2018;**227**:99-110. DOI: 10.1016/J.FUEL.2018.04.082

[62] Fetanat A, Tayebi M. Sustainability and availability prioritization of solar thermal energy technologies for Enhanced Oil Recovery using T-spherical fuzzy set-based ordinal priority approach. Clean Technologies and Environmental Policy. 2024:1-30. DOI: 10.1007/S10098-024-03047-Y

[63] Lourdes RS, Cheng SY, Chew KW, Ma Z, Show PL. Prospects of microbial Enhanced Oil Recovery: Mechanisms and environmental sustainability. Sustainable Energy Technologies and Assessments. 2022;**53**:102527. DOI: 10.1016/J.SETA.2022.102527

[64] Zyrin V, Ilinova A. Ecology safety Technologies of Unconventional oil Reserves Recovery for sustainable oil and gas industry development. Journal of Ecological Engineering. 2016;**17**:35-40. DOI: 10.12911/22998993/64637

[65] Oruwari HO. Green technological approach for sustainable operations of oil and gas industry. In: Soc. Pet. Eng. - SPE Niger. Annu. Int. Conf. Exhib. NAIC 2023. OnePetro; 2023. DOI: 10.2118/217178-MS

[66] Kharazi M, Saien J, Torabi M, Zolfigol MA. Green nano multicationic ionic liquid based surfactants for enhanced oil recovery: A comparative study on design and applications. Journal of Molecular Liquids.

2023;383:122090. DOI: 10.1016/j.molliq.2023.122090

[67] Quadri SMR, Jiran L, Shoaib M, Hashmet MR, Al Sumaiti AM, Alhassan SM. Application of biopolymer to improve oil recovery in high temperature high salinity carbonate reservoirs, in: Soc. pet. Eng. - Abu Dhabi Int. pet. Exhib. Conf. ADIPEC. Society of Petroleum Engineers. 2015:2015. DOI: 10.2118/177915-MS

[68] Gao C. Application of a novel biopolymer to enhance oil recovery. Journal of Petroleum Exploration and Production Technologies. 2016;**6**:749-753. DOI: 10.1007/S13202-015-0213-7

[69] Massarweh O, Abushaikha AS. The use of surfactants in Enhanced Oil Recovery: A review of recent advances. Energy Reports. 2020;**6**:3150-3178. DOI: 10.1016/J.EGYR.2020.11.009

[70] Hu B, Zhao JY, Nie Y, Qin XY, Zhang KD, Xing JM, et al. Bioemulsification and microbial community reconstruction in thermally processed crude oil. Microorganisms. 2021;**9**. DOI: 10.3390/microorganisms9102054

[71] Mathews TA, Cortes J, Hascakir B. Evaluation of environmentally friendly green solvents for the recovery of heavy oils. In: Proc. - SPE Symp. Improv. Oil Recover., Society of Petroleum Engineers (SPE). 2022. DOI: 10.2118/209433-MS

[72] Armenta S, Esteve-Turrillas FA, Garrigues S, de la Guardia M. Alternative green solvents in sample preparation. Green Analytical Chemistry. 2022;**1**:100007. DOI: 10.1016/J.GREEAC.2022.100007

[73] Le Nian Y, Cheng WL. Insights into heat transport for thermal oil recovery. Journal of Petroleum Science and Engineering. 2017;**151**:507-521. DOI: 10.1016/j.petrol.2017.01.026

[74] Ying S, Guan Z, Ofoegbu PC, Clubb P, Rico C, He F, et al. Green synthesis of nanoparticles: Current developments and limitations. Environmental Technology and Innovation. 2022;**26**:102336. DOI: 10.1016/J.ETI.2022.102336

[75] Singh N, Ogunseitan OA, Wong MH, Tang Y. Sustainable materials alternative to petrochemical plastics pollution: A review analysis. Sustainable Horizons. 2022;**2**:100016. DOI: 10.1016/J.HORIZ.2022.100016

[76] Bera A. Chapter 6 - Applications of ionic liquids as green solvents in enhanced oil recovery. In: Inamuddin D, T.B.T.-G.S.P. for C., E.E., Altalhi S, editors. Elsevier; 2023. pp. 125-144. DOI: 10.1016/B978-0-323-95156-2.00016-7

[77] Al-Ghamdi A, Haq B, Al-Shehri D, Muhammed NS, Mahmoud M. Surfactant formulation for green enhanced oil recovery. Energy Reports. 2022;**8**:7800-7813. DOI: 10.1016/J.EGYR.2022.05.293

[78] Pothula GK, Vij RK, Bera A. An overview of chemical Enhanced Oil Recovery and its status in India. Petroleum Science. 2023;**20**:2305-2323. DOI: 10.1016/J.PETSCI.2023.01.001

[79] Speight JG, El-Gendy NS. Microbial enhanced oil recovery. Introduction to Pet Biotechnology. 2018:103-130. DOI: 10.1016/B978-0-12-805151-1.00004-7

[80] Sun J, Liu Y, Zhu X, Hu F, Wang Y, Yi X, et al. Surfactant–polymer flooding: Chemical formula design and evaluation for high-temperature and high-salinity Qinghai Gasi reservoir. PRO. 2024;**12**:1082. DOI: 10.3390/PR12061082

[81] Phukan R, Gogoi SB, Tiwari P. Enhanced oil recovery by alkaline-surfactant-alternated-gas/CO_2 flooding. Journal of Petroleum Exploration and Production Technologies. 2019;**9**:247-260. DOI: 10.1007/S13202-018-0465-0/FIGURES/6

[82] Wei J, Li J, Shi X, Zhou R. Combination of alkali–surfactant–polymer flooding and horizontal wells to maximize the oil recovery for high water cut oil reservoir. Energy Reports. 2021;**7**:5955-5964. DOI: 10.1016/J.EGYR.2021.Ø9.019

[83] Shakeel M, Sagandykova D, Mukhtarov A, Dauyltayeva A, Maratbekkyzy L, Pourafshary P, et al. Maximizing oil recovery: Innovative chemical EOR solutions for residual oil mobilization in Kazakhstan's waterflooded sandstone oilfield. Heliyon. 2024;**10**:e28915. DOI: 10.1016/J.HELIYON.2024.E28915

[84] Ahmadi M, Chen Z. Challenges and future of chemical assisted heavy oil recovery processes. Advances in Colloid and Interface Science. 2020;**275**:102081. DOI: 10.1016/J.CIS.2019.102081

[85] Kamal MS, Hussein IA, Sultan A. Review on Chemical EOR for High Temperature/High Salinity Reservoir Conditions|Request PDF. Int. EOR Heavy Oil Conf; 2013. Available from: https://www.researchgate.net/publication/264337916_Review_on_Chemical_EOR_for_high_temperaturehigh_salinity_reservoir_conditions [Accessed: January 22, 2025]

[86] Uzoho CU, Onyekonwu M, Akaranta O. Formulation of local alkaline-surfactant-polymer (asp) for Enhanced Oil Recovery in Niger delta: A review. In: Soc. Pet. Eng. - SPE Niger. Annu. Int. Conf. Exhib. NAICE 2015. Society of Petroleum Engineers; 2015. DOI: 10.2118/178300-MS

[87] Lu X, Cao B, Xie K, Cao W, Liu Y, Zhang Y, et al. Enhanced oil recovery mechanisms of polymer flooding in a heterogeneous oil reservoir. Petroleum Exploration and Development. 2021;**48**:169-178. DOI: 10.1016/S1876-3804(21)60013-7

[88] Zekri AY, Jerbi KK. Economic evaluation of Enhanced Oil Recovery. Oil & Gas Science and Technology. 2002;**57**:259-267. DOI: 10.2516/OGST:2002018

[89] Al-Murayri MT, Al-Mayyan HE, Moudi K, Al-Ajmi F, Pitts D, Wyatt MJ, et al. Chemical EOR economic evaluation in a low oil price environment: Sabriyah lower burgan reservoir case study. In: Society of Petroleum Engineers (SPE) - SPE EOR Conference at Oil and Gas West Asia 2018. Society of Petroleum Engineers; 2018. DOI: 10.2118/190337-MS

[90] Johnston J, Cushing L. Chemical exposures, health and environmental justice in communities living on the fenceline of industry. Current Environmental Health Reports. 2020;**7**:48. DOI: 10.1007/S40572-020-00263-8

[91] Wang Y, Ren S, Zhang L, Hu C. Energy efficiency and greenhouse gas emissions of current steam injection process and promising steam based techniques for heavy oil reservoirs. Journal of Petroleum Science and Engineering. 2018;**166**:842-849. DOI: 10.1016/J.PETROL.2018.03.094

[92] Rizvi SMH. Nanotechnology applications in enhanced oil recovery (EOR). International Journal of Scientific and Management Research. 2024;**12**:135-143. DOI: 10.18535/IJSRM/V12I06.C03

[93] Pinaeva LG, Noskov AS. Biodegradable biopolymers: Real

impact to environment pollution. Science of the Total Environment. 2024;**947**:174445. DOI: 10.1016/J.SCITOTENV.2024.174445

[94] Bera H, Abbasi YF, Hasnain MS, Nayak AK. Sterculia gum in drug delivery applications. Natural Polysaccharides in Drug Delivery and Biomedical Applications. 2019:223-247. DOI: 10.1016/B978-0-12-817055-7.00009-1

[95] Speight JG, El-Gendy NS. Chapter 4 - Microbial enhanced oil recovery. In: Speight JG, N.S.B.T.-I. to El-Gendy, editors. Boston: Gulf Professional Publishing; 2018. pp. 103-130. DOI: 10.1016/B978-0-12-805151-1.00004-7

[96] Sveistrup M, van Mastrigt F, Norrman J, Picchioni F, Paso K. Viability of biopolymers for enhanced oil recovery. Journal of Dispersion Science and Technology. 2016;**37**:1160-1169. DOI: 10.1080/01932691.2015.1088450

[97] Bordoloi NK, Konwar BK. Bacterial biosurfactant in enhancing solubility and metabolism of petroleum hydrocarbons. Journal of Hazardous Materials. 2009;**170**:495-505. DOI: 10.1016/J.JHAZMAT.2009.04.136

[98] Sagir M, Mushtaq M, Tahir MS, Tahir MB, Shaik AR. Surfactants for Enhanced Oil Recovery applications. Surfactant Enhanced Oil Recovery Application. 2020:1-129. DOI: 10.1007/978-3-030-18785-9

[99] Nagtode VS, Cardoza C, Yasin HKA, Mali SN, Tambe SM, Roy P, et al. Green surfactants (biosurfactants): A petroleum-free substitute for sustainability—Comparison. Applications, Market, and Future Prospects, ACS Omega. 2023;**8**:11674. DOI: 10.1021/ACSOMEGA.3C00591

[100] da Faccioli YES, de Oliveira KW, Campos-Guerra JM, Converti A, de Soares da Silva RCF, Sarubbo LA. Biosurfactants: Chemical properties. Ecofriendly Environmental Applications, and Uses in the Industrial Energy Sector, Energies. 2024;**17**:5042. DOI: 10.3390/EN17205042

[101] Saxena N, Goswami A, Dhodapkar PK, Nihalani MC, Mandal A. Bio-based surfactant for Enhanced Oil Recovery: Interfacial properties, emulsification and rock-fluid interactions. Journal of Petroleum Science and Engineering. 2019;**176**:299-311. DOI: 10.1016/J.PETROL.2019.01.052

[102] Rehman S, Ali M, Zuber M, Zia KM, Iqbal R. Future prospects of algae-based materials. In: Algae Based Polymers, Blends, and Composites: Chemistry, Biotechnology and Material Sciences. Elsevier; 2017. pp. 687-691. DOI: 10.1016/B978-0-12-812360-7.00020-3

[103] Ogunyemi AT, Eluwa TD, Adeyi AA, Giwa A, Igbafe AI, Gbadamosi AO. Enhanced oil recovery using novel bio-based surfactant synthesized from algae. In: Soc. Pet. Eng. - SPE Niger. Annu. Int. Conf. Exhib. NAIC 2024. Society of Petroleum Engineers; 2024. DOI: 10.2118/221609-MS

[104] Tello P, Santos J, Calero N, Trujillo-Cayado LA. Formulation and characterization of sustainable algal-derived Nanoemulgels: A green approach to minimize the dependency on synthetic surfactants. Polymers (Basel). 2024;**16**:194. DOI: 10.3390/POLYM16020194

[105] Khan I, Saeed K, Khan I. Nanoparticles: Properties, applications and toxicities. Arabian Journal of Chemistry. 2019;**12**:908-931. DOI: 10.1016/J.ARABJC.2017.05.011

[106] Ali JA, Kolo K, Manshad AK, Mohammadi AH. Recent advances in application of nanotechnology in chemical Enhanced Oil Recovery: Effects of nanoparticles on wettability alteration, interfacial tension reduction, and flooding. Egyptian Journal of Petroleum. 2018;**27**:1371-1383. DOI: 10.1016/J.EJPE.2018.09.006

[107] Sircar A, Rayavarapu K, Bist N, Yadav K, Singh S. Applications of nanoparticles in Enhanced Oil Recovery. Petroleum Research. 2022;**7**:77-90. DOI: 10.1016/J.PTLRS.2021.08.004

[108] Weiss M, Haufe J, Carus M, Brandão M, Bringezu S, Hermann B, et al. A review of the environmental impacts of biobased materials. Journal of Industrial Ecology. 2012;**16**. DOI: 10.1111/J.1530-9290.2012.00468.X

[109] Sapuan SM. Materials selection for composites: Concurrent engineering perspective. In: Compos. Mater. Elsevier; 2017. pp. 209-271. DOI: 10.1016/B978-0-12-802507-9.00006-4

[110] Mehmood S, Zaman K, Khan S, Ali Z, Khan HURR. The role of green industrial transformation in mitigating carbon emissions: Exploring the channels of technological innovation and environmental regulation. Energy and Built Environment. 2024;**5**:464-479. DOI: 10.1016/J.ENBENV.2023.03.001

[111] Belhaj H, Abukhalifeh H, Javid K. Miscible oil recovery utilizing N2 and/or HC gases in CO2 injection. Journal of Petroleum Science and Engineering. 2013;**111**:144-152. DOI: 10.1016/J.PETROL.2013.08.030

[112] Cleveland CJ, Morris C. Miscible gas injection. In: Handb. Energy Chronol. Top Ten List. Word Clouds. Elsevier; 2014. pp. 85-131. DOI: 10.1016/B978-0-12-417013-1.00006-6

[113] Enab K, Elizondo T, Elmasry Y, Flores L, Addo-Mensah A. Experimental investigation of liquid, supercritical CO2, CH4, and CO2/CH4 mixture to improve oil recovery. In: International PET Technologies Conference IPTC 2024. International Petroleum Technology Conference (IPTC); 2024. DOI: 10.2523/IPTC-24523-MS

[114] Al Hinai NM, Saeedi A. Miscible displacement oil recovery. In: Enhanced Oil Recovery. IntechOpen; 2022. DOI: 10.5772/INTECHOPEN.105757

[115] Vishnyakov V, Suleimanov B, Salmanov A, Zeynalov E. Oil recovery stages and methods. In: Primer on Enhanced Oil Recovery. Elsevier; 2020. pp. 53-63. DOI: 10.1016/B978-0-12-817632-0.00007-4

[116] Wang L, Mohanty K. Enhanced oil recovery in gas-flooded, carbonate reservoirs by wettability-altering surfactants. In: Proceedings - SPE Annual Technical Conference and Exhibition. Society of Petroleum Engineers (SPE); 2013. pp. 2448-2464. DOI: 10.2118/166283-MS

[117] Zendehboudi S, Bahadori A. Production methods in shale oil reservoirs. In: Shale Oil Gas Handbook. Elsevier; 2017. pp. 285-319. DOI: 10.1016/B978-0-12-802100-2.00008-3

[118] Alvarado V, Manrique E. Enhanced oil recovery concepts. In: Enhanced Oil Recover. Elsevier; 2010. pp. 7-16. DOI: 10.1016/B978-1-85617-855-6.00008-5

[119] Anikin OV, Derevyanko VK, Bolotov AV, Minkhanov IF, Varfolomeev MA, Konstantinovich EA, et al. Gas Enhanced Oil Recovery methods for offshore oilfields: Features, implementation, operational status.

Petroleum Research. 2024) In Press. DOI: 10.1016/J.PTLRS.2024.12.003

[120] Cleveland CJ, Morris C. Section 6 – Oil. In: Cleveland CJ, C.B.T.-H. of Morris E, editors. Boston: Elsevier; 2014. pp. 85-131. DOI: 10.1016/B978-0-12-417013-1.00006-6

[121] Santos R, Sgouridis S, Alhajaj A. Potential of CO2-Enhanced Oil Recovery coupled with carbon capture and storage in mitigating greenhouse gas emissions in the UAE. International Journal of Greenhouse Gas Control. 2021;**111**:103485. DOI: 10.1016/J.IJGGC.2021.103485

[122] Khoshsima A, Sedighi M, Mohammadi M. Enhanced oil recovery by water alternating gas injection. In: Gas Inject. Methods a Vol. Enhanc. Oil Recover. Ser. Elsevier; 2022. pp. 295-316. DOI: 10.1016/B978-0-12-822302-4.00001-6

[123] Ahmed T. Principles of Waterflooding. In: Reservation English Handbook. Elsevier; 2010. pp. 909-1095. DOI: 10.1016/B978-1-85617-803-7.50022-5

[124] Ozowe W, Daramola GO, Ekemezie IO. Recent advances and challenges in gas injection techniques for Enhanced Oil Recovery. Magna Scientia Advanced Research and Reviews. 2023;**9**:168-178. DOI: 10.30574/MSARR.2023.9.2.0180

[125] McCoy ST, Rubin ES. The effect of high oil prices on EOR project economics. Energy Procedia. 2009;**1**:4143-4150. DOI: 10.1016/J.EGYPRO.2009.02.223

[126] Alam MMM, Hassan A, Mahmoud M, Sibaweihi N, Patil S. Dual benefits of enhanced oil recovery and CO2 sequestration: The impact of CO2 injection approach on oil recovery. Frontiers in Energy Research. 2022;**10**:877212. DOI: 10.3389/FENRG.2022.877212/BIBTEX

[127] Mortezaei K, Amirlatifi A, Ghazanfari E, Vahedifard F. Potential CO2 leakage from geological storage sites: Advances and challenges. Environmental Geotechnics. 2017;**8**:3-27. DOI: 10.1680/JENGE.18.00041

[128] Diaz OC, Katiyar A, Hassanzadeh A, Crosley M, Knight T, Rozowski P. Evaluation of carbon footprint for a hydrocarbon foam EOR field pilot. In: Proc. - SPE Symp. Improv. Oil Recover. OnePetro; 2022. DOI: 10.2118/209366-MS

[129] Alsheyab MAT. Recycling of construction and demolition waste and its impact on climate change and sustainable development. International journal of Environmental Science and Technology. 2022;**19**:2129-2138. DOI: 10.1007/S13762-021-03217-1

[130] He M, Luis S, Rita S, Ana G, Euripedes V, Zhang N. Risk assessment of CO2 injection processes and storage in carboniferous formations: A review. Journal of Rock Mechanics and Geotechnical Engineering. 2011;**3**:39-56. DOI: 10.3724/SP.J.1235.2011.00039

[131] Maleki M, Mehrjoo H, Kazemzadeh Y, Mohmmadinia F, Ranjbar A. Green innovation in Enhanced Oil Recovery: Pioneering sustainable solution for a cleaner future. Environment and Earth Science. 2025;**84**:1-26. DOI: 10.1007/S12665-024-12061-8

[132] Zuiderveen EAR, Kuipers KJJ, Caldeira C, Hanssen SV, van der Hulst MK, de Jonge MMJ, et al. The potential of emerging bio-based products to reduce environmental impacts. Nature Communications.

2023;**14**:1-7. DOI: 10.1038/s41467-023-43797-9

[133] Yasemi S, Khalili Y, Sanati A, Bagheri M. Carbon capture and storage: Application in the oil and gas industry. Sustainability. 2023;**15**:14486. DOI: 10.3390/SU151914486

[134] Saravanan A, Kumar PS, Vardhan KH, Jeevanantham S, Karishma SB, Yaashikaa PR, et al. A review on systematic approach for microbial Enhanced Oil Recovery technologies: Opportunities and challenges. Journal of Cleaner Production. 2020;**258**:120777. DOI: 10.1016/J.JCLEPRO.2020.120777

[135] Goren AY, Dincer I, Gogoi SB, Boral P, Patel D. Recent developments on carbon neutrality through carbon dioxide capture and utilization with clean hydrogen for production of alternative fuels for smart cities. International Journal of Hydrogen Energy. 2024;**79**:551-578. DOI: 10.1016/J.IJHYDENE.2024.06.421

[136] Sikiru S, Soleimani H, Yusuf JY, Hassan YM, Hamza MF, Singh R. Recent advance and Prospect of enhanced oil recovery mechanisms in reservoir sandstone. Energy and Fuel Technology. 2023. DOI: 10.20944/PREPRINTS202309.1654.V2

[137] Tileuberdi N, Gussenov I. Review on miscible, immiscible, and progressive nitrogen injection for Enhanced Oil Recovery. Energy Reports. 2024;**12**:360-367. DOI: 10.1016/J.EGYR.2024.06.004

[138] Singh N, Farina I, Petrillo A, Colangelo F, De Felice F. Carbon capture, sequestration, and usage for clean and green environment: Challenges and opportunities. International Journal of Sustainable Engineering. 2023;**16**:248-268. DOI: 10.1080/19397038.2023.2256379

[139] Purwasena IA, Amaniyah M, Astuti DI, Firmansyah Y, Sugai Y. Production, characterization, and application of Pseudoxanthomonas taiwanensis biosurfactant: A green chemical for microbial Enhanced Oil Recovery (MEOR). Scientific Reports. 2024;**14**:1-13. DOI: 10.1038/s41598-024-61096-1

[140] Ismail NL, Shahruddin S, Othman J, Ismail NL, Shahruddin S, Othman J. Perspective chapter: Overview of bio-based surfactant – Recent development, industrial challenge, and future outlook. In: Surfactants Deterg. - Updat. New Insights. London, UK: IntechOpen; 2022. DOI: 10.5772/INTECHOPEN.100542

[141] Panchal H, Patel H, Patel J, Shah M. A systematic review on nanotechnology in Enhanced Oil Recovery. Petroleum Research. 2021;**6**:204-212. DOI: 10.1016/J.PTLRS.2021.03.003

[142] Hu X, Li M, Peng C, Davarpanah A. Hybrid thermal-chemical enhanced oil recovery methods; an experimental study for tight reservoirs. Symmetry. 2020;**12**:947. DOI: 10.3390/SYM12060947

[143] Calderón AJ, Pekney NJ. Optimization of Enhanced Oil Recovery operations in unconventional reservoirs. Applied Energy. 2020;**258**:114072. DOI: 10.1016/J.APENERGY.2019.114072

[144] Ahmed A. Geological and Environmental Implications of Enhanced Oil Recovery Techniques: Balancing Efficiency with Sustainability, EaArX; 2024, X5M69G. DOI: 10.31223/X5M69G

[145] Kobayashi T, Nakajima L. Sustainable development goals for advanced materials provided by industrial wastes and biomass sources. Current Opinion in Green and Sustainable Chemistry. 2021;**28**:100439. DOI: 10.1016/J.COGSC.2020.100439

[146] Mapping the Oil and Gas Industry to the Sustainable Development Goals: An Atlas – SDGs. Department of Economic and Social Affairs; n.d. pp. 1-17

[147] Omer MAB, Noguchi T. A conceptual framework for understanding the contribution of building materials in the achievement of sustainable development goals (SDGs). Sustainable Cities and Society. 2020;**52**:101869. DOI: 10.1016/J.SCS.2019.101869

[148] Annual report. Educational Research. 2003;**32**:44-54. DOI: 10.3102/0013189x032009044

Chapter 3

Innovation Technique for Enhanced Oil Recovery: Sustainable Approaches and Environmental Considerations

Bongiwe Ladyfair Hem

Abstract

Enhance oil recovery development is under the intervention of rapid change in the increase of sustainability efforts and technological innovation. Data analytics and nanotechnology are being used to optimize enhanced oil recovery (EOR) processes, enhancing recovery efficiency and resource efficiency. Renewable energy sources, such as solar thermal, are being integrated into EOR operations, complementing global decarbonization efforts. The policy and regulatory support are also important for sustainable EOR, including water treatment and emission reductions. While innovation is ongoing, EOR could also bridge conventional fossil fuel production and clean energy sources.

Keywords: enhance, oil, recovery, sustainability, environmental

1. Introduction

Advanced extraction method applied to recover oil that cannot be retrieved through primary and secondary recovery processes, but tertiary recovery is called enhanced oil recovery (EOR) [1]. EOR has become a foundation of the global oil industry, driven by its capacity to meet rising energy demands in the face of reducing conventional resources. While traditional extraction methods leverage natural reservoir pressure and water injection to push oil to the surface, EOR techniques aim to maximize oil extraction by altering the properties of the reservoir fluid. Utilizing advanced recovery techniques, EOR can boost oil recovery rates by up to 30–60%, making it an essential tool for stabilizing supply in a competitive and fluctuating energy market [2].

The risks of natural ecosystems that are linked to standard EOR methods, which result in environmental impacts, such as elevated carbon emissions, increased water consumption, and the risk of chemical contamination, need consideration [3]. The noted challenges require sustainable and environmentally friendly methods that can balance a long-term oil recovery with long-term ecological stability. Low-carbon thermal recovery, biodegradable chemical injections, and renewable energy aim to minimize environmental impact and reduce greenhouse emission while maintaining

EOR extraction methods [4]. The development of minimizing environmental impact is important for also compliance, cost reduction, and feasibility of oil production [5]. This chapter examines recent advancements in enhanced oil recovery (EOR), focusing on sustainable methods and environmental factors crucial for reducing impact while addressing global energy requirements.

1.1 Background

Reducing oil viscosity and interfacial tension after primary and secondary recovery stages, while improving oil efficiency is what enhanced oil recovery (EOR) intends to achieve [6]. EOR research has developed over the years with advanced process that requires compliance standards due to population growth. Forty years ago, carbon dioxide EOR was developed to increase the percentage of oil that's within the reservoir [7]. Nonetheless, the 1950s witnessed notable developments in chemical injection and thermal recovery, establishing the groundwork for modern OER, aimed at maximizing recovery and managing declining production rates [8].

1.2 Fundamentals of enhanced oil recovery

The following summarizes the stages of oil production that lead to improved oil recovery [9]:

a. Initial recovery: this is the first phase of oil production from a reservoir, during which natural processes such as drainage, water migration, gas expansion, and oil expansion push oil to the surface [10].

b. Secondary recovery: recovers up to 40% of oil by employing surface and submerged pumps to mechanically lift fluids in the reservoir, as well as water injection, natural gas reinjection, and gas lift [10].

c. Third recovery, also known as improved oil recovery, is a method of boosting production and field life by injecting CO2 into exhausted oil or gas reservoirs. Accordingly, the recovery is increased to 30–60% [11].

When enhanced oil recovery (EOR) techniques are used to increase both the amount of oil recovered and the pace of recovery, two parts can be identified (**Figure 1**): both traditional and non-traditional [13].

1.3 Conventional

Conventional oil recovery involves natural flow, artificial lift, water floods, and pressure maintenance, as illustrated in **Figure 1** [12]. Traditional conventional oil recovery often utilizes well drilling in those fields where a gas drive mechanism pushes the oil into production wells. However, the problem of pressure depletion can easily arise much earlier in the life of such a reservoir. In order to delay this process, reservoir engineers generally reinject the produced gas to maintain pressure.

According to Geologist skills [14], when a strong water drive exists, water flooding is commonly used once oil production slows. The objective is to push the remaining oil into production wells. Dry gas injection is done into the same types of porous (**Figure 2**).

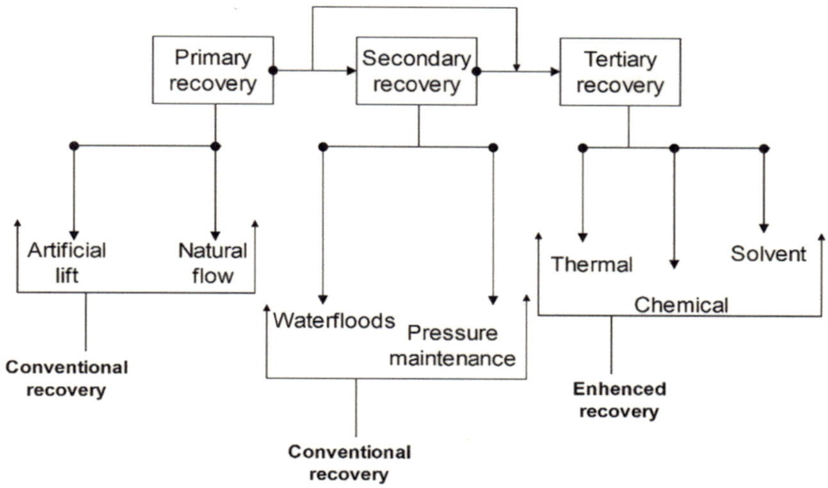

Figure 1.
Fundamental of recovery [12].

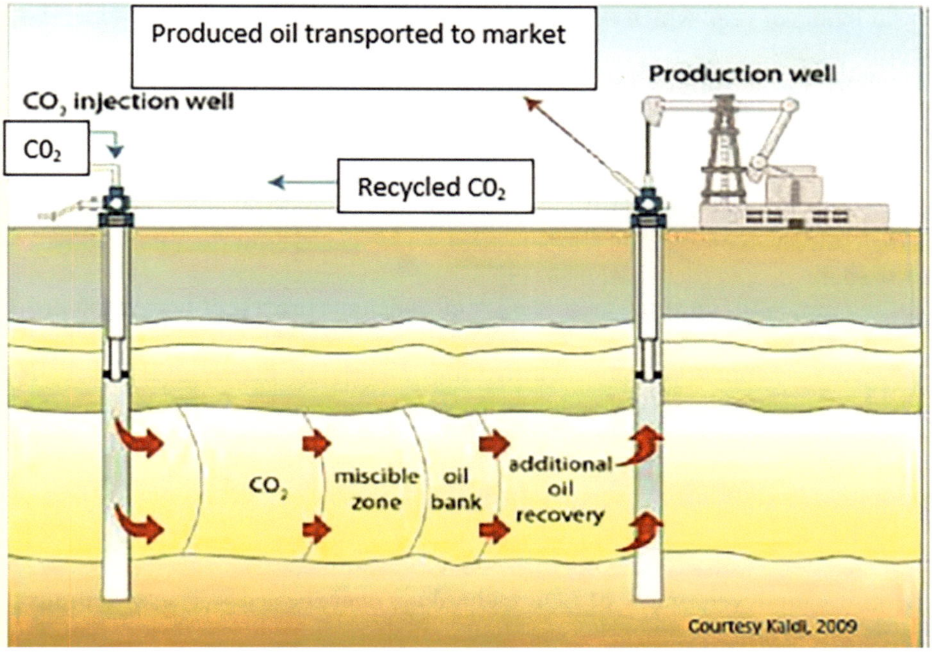

Figure 2.
Traditional conventional oil recovery [15].

Conventional EOR methods can be categorized into the following groups:

Thermal EOR: By heating the reservoir, the oil's viscosity is decreased, and its flow is enhanced with two popular techniques, which are combustion and steam injection [16].

Gas Injection EOR: Gases used in this process include CO_2, nitrogen, or natural gas, which can either maintain reservoir pressure or reduce oil viscosity, facilitating improved recovery [8].

1.4 Unconventional EOR

Unconventional EOR methods are those using innovative technologies and sustainability-driven approaches to extract oil more effectively. Examples include low-salinity water flooding, which lowers the interfacial tension between water and oil; nanotechnology-based EOR, in which nanoparticles change the ability to absorb water within the rock elements; and microbial EOR, which uses bacteria to improve oil recovery by producing surfactants or gases [17]. In recent years, such unconventional approaches have found wider application as sustainable and potentially cost-effective methods, especially in environmentally sensitive or complex reservoirs [18].

2. Economic and technical challenges in EOR

EOR presents great economic and technical challenges, along with its potential. This is due to most of the EOR projects are capital-intensive, they requiring huge upfront costs in deploying technology, chemicals, and infrastructure, which may not be economically viable in a low oil price environment. In addition, EOR is usually energy-intensive, with thermal methods such as steam injection requiring huge amounts of fuel, which contributes to greenhouse gas emissions and increases operational costs.

Technically, the different EOR techniques need to be tailored with care for a reservoir whose temperature, pressure, rock composition, and fluid properties may be different from each other. This requires thorough reservoir characterization and modeling. The main reasons are therefore complexity and high implementation cost. Additionally, the majority of EOR compounds are vulnerable to harsh tank conditions like high salt levels or elevated temperatures, which could limit their use and necessitate a sophisticated formulation[8, 19].

In general, EOR is one of the necessary means in the process of increasing production from mature and complex reservoirs; it has constant needs for innovation to mitigate economic and technical difficulties. Developments in sustainable EOR methods are needed to mitigate environmental impact and minimize costs for the long-term sustainability of oil fields worldwide.

2.1 Innovative methods of enhanced oil recovery

The need to produce more oil with fewer environmental impacts has resulted in a significant development of EOR technology over the years. These methods are very important in recovering more oil in mature or problematic reservoirs and include a wide range of approaches tailored to specific reservoir conditions. This section highlights the latest innovations across chemical, thermal, gas injection, and microbial EOR techniques, with a focus on sustainable and eco-friendly practices.

2.2 Chemical EOR techniques

To improve the process of oil movement or recovery, fluids containing active agents are injected into the formation using physical and chemical EOR techniques. Surfactants, polymers, alkalis, sulfuric acid, carbon dioxide, and micellar solutions are examples of active agents [20]. However, solvents improve mobility by lowering the interfacial tension between water and oil. Polymers improve the flooding process's sweep efficiency by making water more viscous [18].

While carbon dioxide dissolves in the oil, decreasing its viscosity and causing it to swell for easier recovery, alkaline agents react with some crudes to create *in situ* surfactants [19]. Micellar solutions blend surfactants and co-surfactants to make microemulsions that displace trapped oil with high efficiency [21]. These methods are tailored to the reservoir characteristics in order to maximize recovery.

2.3 Flooding with polymers

In order to improve the efficiency of oil displacement by decreasing the mobility ratio between water and oil, polymer flooding involves injecting water-soluble polymers, usually polyacrylamides, which raise the viscosity of the injected water [8]. Although polymers increase the viscosity of injected water, they improve sweep efficiency and decrease water channeling, which promotes the mobilization and recovery of oil that would otherwise be trapped (**Figure 3**) [21].

The reason for the addition of polymers in water is to increase viscosity, reducing permeability due to mechanical entrapment, which will result in decreased mobility in water [22]. Recently, there has been progress in developing high levels of salt alongside elevated heat-resistant polymers in addressing challenging reservoir conditions [18].

2.3.1 Surfactant and alkaline flooding

According to references [23, 24], surfactant and alkaline flooding is the method of delivering alkali, surfactant, and polymer onto the same drum. Surfactants improve the mobilization of oil droplets by reducing the interfacial tension between water and oil [25].

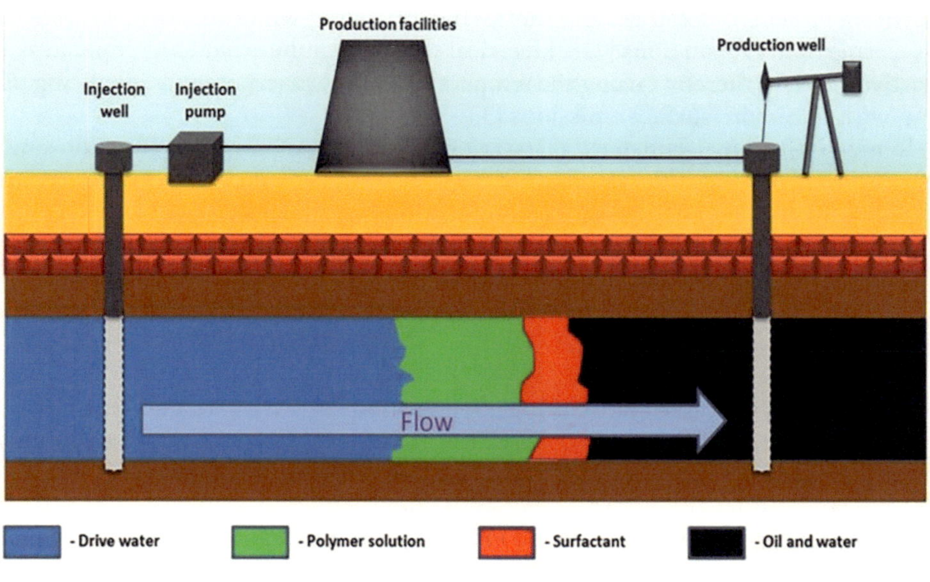

Figure 3.
Polymer flooding [22].

According to reference [8], this combination can greatly improve oil recovery, especially in reservoirs with highly salinized water.

To address environmental concerns, much research has been conducted on the development of biodegradable and eco-friendly surfactants and polymers from natural sources, such as lignin, cellulose, and starch-based materials [26]. These green chemicals aim to reduce the environmental footprint of chemical EOR and lower the risk of contamination of water sources [18].

2.3.2 Thermal EOR techniques

Conventional thermal methods, such as ordinary steam injection, do not apply for outsized oils from underground storage with a light coat [27]. To address this challenge, the following we looked into expect were looked into details:

2.3.2.1 Steam injection and its environmental impact

One of the most often used thermal EOR techniques is vapor infusion, primarily in the form of cyclic vapor stimulation and steam flooding [28]. In order to heat the reservoir and, as a result, reduce the viscosity of the oil for easier flow toward the production wells, steam at high temperatures is introduced into it [29]. Active agents and gases have been put in place to strategize how the optimization of steam injection can be improved; various works have explored the use of steam in combination with carbonate minerals as catalysts in improving combustion processes for improved recovery rates of heavy oil [30].

Steam injection may be a good way, but it does need massive energy investments and gives rise to carbon emissions during the combustion process for steam generation, thereby posing colossal environmental concerns [19]. Thus, in response to the environmental challenges of steam injection, microwave-assisted and electrical heating techniques have certainly emerged as possible alternatives [31]. Microwave heating produces heat right within the reservoir, thus reducing the energy required to achieve target temperatures and improving efficiency, while it has a major role in low-permeability formations [32]. Electrical heating, on the other hand, operates on resistive heating, thereby raising the temperature of the reservoir and enhancing oil flow, with minimum surface emissions [33].

Innovative thermal-enhanced oil recovery solutions for emission reduction consist of incorporating renewable energy sources such as solar steam generation, where solar energy is used to produce steam for injection [34]. Solar-assisted EOR has shown promise in reducing carbon footprints from steam injection, especially with respect to sunny areas while providing cheaper fuel use [4].

2.3.2.2 Gas injection EOR techniques

In particular, the three main gas injection techniques, CO_2, nitrogen, and hydrocarbon gas injection, have revolutionized the oil business by increasing the recovery rate and production life cycle [35]. According to reference [36], carbon dioxide (CO_2) is a renewable, inexpensive, abundant, and harmless material used in organic synthesis. Coinjection is the term used in the context of enhanced oil recovery (EOR) to describe the injection of CO_2 along with certain additives that create new ways to maximize recovery, reduce costs, and increase environmental benefits by utilizing the agents' synergies [17].

2.3.2.2.1 Water-substituting gas infusion using CO_2

Substituting gas infusion is a popular EOR technique that uses alternate gas and water injection to increase oil recovery and sweep efficiency (**Figure 4**) [38]:

- *Benefits:* Combining the advantages of water-substituting gas infusion with CO_2 helps regulate movement and reduces the influence of viscosity, increasing the extraction of oil [38].

- *Innovation*: The optimum well controls the ratios of CO_2 and water injection [39].

- *Studies:* An investigation titled water-alternating-gas injection with chemical assistance for increased sweep efficiency in CO_2-EOR was performed by Fang et al. [37]. Chemical-assisted CO_2-WAG has various advantages relative to conventional methods, including an enhanced mobility ratio, sealing off fractures that may otherwise extend the production life in some cases, and easier access to production zones tightly connected by permeable channels. The technologies are shown to be critical in developing heterogeneous reservoirs economically, as proven by the recovery factors above 5% beyond conventional methods.

2.3.2.2.2 CO_2-polymer coinjection

The use of CO_2 and polymers for coinjection into the reservoir was recently developed as an EOR alternative that involves using CO_2 with polymers for mobility control of CO_2 to enhance the sweep efficiency and maximize the recovery [40].

- *Benefits:* The CO_2-polymer coinjection has several advantages, which include good operational performance, good behavior of the esthetic, cost-effective, inclusion of other types of polymers with unique characteristics, enhanced

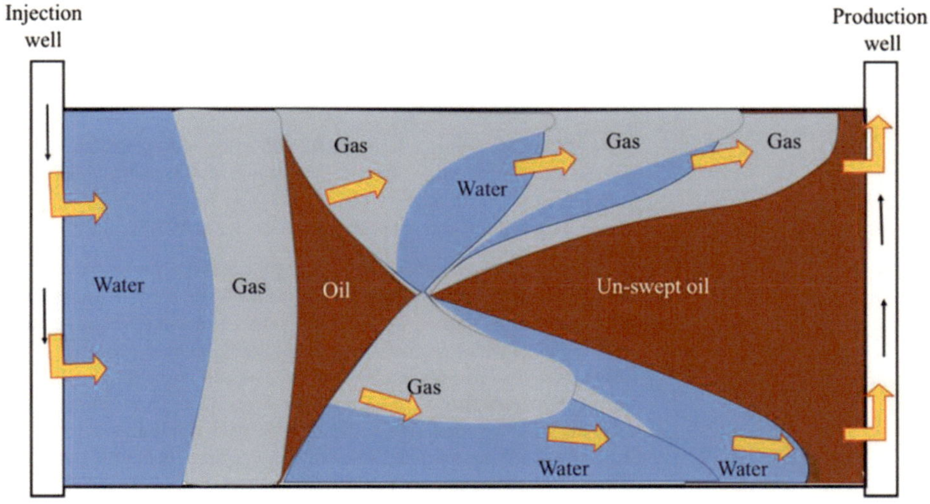

Figure 4.
Water injection to increase oil recovery [37].

control over mobility and volumetric sweep efficiency, and advances in special physical development properties, according to reference [41].

- *Innovation:* Introduction of environmentally friendly polymers and Nano polymers. Running along is the rising tide of CO_2-based polymerization of CO_2/olefins, CO_2/alkyne, CO_2/isocyanides, and CO_2/FLP polymerizations [17].

- *Studies*: Oil recovery enhancement by about 30% has been shown by laboratory tests in the case of CO_2-only injection (**Table 1**) [18].

2.3.2.2.3 Coinjection of CO_2-nitrogen (N_2)

The method for recovering methane from hydrate structures is coinjection of CO_2-nitrogen (N_2) [42].

- *Benefits:* Reduction of CO_2 in coal seams, improved gas drainage efficiency, improved geological underground storage of CO_2, improved restoration, and increased CH_4 diffusion [43–46].

Year of study	Research and findings
2008	Thomas [19], conducted a study using a mechanism of reduced oil viscosity by injecting CO_2, which expanded the oil recovery. • Polymer improves mobility control by increasing water viscosity and reducing fingering. The findings were as follows: • Polymer improves mobility control by increasing water viscosity and reducing fingering, with a range of concentration of polymers within 500 to 2000 ppm. • CO_2 injection rates: 0.5–1.5 MSCF/day.
2010	Research to study the improvement of oil recovery by integrating CO_2 injection with polymer flooding in order to enhance the sweeping efficiency while reducing the mobility of CO_2 was conducted by Alvarado and Manrique [17]. Some of the following results were found: • increase in oil recovery of 10–20% and • Reduction in the mobility of CO_2 by 30–60%. Alvarado and Manrique [17], also performed an experiment that proved that the use of the introduction of environmentally friendly polymer Reduced CO_2 recycling costs. Cost-saving contribution in operational efficiency at 15–30%.
2014	A discovery was made by Sheng [18] that while the area of polymers has been one of the most interesting research fields in the years, there are challenges yet associated with the degrading of polymers at high temperatures, compatibility of the polymer with reservoir conditions, salinity, hardness. It was found that hydrolyzed polyacrylamides were susceptible to degradation under harsh conditions; therefore, much effort has been done toward developing sulfonated and thermally stable polymers [18]. Advancement in cross-linked and modified polymers enhanced performance in saline reservoirs and achieved 10–15% incremental oil recovery [18].

Table 1.
Summarized work done on CO_2-polymer coinjection.

- *Innovation:* Gas mixtures can be optimized for particular reservoir conditions by real-time monitoring and simulation [47].

- *Studies:* McLaughlin et al. [41] conducted an investigation on the recovery of methane from hydrate structures using a binary CO_2/N_2 gaseous mixture, where they compared to pure carbon dioxide, and the proportion of methane recovered was low, staying below 26%, the mixture, which is 50/50 volume % carbon dioxide and nitrogen, was found to be less expensive.

2.3.2.2.4 CO_2 with surfactants

The choice of the CO_2-foam descending method depends on the selection of surfactant, which should have low adsorption, low oil phase partitioning, low adsorption, poor sodium tolerance, and low heat stability (**Figures 5** and **6**) [50].

- *Benefits*: EOR efficiency is increased by mixing carbonated water with chemical additions like surfactants and nanoparticles [51]. There are some open issues regarding early breakout problems, limited mobility control and gravity segregation that were reported by Peach and Eastoe [52]]. Many surfactants with higher partition coefficients are much more efficient at improving oil recovery [53]. Although mobility control is an issue still to be overcome, CO_2 injection projects can contribute to climate change mitigation efforts through underground CO_2 storage [47].

- *Innovation:* Recent advancements in surfactant compositions, foam fluid movement, fluid imaging, modeling tools, process design, and reservoir performance

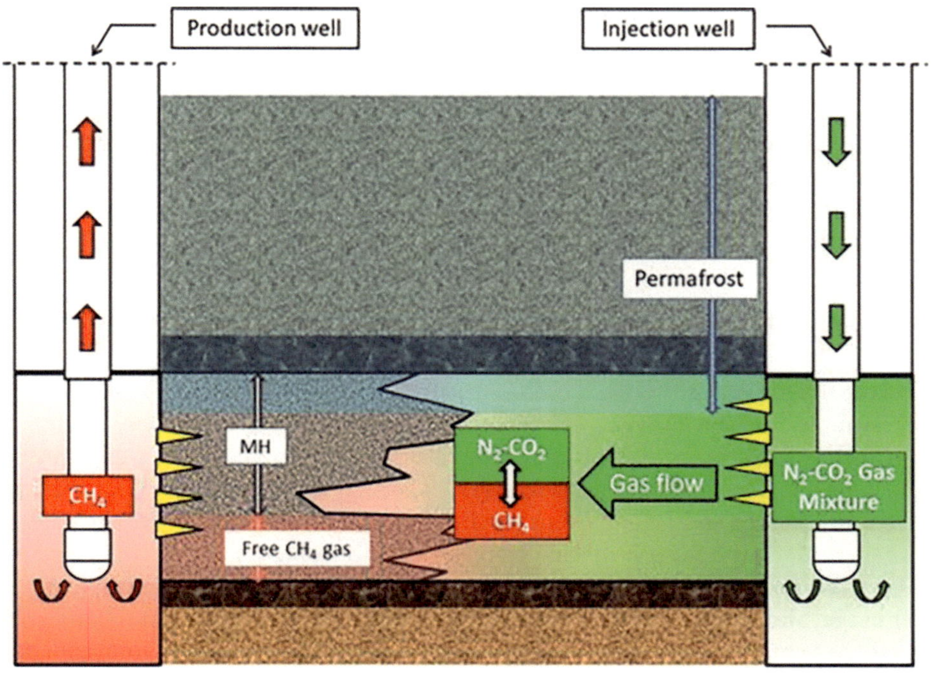

Figure 5.
Representation of CO_2-nitrogen (N_2) to improved restoration, and increased CH_4 diffusion [48].

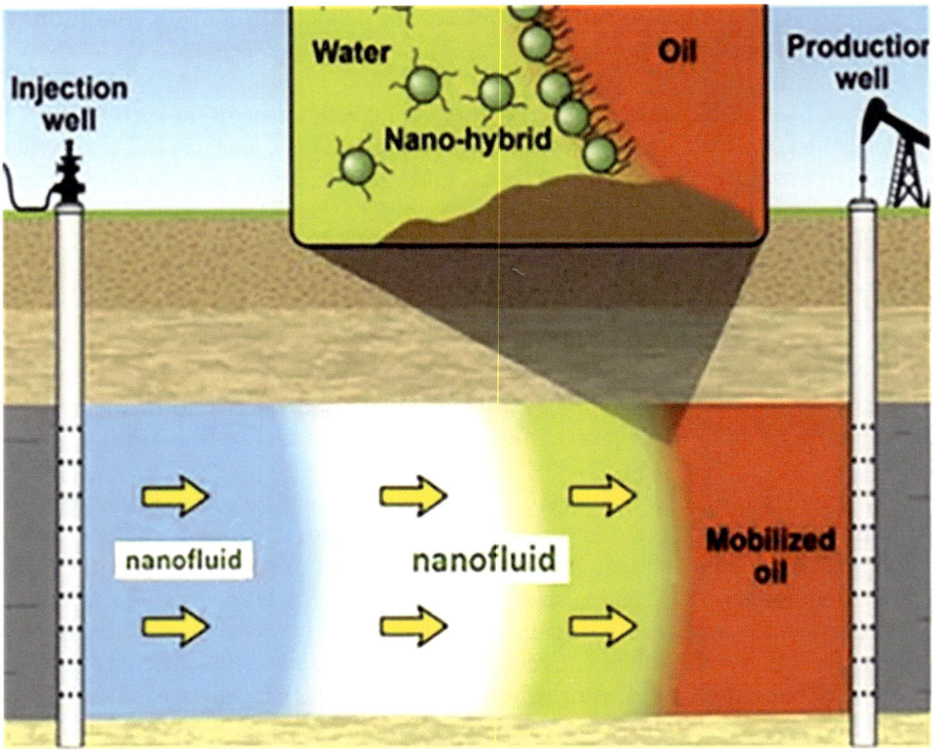

Figure 6.
Addition of nanoparticles to mobilized oil [49].

evaluation issues are all explored, along with preliminary experiments including the deployment of CO_2 foam in carbonate reservoirs [47].

- *Studies:* Zhang et al. [53], assessed surfactants and nanoparticles in oil recovery foam. Their main focus was to find conditions under which the process had increased stability and strength for difficult reservoir conditions. Oil recovery addition increased nanoparticle addition substantially; the addition may act in ways that could stabilize CO_2/Brine foam.

3. Impact on the environment and risk reduction

On the other hand, although these techniques are viewed as enhancing the oil recovery rates, they have adverse effects on the environment with emissions, water usage, and land disturbance. Thus, effective risk management techniques are significantly needed, in order to weed these out [18]. Thermal and other such EOR techniques warrant monitoring and management for their polluting nature-emissions due to fuel burning, local water bodies withering away, and land subsidence [19].

This is why a complex regulatory framework – prescripts regarding emissions, water usage, and disposal of waste – must be established for any given EOR project to manage these matters, to ensure compliance with environmental standards, and, at the same time, reduce ecological harm. For example, CO_2 injection for EOR must conform to CO_2 storage regulations to prevent leaks and protect groundwater quality [3]. The sequential order of effective risk management involves the assessing of risks, which aptly fall within the frames of certain EOR methodologies. Hence, reservoir characterization, environmental impact analyses, and groundwater modeling become vital instruments for understanding probable hazards and formulating mitigation plans.

Gbadamosi et al. [5] mentioned that such evaluations assist operators in predicting the possible effects of EOR operations on the ecosystem and community resources. Operators can often employ best practices in monitoring and clean-up so as to minimize environmental issues. Environmental concerns can be ameliorated in EOR projects through timely remediation and routine monitoring of emissions and water characteristics. Environmental condition-monitoring technologies include automated leak detection systems and real-time monitoring systems to ensure that any problem becomes an issue of the past [19].

4. Conclusion

EOR development has reached an important turning point as rising sustainability efforts are becoming part of technological innovations and an even stronger commitment to environmental responsibility. Insights from renewable energy integration always suggest that EOR remains one of the best solutions to meet the global energy demand and avoid threatening the environment.

4.1 An overview of the main conclusions

- Technology innovations: Uses of data analytics for optimizing the EOR process will contribute to online decision-making that enhances recovery efficiency while using fewer resources and nanotechnology to enhance mobility that makes recovery viable in low-permeability formations, especially when mixing eco-friendly and biodegradable materials.

- The environmental benefits of renewable energy and CO_2-EOR: The incorporation of renewable energy sources like solar thermal into EOR operations allows them to become greener and complement global decarbonization efforts. CO_2-EOR fits into dual-stage procedures: stimulating oil production and sequestration of CO2 as a very strong enabling technique for sustainable energy production. $_2$.

- Policy and regulatory support: These include regulations around water treatment and emission reductions, which are examples of how regulation is among initiatives that the realization of sustainable EOR will need. Such support will thus ensure environmental compliance as well as cheap and resource-efficient recovery.

- As innovation progresses, EOR could serve as a go-between for traditional fossil fuel production on one hand and clean energy sources on the other. The future of EOR will begin tagging along with its responsiveness, as it gradually turns into an active technique of recovery into a New World that is bound to be carbon-conscious.

4.2 Recommendation

- Environmentally-safe materials: Develop environmentally safe materials that would minimize environmental pollution while meeting environmental sustainability standards. Utilizing renewable energy for thermal EOR, instead of using steam injection techniques, the technique would apply solar-assisted steam generation. These developments, especially in very dry regions, would reduce the energy used and carbon emitted per barrel of oil while enhancing the recovery efficiency.

- Apply advanced techniques of CO2 injection with chemical assistance, and this could enhance recovery by improving the efficiency of techniques to increase oil recovery. Regarding certain reservoirs, both simulation and real-time optimization would identify the gas-to-water ratio.

- Enhancing gas injection methodology by the use of nanotechnology. This would also improve foam stability, mobility control, and oil recovery in challenging reservoir conditions, as well as assist in carbon separation.

- Improve the level of risk mitigation and environmental monitoring through the development of comprehensive regulatory frameworks and the deployment of technologies such as real-time monitoring and automatic leak detection. Conduct detailed reservoir characterization, environmental impact studies, and groundwater modeling in order to reduce potential risks and ensure sustainability.

Author details

Bongiwe Ladyfair Hem
Tshwane University of Technology, Pretoria, South Africa

*Address all correspondence to: bongiwehemm@gmail.com

References

[1] Liang F, Wang W, Zhu S, Hu Y, Zhao Z, Tan Y, et al. Nanofluids application in enhanced oil recovery process-opportunities and challenges. Arabian Journal of Chemistry. 2024;**18**:106053-106053. DOI: 10.1016/j.arabjc.2024.106053

[2] Ahmed A. Geological and Environmental Implications of Enhanced Oil Recovery Techniques: Balancing Efficiency with Sustainability. Egypt: California Digital Library, Department of Geology, Faculty of Science, Minia University; 2024. DOI: 10.31223/X5M69G

[3] Lake LW, Johns RT, Rossen WR, Pope GA. Fundamentals of Enhanced Oil Recovery. Taxas: Society of Petroleum Engineers; 2014. DOI: 10.2118/9781613993286

[4] Kokal S, Al-Kaabi A. Enhanced Oil Recovery: Challenges and Opportunities. Saudi Aramco: World Petroleum Congress, Expect Advanced Research Centre; 2010. Available at: https://firstforum.org/wp-content/uploads/2021/05/Publication_00466.pdf

[5] Gbadamosi AO, Junin R, Manan MA, Agi A, Yusuff AS. An overview of chemical enhanced oil recovery: Recent advances and prospects. International Nano Letters. 2019;**9**(3):171-202. DOI: 10.5772/intechopen.88140

[6] Ojha D. Enhanced oil recovery (EOR). India: Cern European Organization for Nuclear Research; 2022. DOI: 10.5281/zenodo.6816528

[7] Conjunction Carbonate Capture Utility and Storage (CCUS). A Brief History of New Development and Reservoir Technologies for CO2 EOR. 2020. Available from: https://www.co2conference.net/wp-content/uploads/2021/01/Melzer-CO2-EOR-History-New-Developments-forweb.pdf.

[8] Green DW, Willhite GP. Enhanced oil recovery. 2nd ed. Lawrence Kansas: Society of Petroleum Engineers; 2018. pp. 102-320

[9] Alagorni AH, Yaacob ZB, Nour AH. An overview of oil production stages: Enhanced oil recovery techniques and nitrogen injection. International Journal of Environmental Science and Development. 2015;**6**:693-701. DOI: 10.7763/ijesd

[10] Speight JG. General Methods of Oil Recovery. Laramie, Wyoming, USA: Elsevier eBooks, CD&W Inc.; 2016. pp. 253-322. DOI: 10.1016/b978-0-12-849906-1.00006-0

[11] Thomas S. Enhanced oil recovery –An overview. Oil & Gas Science and Technology. 2007;**63**:9-19. DOI: 10.2516/ogst:2007060

[12] Lopez VN. Webinar: Fundamentals of CO2 Enhanced Oil Recovery (English). 2014. Available from: https://www.slideshare.net/slideshow/webinar-fundamentals-of-co2-enhanced-oil-recovery-english/36823823#3 [Accessed: December 1, 2024].

[13] Enhanced Oil Recovery. Petroleum Engineering Explained: Basic Concepts for Novices. 2020. pp. 229-265. DOI: 10.1039/9781788016681-00229.

[14] Geologist Skills. Geological Considerations and Types of Wells designed in Oilfields. Geo-skill.com. Algeria. 2020 Available from: https://www.geo-skill.com/2020/08/

geological-considerations-and-types-of.html [Accessed: December 1, 2024]

[15] Kaldi, C. Geological Considerations and Types of Wells Designed in Oilfields. 2009. Available from: https://www.geo-skill.com/2020/08/geological-considerations-and-types-of.html [Accessed: December 3, 2024]

[16] Dong X, Liu H, Chen Z, Wu K, Lu N, Zhang Q. Enhanced oil recovery techniques for heavy oil and oilsands reservoirs after steam injection. Applied Energy. 2019;**239**:1190-1211. DOI: 10.1016/j.apenergy.2019.01.244

[17] Alvarado V, Manrique E. Enhanced Oil Recovery: An Update Review. Energies. 2010;**3**(9):1529-1575

[18] Sheng JJ. Modern Chemical Enhanced Oil Recovery: Theory and Practice. Gulf Professional Publishing; 2011

[19] Thomas S. Enhanced oil recovery an overview. Oil & Gas Science and Technology. 2008;**63**:9-19. DOI: 10.2516/OGST%3A2007060

[20] Malozyomov BV, Martyushev NV, Kukartsev VV, Tynchenko VS, Bkhtoyarov VV, Wu X, et al. Overview of methods for enhanced oil recovery from conventional and unconventional reservoirs. Energies. 2023;**16**:1-41. DOI: 10.3390/en16134907

[21] Lake LW. Enhanced Oil Recovery. Prentice Hall; 1989. Available from: https://www.scribd.com/document/332509420/Larry-W-Lake-Enhanced-oil-recovery-Prentice-Hall-1989-pdf

[22] PPT Online Organization. Polymer Flooding for Enhanced Oil Recovery-Presentation. 2024. Available from: https://en.ppt-online.org/670858 [Accessed: December 1, 2024]

[23] Sheng JJ. A comprehensive review of alkaline surfactant polymer (ASP) flooding. Journal of Chemical Engineering. 2014;**9**:471-489. DOI: 10.1002/apj.1824

[24] Hirasaki GJ, Toksoz MN, Sharma MM. Enhanced oil recovery: Field case studies. In: Symposium on Enhanced Oil Recovery, Society of Petroleum Engineers. USA: Texas Tech University; 2011

[25] Morrow N. Influence of Electrical Surface Charges on the Wetting Properties of Crude Oil. Vol. 4. 1989. DOI: 10.2118/16964-PA

[26] Rai P, Mehrotra S, Priya S, Gnansounou E, Sharma SK. Recent advances in the sustainable design and applications of biodegradable polymers. Bioresource Technology. 2021;**325**:124-739. DOI: 10.1016/j.biortech.2021.124739

[27] Sun H, Wang H, Cao X, Shu Q, Fan Z, Wu G, et al. Innovations and applications of thermal recovery techniques for heavy oil. Energy Geology Science. 2024;**5**:100332-100332. DOI: 10.1016/j.engeos.2024.100332

[28] Butler RM. Thermal Recovery of Oil and Bitumen. Prentice-Hall; 1991. pp. 12-20. Available from: https://www.scribd.com/document/378721955/Thermal-Recovery-of-Oil-Butler-1991

[29] Thomas T, Sharma P, Gupta DK. Understanding & screening of DCW through application of data analysis of experiments and ML/AI. Energies. 2023;**16**:3376. DOI: 10.3390/en16083376

[30] Ebaid H, Gasser M, Yehia T, Okoroafor E. Carbon Capture, Utilization and Storage Conference. 2024. DOI: 10.15530/ccus-2024-4015303

[31] PETROWIKI. Electrical Heating Petro-Wiki. 2024. Available from: https://

petrowiki.spe.org/Electrical_heating [Accessed: December 1, 2024]

[32] Meloni E, Ievolino G, Palma V. Basics of microwave heating and recent advances. Book.rsc.org. 2023;**45**:1-24. DOI: 10.1039/BK9781837670277-00001

[33] Williams O, Gideon OD, Ifeanyi OE. Innovative approaches in enhanced oil recovery: A focus on gas injection synergies with other EOR methods. Magna Scientia Advanced Research and Reviews. 2024;**13**:311-324. DOI: 10.30574/msarr.2024.11.1.0095

[34] Ericson SJ, Engel-Cox J, Arent DJ. Approaches for Integrating Renewable Energy Technologies in oil and gas Operations. OSTI OAI (U.S. Department of Energy Office of Scientific and Technical Information); 2019. DOI: 10.2172/1491378

[35] Ozowe W, Oluseyi GO, Ifeanyi OE. Innovative approaches in enhanced oil recovery: A focus on gas injection synergies with other EOR methods. Magna Scientia Advanced Research and Reviews. 2024;**11**(1):311-324. DOI: 10.30574/msarr.2024.11.1.0095

[36] Song B, Qin A, Tang BZ. Syntheses, properties, and applications of CO_2-based functional polymers. Cell Reports Physical Science. 2022;**3**:1-28. DOI: 10.1016/j.xcrp.2021.100719

[37] Fang P, Zhang Q, Zhou C, Yang Z, Yu H, Du M, et al. Chemical assisted Co_2 water alternative gas injection for enhanced sweep efficiency in co_2 EOR. Molecules. 2024;**29**:3978. DOI: 10.3390/molecules29163978

[38] Afzali S, Rezaei N, Zendehboudi S. A comprehensive review on enhanced oil recovery by water alternating gas (WAG) injection. Fuel. 2018;**227**:218-246. DOI: 10.1016/j.fuel.2018.04.015

[39] Chen B, Reynolds AC. Co_2 water alternating gas injection for enhances oil recovery: optimal well controls and half cycle lengths. Computer and Chemical Engineering. 2018:44-56. DOI: 10.1016/j.compchemeng.2018.03.006

[40] Padmasri B, Nagaraju R, Prasanth D. A comprehensive review on in situ gels. International Journal of Applied Pharmaceutics. 2020;**12**:24-33. DOI: 10.22159/Ijap.2020v12i6.38918

[41] McLaughlin H, Littlefield AA, Menefee M, Kinzer A, Hull T, Sovacool BK, et al. Carbon capture utilization and storage in review: Sociotechnical implications for a carbon reliant world. Renewable and Sustainable Energy Reviews. 2023;**177**:1-41. DOI: 10.1016/j.rser.2023.113215

[42] Haq B, Muhammed NS, Liu J, Hui TC. Enhanced natural gas production using co2 injection: Application to sustainable hydrogen production. Fuel. 2023;**347**:128474-1284474. DOI: 10.1016/j.fuel.2023.128474

[43] Lin J, Ren T, Wang G, Booth P, Nemcik J. Experimental investigation of N2 injection to enhance gas drainage in co2 rich low permeable seam. Fuel. 2018;**215**:665-674. DOI: 10.1016/j.fuel.2017.11.129

[44] Zhou X, Liang D, Liang S, Yi L, Lin F. Recovering CH4 from natural gas hydrates with the injection of co_2-n_2 gas mixtures. Energy and Fuel. 2015;**29**:1099-1106. DOI: 10.1021/ef5025824

[45] Wang Z, Fu X, Pan J, Deng Z. Effect of N2/CO2 injection and alternate injection on volume swelling/shrinkage strain of coal. Energy. 2023;**275**:127377-127377. DOI: 10.1016/j.energy.2023.127377

[46] Li Z, Hu H, Wang Y, Yan F, Bai Y, Gao Y, et al. Molecular simulation of CO2/N2 injection on CH4 adsorption and diffusion. Scientific Reports. 2024;**14**:1-14. DOI: 10.1038/s41598-024-71382-7

[47] Tupsakhare SS, Castaldi MJ. Efficiency enhancements in methane recovery from natural gas hydrates using injection of CO2/N2 gas mixture simulating in-situ combustion. Applied Energy. 2019;**236**:825-836. DOI: 10.1016/j.apenergy.2018.12.023

[48] Environment Guru. Greater than. 2020. Available from: https://za.pinterest.com/pin/333547916154682462/ [Accessed: January 2, 2025]

[49] University of Houston: Department of Physics. UH's Zero-Emissions Enhanced Oil Recovery Nanotechnology Scales up for Commercial Production. 2022. Available from: https://www.uh.edu/nsm/physics/news-events/stories/2021/0423-oil-recovery.php [Accessed: January 2, 2025]

[50] Jian G, Fernandez C, Puerto M, Ramesh S, Bonneville A, Sibani LB. Advances and Challenges in CO2 Foam Technologies for Enhanced Oil Recovery in Carbonate Reservoirs, vol. 202. 2021. pp. 108447-108447. DOI: 10.1016/j.petrol.2021.108447

[51] Halari D, Yadav S, Kesarwani H, Saxena A, Sharma S. Nanoparticle and surfactant stabilized carbonated water induced in-situ CO_2 foam: An improved oil recovery approach. Energy & fuels. 2024;**38**:3515-4768. DOI: 10.1021/acs.energyfuels.3c04232

[52] Peach J, Eastoe J. Supercritical carbon dioxide: A solvent like no other. Beilstein Journal of Organic Chemistry. 2014;**10**:1878-1895. DOI: 10.3762/bjoc.10.196

[53] Zhang F, Wang X, Wang B, Lou X, Lipiński W. Effects of silica gel nanopores and surfactants on CO_2 hydrate formation kinetics an experimental and modelling study. Chemical Engineering Science. 2022;**262**:118002. DOI: 10.1016/j.ces.2022.118002

Section 2

Emerging Technologies and Smart Solutions

Chapter 4

Perspective Chapter: Using Artificial Intelligence Tools to Determine the Status of an Oil Well Prior to Applying Enhanced Oil Recovery Techniques

Esbeydi Villicaña-García, Arturo Palacio-Pérez and Silvia Raquel García-Benítez

Abstract

As the global demand for fossil fuels is increasing, and access to them is becoming more complicated and expensive, it has been important to find alternatives where existing resources are better used. For this reason, this work proposes a methodology to detect areas where oil production is low compared to its peak but still has proven available crude resources. Then, enhanced recovery techniques can be practiced, diminishing the risk of failure. Using artificial intelligence tools such as fuzzy logic, the aim is to categorize oil wells according to geophysical variables pertaining to crude oil such as pressure, production, saturation, viscosity, API gravity, compressibility and volume factor, highlighting that historical data is not required to determine if the well can be subject to recovery. The results show that the values obtained have a good margin of coincidence with the interpretation that is proposed.

Keywords: exhausted well, oil wells, artificial intelligence, fuzzy logic, enhanced oil recovery

1. Introduction

Fossil fuels account for 80% of current global primary energy demand [1]. Fossil fuel consumption has increased significantly over the past half-century, approximately eightfold from 1950 to 1980. The fuel type shifted from coal alone to a combination of oil and gas [2]. Today, fossil fuels are necessary for human survival and daily life and are the world's primary source of energy. Most countries with large fossil fuel deposits have economies dependent on fossil fuel extraction [3]. In 2023, proven crude oil reserves in Mexico reached 6.154 billion barrels (**Figure 1**), which represented an increase of 1.58% compared to proven reserves in 2022 [4]. Proven reserves are those reserves which, from geoscience and engineering data, are

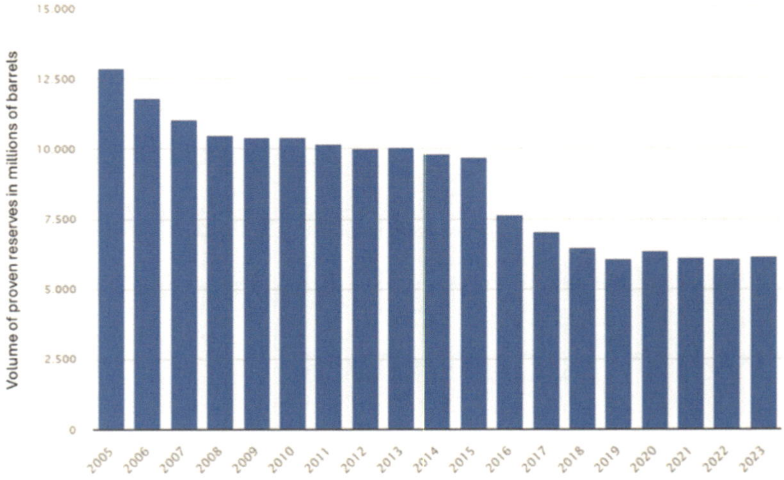

Figure 1.
Proven oil reserves in Mexico [4].

estimated with reasonable certainty to be commercially recoverable from a given date onwards from known reservoirs under defined economic conditions, operating methods and governmental regulation [5].

Oil-producing wells are abandoned when the reservoir is depleted, that is, when the amount of oil produced is not sufficient to make hydrocarbon extraction profitable for the company. The United States has more than a century of oil production history, and in places such as Pennsylvania, Texas or California it is estimated that there are millions of abandoned installations, many of which are poorly documented or unrecognized [6]. The cost of a failed exploration, such as seismic surveys and drilling a dry hole, can cost between 5 and 20 million dollars per exploration site and, in some cases, much more. However, when an exploration site is successful and oil and gas extraction is productive, exploration costs are recovered and are significantly lower compared to other production costs [7]. Applications have been shown for the use of abandoned horizontal wells for underground storage of natural gas [8]. The methodology for fuzzy integral multi-criteria decision-making is proposed to select the most suitable sites that have the role of storage, where the solutions are based on a set of criteria and linguistic variables to be used in the model [9].

It is important to find techniques to predict descriptive parameters of oil-producing wells, such as production rate, with the help of machine learning methodology that speeds up calculations, since previously to predict detailed and accurate reservoir information such as porosity and permeability that are used to determine the estimated final recovery could take hours or days making it computationally expensive [10]. Likewise, machine learning techniques have significantly solved huge problems related to real-time data anomalies and the decrease in interaction processes of data engineers [11]. In addition, the combination of several techniques such as convolutional neural network, long-term memory neural network and self-attention mechanism can be used to predict oil well production, achieving improved performance in terms of adaptation to the basic trend of oil well production [12].

Fuzzy logic is an extension of conventional Boolean logic developed to handle the concept of partial truth – truth values between "completely true" and "completely false" [13]. Fuzzy logic has presented the industry with a new interpretation technique that fully takes into account uncertainty to predict reservoir parameters such as permeability [14]. The advantage of fuzzy logic over other techniques in reservoir characterization is the fact that it is not affected by data deficits, which is a serious challenge in the early stages of reservoir exploitation [15].

However, in the reviewed literature, the use of fuzzy logic has not been considered to categorize an oil well as "subject to recovery", which means that its production is very low compared to its maximum peak, but it still has proven reserves, making it necessary to use enhanced recovery techniques for its exploitation. For the evaluation of oil wells, it is necessary to know geophysical variables such as operating pressure, bubble pressure, pressure gradient, hydrostatic pressure and formation pressure, highlighting that this way of approaching the problem facilitates analysis and decision-making despite the low availability of related historical data concerning wells production.

2. Problem statement

Nearly 30 million oil and gas wells around the world have been abandoned in recent decades by the oil industry [16]. Three out of five oil wells are inactive, and only one-third of these have been permanently closed [17]. Due to the high energy consumption and the great need to meet the associated demands today, it is vitally important to find better alternatives for the use of existing fossil fuels. For this reason, a new methodology is proposed to detect areas that were considered lost for obtaining fossil resources and determine if they can be categorized as subject to recovery [Recoverable (RE) or Barely recoverable (BRE)], that is, where production is very low compared to its maximum production point but proven resources continue to exist in the subsoil, making it necessary to use enhanced oil recovery (EOR) techniques. If the well can carry on producing autonomously, that is, without the need of any EOR, then it will be considered as No Recoverable (NRE).

3. Methodology

To establish if a well may be categorized as "subject to recovery", the use of fuzzy logic tools is proposed, which, through membership functions and the construction of "if-then" rule blocks, allow determining the level of membership to a certain category and consequently choosing the wells that meet specifications such as operating pressure, bubble pressure, pressure gradient, hydrostatic pressure and formation pressure (**Figure 2**). As a following stage of this work, other parameters such as saturation, ° API, viscosity, compressibility and volume factor will be considered.

The concept of fuzzy logic is related to the way people perceive their surroundings. If someone says "it's hot today" it is interesting to know to what extent this is true. Fuzzy sets define such ambiguities and are an extension of classical set theory, where an element belongs or does not belong to a set, but ambiguities are not defined. Fuzzy sets are intended to model the ambiguity with which a variable is perceived [18]. Decision-making based on non-specific information is considered an everyday procedure. This is what fuzzy logic attempts to emulate based on observation of the environment and formulation of logical rules and mechanisms for decision-making. This topic is specific to artificial intelligence, where the aim is to emulate human thought.

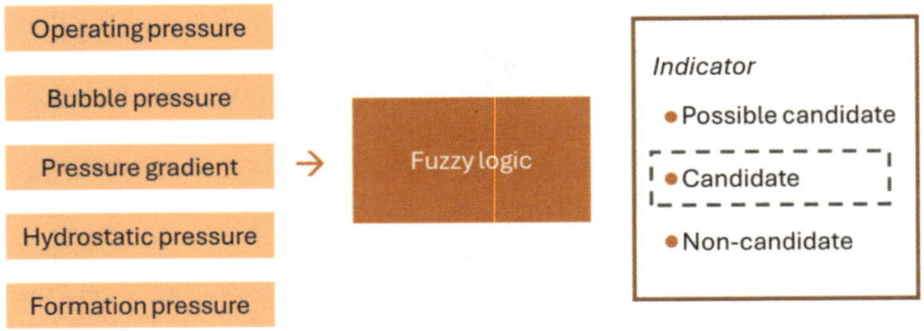

Figure 2.
Fuzzy logic system for categorizing oil wells.

3.1 Fuzzy logic basics

Fuzzy logic (FL) is a mathematical tool that aims to imitate the way human beings reason and process information. It is a simple method that is ideal for information processing, decision-making and process control [19]. Zadeh [20], its creator, concluded that traditional techniques were too rigid to model most complex problems. The central idea is that data have a partial membership value associated with sets that make up a universe of analysis. Among the most successful applications of FL are control systems (the most documented case is the subway trains of Sendai, Japan [21]), but its field of application is very extensive: any process that is associated with knowledge (experience).

The concept of a set is intuitively understood as a perfectly defined grouping of elements in which it is possible to determine for any element, in a given universe, whether it belongs or not to the set (classical or Boolean sets). In fuzzy sets, a value, degree of belonging or membership (ranging from 0 to 1) is assigned to the element with respect to a set, so it is understood that classical sets are contained in fuzzy sets. Each fuzzy set, comprised by a universe of discourse U, which is the aggregate of objects that are spoken of in a given context or situation, is associated with a linguistic value that can be a word or a label to associate it with the name of the set. The fuzzy set is characterized by a membership function called μA(x) that indicates the degree to which the variable x is included in the set represented by the label A ($0 \leq \mu A(x) \leq 1$). If this function takes the value 0, it means that such value of x is not included in A and if it takes the value 1, it is absolutely included in A [22]. All elements of the universe of discourse belong to a certain degree to all sets of the universe of discourse. In this way, a fuzzy set in a universe of discourse is defined as:

$$A = \{(x, \mu_A(x)) | x \in U\} \tag{1}$$

An example of this is shown in **Figure 3**, where the universe of discourse is the heights in the interval 140 to 200 centimeters, that is: U[140, 200 cm]. Three fuzzy sets were chosen whose linguistic values are [Short, Average, Tall]. In this way, an individual who is 180 centimeters tall belongs to the three sets as follows: 0 in Short, 0.5 in Average and 0.5 in Tall.

A linguistic variable is used to represent any element that is too complex, for which there is no concrete definition, or whose numerical conception is uncertain (its obtention is susceptible to contamination by epistemic or random sources). Each variable X is associated with T(X), which is its set of terms (or linguistic labels), U is the universe of discourse (or underlying domain), G is a context-free grammar

	Intersection (AND): $\mu_{A \cap B}$	Union (OR): $\mu_{A \cup B}$	Complement (NOT): $\mu_{\overline{A}}(x)$
Zadeh operators MIN/MAX	$\min(\mu_A(x), \mu_B(x))$	$\max(\mu_A(x), \mu_B(x))$	$1 - \mu_A(x)$
Probabilistic PROD/PROBOR	$\mu_A(x) \times \mu_B(x)$	$\mu_A(x) + \mu_B(x) - \mu_A(x) \times \mu_B(x)$	$1 - \mu_A(x)$

Table 1.
Operators for fuzzy sets.

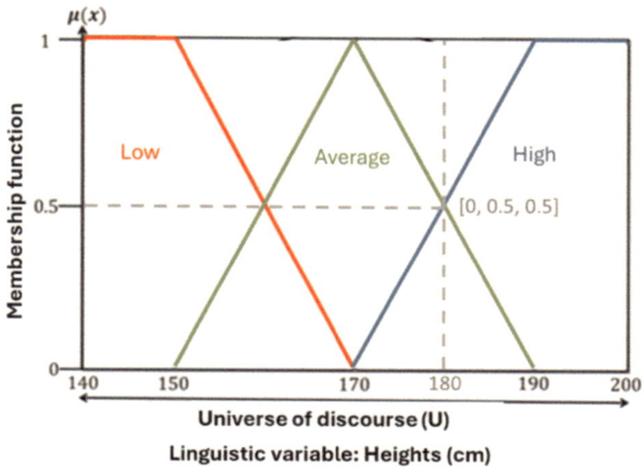

Figure 3.
Elements of fuzzy sets.

through which terms in T(X) are generated (in the case of the example it could be "very high", "not very low"), and M is a semantic rule that associates each linguistic value of X with a meaning M(X).

The choice of membership functions is fundamental to the creation of a system and depends on the principle that it applies to solve a problem (reasoning). The only condition that a membership function must fulfill is that its values oscillate between 0 and 1 continuously. Identifying the variables that are determinants for the categorization of a well as "subject to recovery" is the starting point for being able to qualify a well that will require enhanced recovery techniques. The proposal is to determine the ranges of values required for operating pressure, bubble pressure, pressure gradient, hydrostatic pressure and formation pressure. With this information, the membership functions that will be used to create a block of rules that determine the level of membership of the well to the "subject to recovery" category are established, this being the main objective of the study.

Three basic operations of conventional sets can be generalized to fuzzy sets: complement, union and intersection. Such operations for fuzzy sets can be analyzed from the point of view of Zadeh operators [23], as observed in **Table 1**. **Figure 4** graphically illustrates the operations based on Zadeh operators.

Fuzzy controllers (FC) were introduced by Mamdani [24] and Mamdani and Assilian [25] for the control of complex processes or those whose mathematical models involve complicated knowledge. FC works in a very different way from

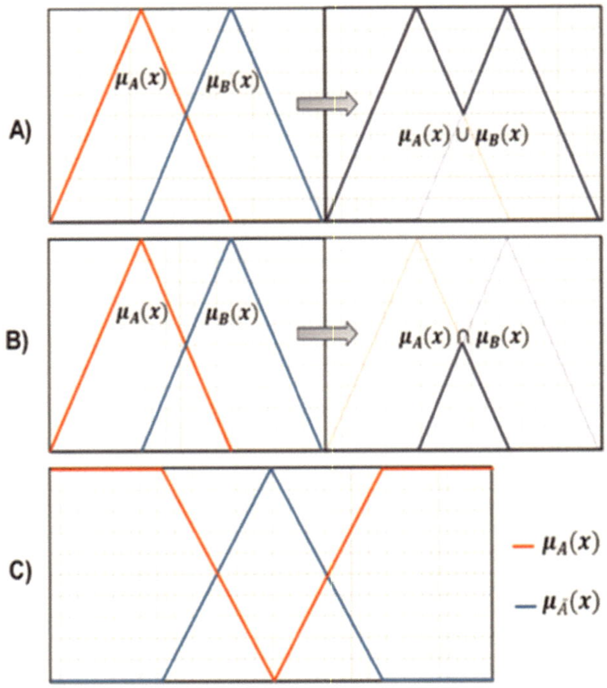

Figure 4.
Zadeh operators for (A) union, (B) intersection and (C) complement.

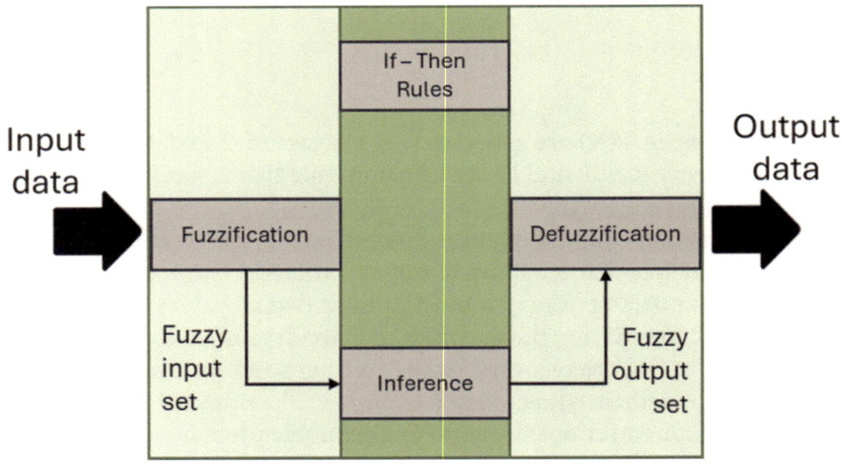

Figure 5.
Structure of a fuzzy controller [26].

conventional control systems because it uses expert reasoning to generate a knowledge base that will give the system the ability to discern certain actions. These systems use a set of rules similar to those that a person would use to control a process and take action (**Figure 5**).

Fuzzification comes from the term fuzzy logic and deals with the transformation of an element of the universe of discourse (measured variable of the process) into a value (of membership) for each membership function. Inference refers to the block with which the input and output fuzzy sets are related, simulating the human decision process. The difference between inference systems is the way in which the membership degrees influence the rules; the two most common inference models are Mamdani and TSK (Takagi, Sugeno and Kang) [27]. In this research, the first of them is used, since TSK is also a method based on fuzzy rules but in which the consequent does not generate a fuzzy set but a series of linear functions. In Mamdani inference, fuzzification is executed and then the rules are evaluated. The rules are composed of an IF part (or antecedents) and a THEN part (consequence or result). If a rule has multiple antecedents, the AND or operators are used to obtain a single value or the result of the evaluation. This number (the truth value) is applied to the consequent with a scaling or trimming derived from the truth value of the antecedent. Details of the process are given in [28].

At the end of the evaluation of all the rules, the aggregation is performed, in which the outputs of all the rules are added, that is, the membership functions of all the previously trimmed or scaled consequences are combined to obtain a single fuzzy set for each output variable. This result is used in the defuzzification stage. In this research, the Center of Gravity (CoG) is calculated to determine the representative value of the fuzzy set, this being the categorization to which the well belongs, that is, the output variable obtained.

3.2 Membership functions

Figure 6 shows the variables that may be considered for further analysis, even though as stated above, in this stage of the research pressure will be used as the main variable in its diverse forms (operating pressure, bubble pressure, pressure gradient, hydrostatic pressure and formation pressure). The pressure value should not be appropriate for a natural flow to exist, and in this way, recovery techniques are used, therefore, the labels for the pressure conditions are *subnormal* in the case of the

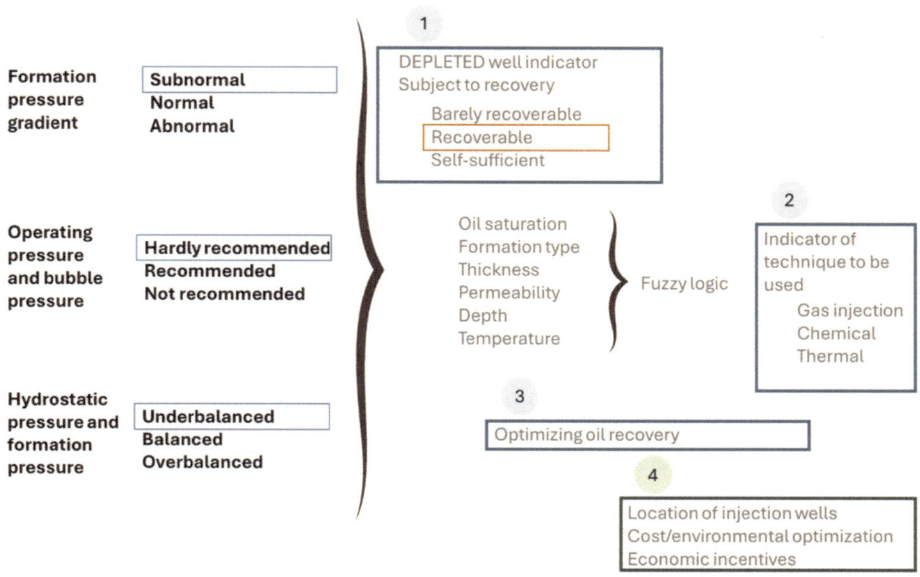

Figure 6.
Work flowchart. The selected variables are framed within blue squares in such a way that in the resulting output variable, the depleted well indicator may lead to "recoverable" (orange square).

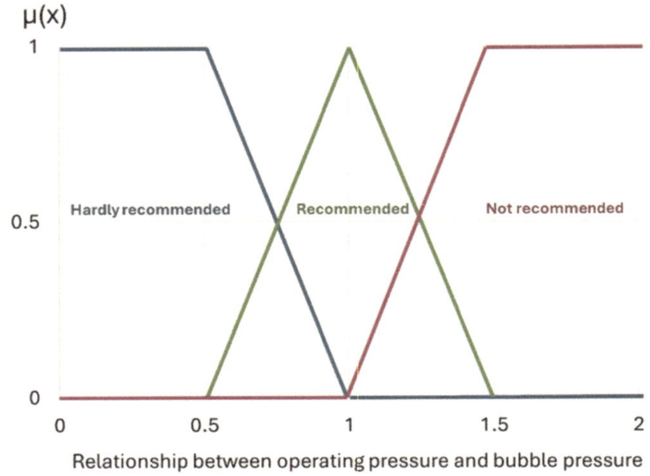

Figure 7.
Membership function for the ratio of operating and bubble pressure.

formation pressure gradient, *hardly recommended* for the relationship between the operating pressure and bubble pressure and *underbalanced* for the relationship between the hydrostatic pressure and formation pressure. This implies that the production level is much lower than the maximum production peak that was ever reached and therefore the use of enhanced recovery techniques is necessary for the exploitation of the resource. The intention of highlighting the labels selected with blue boxes in **Figure 7** for each pressure condition (formation pressure gradient, relationship between operating pressure and bubble pressure, relationship between hydrostatic pressure and formation pressure) is that in the output variable, the depleted well indicator may lead to "recoverable" (orange box). It is important to reiterate that as a first approximation, only pressure is considered. For future work, other variables will be considered such as production, saturation, viscosity, API gravity, compressibility and formation volume. In this way, in the first instance, the well must categorized as recoverable (step 1), only the well categorized as *recoverable* should continue the process. Afterward, according to other different parameters, the best-enhanced oil recovery method will be determined (step 2) in order to optimize the process (step 3) and finally select the best location of injection wells (step 4).

3.3 Pressure-related variables

Bubble pressure is an important oil property. The bubble point implies the pressure at which the most volatile components present in the oil begin to bubble. The reservoir performance changes significantly when producing with pressure below the bubble point, which corresponds to the oil and gas flows in producing wells. The pressure at the bubble point in conventional oil reservoirs varies between 1800 and 2600 psi [29]. That is, if the operating pressure is less than 1800 psi, a technique is sought to increase the pressure or to cause the oil still trapped in the subsoil to flow to the surface and maintain production. The way to consider the bubble pressure is by defining the relationship between the well operating pressure and the bubble pressure (Poperation/Pbubble). The well is categorized as "recoverable" when the well operation pressure is less than the bubble pressure, that is, Poperation/Pbubble <1, which implies using an external agent to improve the pressure and that the relationship between operating pressure and bubble pressure is at least equal to 1, therefore the label selected when the relationship is less than 1 is "hardly recommended" (**Figure** 7).

Another pressure that is related to extraction is the formation pressure, which is the pressure exerted by the liquid or gas content in the pore spaces of the formation. The formation pressure can be classified according to the range of values that it can achieve.

Normal Formation: Normal pressure formations exert a pressure equal to the column of native fluid in that formation up to the surface. The pressure gradient typically ranges from 0.443 to 0.465 psi/ft [30].

Abnormal formation: Abnormally pressured formations exert pressure greater than the hydrostatic pressure of the fluids contained in the formation. Pore pressure increases, generally exceeding 0.465 psi/ft. The resulting increase in overload causes the latter to be partially supported by pore fluids rather than by the rock grains.

Subnormal formation: Formations with subnormal pressures have gradients lower than those of fresh water, that is, less than 0.433 psi/ft. Formations with subnormal pressures may be developed when overload has been eroded, leaving the formation exposed to the surface.

For the purposes of this study, the subnormal pressure is chosen, since it implies having a pressure lower than that required for the well to perform in a "natural" way, and based on that, the well can be classified as "recoverable" (**Figure 8**) and therefore an EOR technique would be required for improvement of the well production. It is important to highlight that when the pressure gradient is 0.454 psi/ft. it implies 100% a normal well ($\mu = 1$) if the pressure decreases, it stops being normal and starts being subnormal (blue line) and if it starts to increase from 0.454 psi/ft. it stops being normal and starts being abnormal (pink line).

In the oil industry, another pressure related to the extraction of crude oil is the hydrostatic pressure [30], which is the pressure exerted at a given point in the well by a fluid column due to its density (ρ) and its height (h). The difference between the formation pressure (FP) and the downhole hydrostatic pressure (HP) is the differential pressure. There are three different cases: Overbalance: HP > FP, Balance: HP = FP, Underbalance: HP < FP.

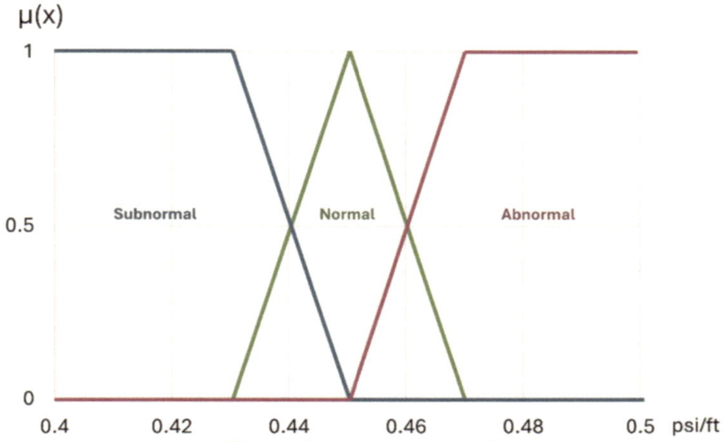

Figure 8.
Membership function for the formation pressure gradient.

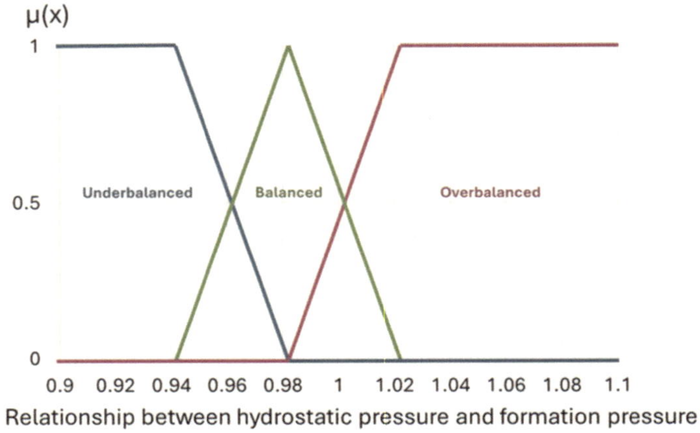

Figure 9.
Membership function representation for the ratio of hydrostatic pressure and formation pressure.

To consider these relationships in the membership functions, it is approached as follows: the relationship between hydrostatic pressure and formation pressure as a balanced well varies between 0.95 and 1, if this relationship is greater than one it is considered overbalanced and if it is less than 0.95 it is considered underbalanced (**Figure 9**) [30].

Therefore, if the different types of pressure that have been discussed are considered, the membership functions would imply that the formation pressure gradient is less than 0.433 psi/ft. (*subnormal*), that the ratio of operation pressure to bubble pressure is less than 1 (*hardly recommended*), and also that the ratio of hydrostatic and formation pressure is less than 0.95 (*underbalanced*) to qualify the oil well as depleted or subject to recovery.

4. Case study

The oil zones of Mexico are distributed mainly in the east of the country and correspond to large sedimentary basins where the accumulation of organic matter has the potential to generate hydrocarbons. In Mexico, 12 oil provinces are recognized (**Figure 10**) [31], six producing provinces: Sabinas-Burro-Picachos, Burgos, Tampico-Misantla, Veracruz, South and Deep Gulf of Mexico, and six provinces with medium-low potential: Yucatan Platforms, Chiapas Fold Belt, Sierra Madre Oriental Fold Belt, Chihuahua, Gulf of California and Vizcaíno-LaPurísima -Iray.

Pemex exploration and production considers that the deepest part of the Gulf of Mexico Basin has the greatest oil potential, with a prospective resource of 29.5 billion barrels of crude oil equivalent, which represents 56% of the country's total resource [32]. As of 2015, there are great expectations of incorporating production from the Gulf of Mexico, neglecting the remaining resources in the land area of the country. That is why this work focuses on the land area of the country, mainly in the basins: Veracruz, Southeast Basin, Tampico-Misantla and Burgos (**Figure 11**). Wells located in that area suitable to be evaluated with the proposed methodology to determine their status for recovery of remaining oil.

A representative and trivial example is herein discussed, where the three most important steps in the use of fuzzy logic are used and described: Fuzzify, infer and

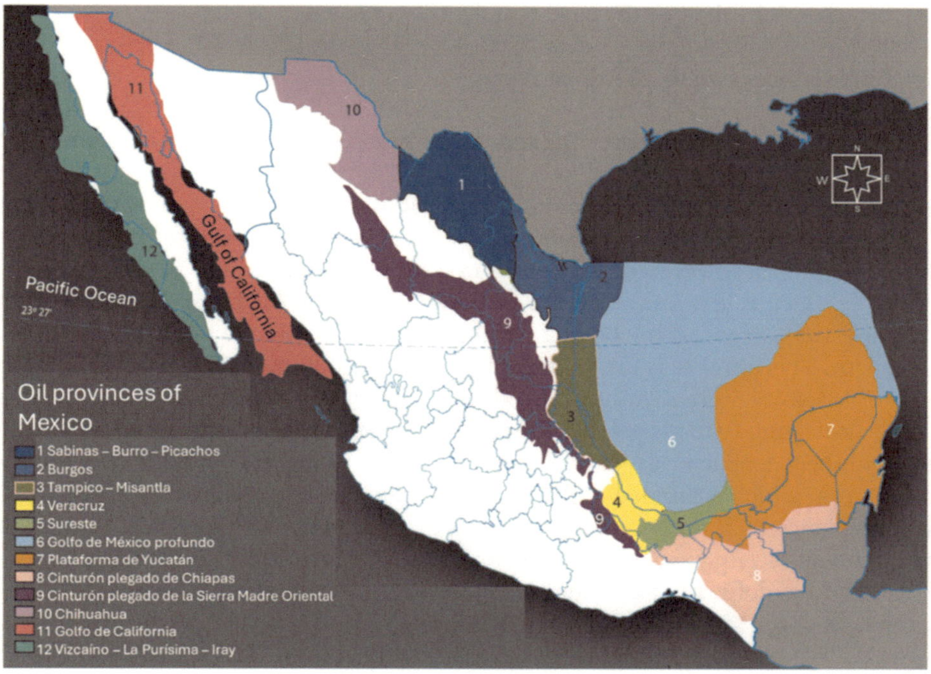

Figure 10.
Oil provinces of Mexico.

Figure 11.
Remaining onshore petroleum resources in Mexico.

aggregate (**Figure 12**). In this example, the fuzzy inference system for the problem takes as inputs the service and the quality of the food and calculates a tip percentage using the following rules:

- If the service is bad or the food is stale, the tip is low.
- If the service is good, the tip is average.
- If the service is excellent or the food is delicious, then the tip is generous.

The first step is to take the inputs and determine the degree to which they belong to each of the appropriate fuzzy sets using membership functions (fuzzification). Once the inputs are fuzzified, the degree to which each part of the antecedent is fulfilled for each rule is known. If the antecedent of a rule has more than one part, the fuzzy operator is applied to obtain a number that represents the result of the rule's antecedent. This number is then applied to the output function. The input to the fuzzy operator is two or more membership values of fuzzy input variables. The result is a single truth value. Before applying the implication method, it must be determined the weight of the rule. Each rule has a weight or a degree of membership (DoS) (a number from 0 to 1), which is applied to the number given by the antecedent. Aggregation is the process by which fuzzy sets representing the results of each rule are combined into a single fuzzy set. Aggregation only occurs once for each output variable, which is before the final defuzzification step. The output of the aggregation process is a fuzzy set for each output variable. For more details see [33].

FuzzyTech software is used to analyze the study system. The input variables are specified: Pressure gradient, operating and bubble pressure ratio and the hydrostatic and formation pressure ratio with their respective ranges of typical values as shown in **Figures 7–9**. Then the output variable is established as the well qualification, where the range of values goes from 0 to 100%, 50% the well is fully recoverable, less than

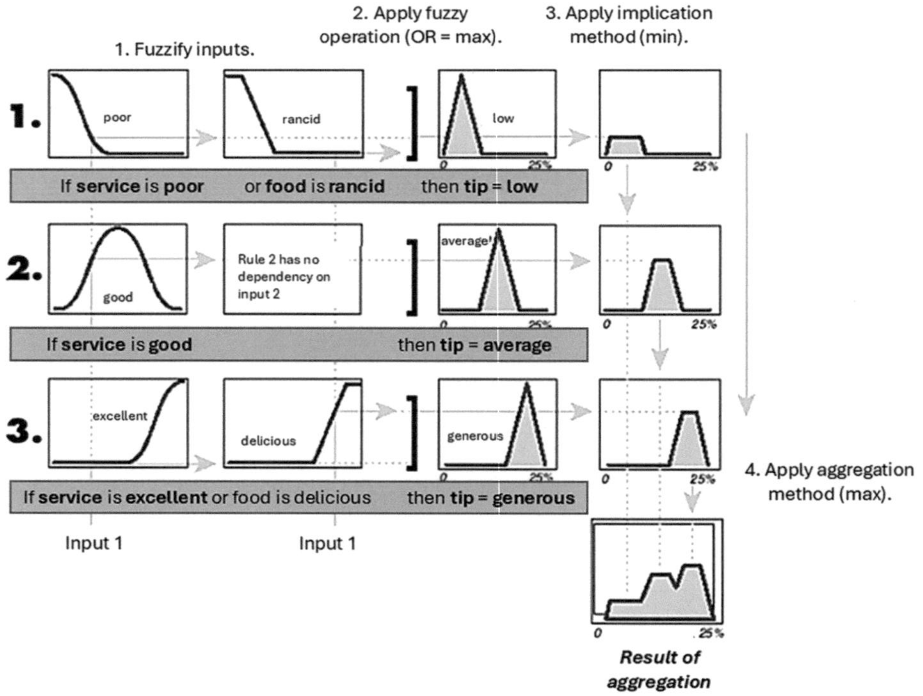

Figure 12.
Evaluation of the three rules: Fuzzification, inference and aggregation.

50% is barely recoverable and greater than 50% is self-sufficient. It should be noted that an oil well is being qualified according to the different pressures already defined in previous paragraphs, to determine if it can be subject to recovery with techniques to modify its conditions, whether primary, secondary or enhanced. Since the variables (input and output) are available, the block of rules with the 81 generated rules and their degrees of belonging must be specified. These rules are generated from the combination of the three variables with their three labels. In this way, the process carried out is the fuzzification and defuzzification of the variables, that is, having a fixed input value, an output value with a well rating can be obtained. The process is shown in **Figure 13**, where it starts with input data that measures the process variable, which may present noise and deviations with respect to the real value. Then, in the fuzzification, a number of values corresponding to the membership functions to which it belongs are converted, in order to subsequently carry out the evaluation of the rules, which define the control or knowledge strategy, and operations are carried out between the sets, to later move to the inference, which involves determining the output set of each rule. Likewise, the aggregate follows, where the membership function of the output variable is obtained from some operation between all the output sets of the inference stage. In the defuzzification, the membership function of the output is defined, and it is determined which data is the most representative of the total output set. Finally, the output data is obtained, which is the variable that will be taken to modify the state of the process.

The values of the degree of support range from 0 to 1, with 0 being totally discarded the rule and 1 implying that it is totally confirmed. The combination of the three variables being considered (pressure gradient, the relationship of operating pressure and bubble pressure, and the relationship of hydrostatic pressure and formation pressure) with their respective labels that were previously analyzed (**Figures 7–9**) results in 81 rules (Appendix A). It is important to mention that all rules were evaluated with the degree of support (DoS) to determine the degree of confirmation.

For example, **Table 2** shows the combination of some labels for the three variables with their respective degrees of support (DoS) in the resulting category “status well”. If you are 100% sure that the resulting category is met (barely recoverable, recoverable and self-sufficient), DoS = 1. On the other hand, if you are 100% sure that this category will not occur DoS = 0. In this way, rules 1, 2 and 3 the pressure gradient is

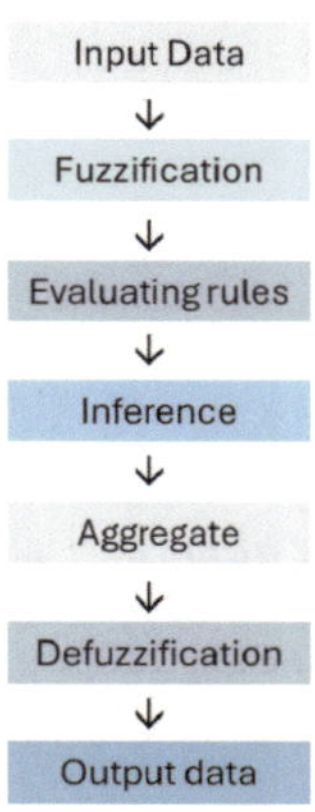

Figure 13.
Stages for decision making.

Rule		Pressure gradient		Operating pressure/bubble pressure		Hydrostatic pressure/formation pressure		DoS	Status well
1	If	Subnormal	And	Hardly recommended	And	Underbalanced	Then	0	Barely recoverable
2	If	Subnormal	And	Hardly recommended	And	Underbalanced	Then	1	Recoverable
3	If	Subnormal	And	Hardly recommended	And	Underbalanced	Then	0	Self-sufficient
55	If	Abnormal	And	Hardly recommended	And	Underbalanced	Then	0.33	Barely recoverable
56	If	Abnormal	And	Hardly recommended	And	Underbalanced	Then	0.66	Recoverable
57	If	Abnormal	And	Hardly recommended	And	Underbalanced	Then	0	Self-sufficient

Table 2.
Block of rules where the type of well is qualified, where DoS = 1 is fully confirmed, DoS = 0 is fully ruled out.

subnormal, the relationship between the operating pressure and the bubble pressure is hardly recommended and the relationship between the hydrostatic pressure and the formation pressure is underbalanced, the combination of these three labels imply that there is a 100% certainty that the well can be recovered, therefore DoS = 1 for recoverable, DoS = 0 for barely recoverable and DoS = 0 for self-sufficient. In case there is no complete certainty what will happen with the combination of labels, for example in rules 55–57, the pressure gradient is abnormal, the ratio of operating pressure to bubble pressure is hardly recommended and the ratio of hydrostatic pressure to formation pressure is underbalanced, two labels meet "hardly recommended and underbalanced" for the recoverable category, that is the reason of DoS = 0.66, "abnormal" is a label for "barely recoverable" and none of the categories would give the conditions for the well to be "self-sufficient", because of that, DoS = 0.

The three oil well pressure conditions discussed above are related:

1. Formation pressure gradient.

2. Relationship between operating pressure and bubble pressure.

3. Relationship between hydrostatic pressure and formation pressure.

In this way, the well can be categorized as depleted, which implies that it is "recoverable" with enhanced recovery (EOR) techniques, as shown in **Figure 14**.

Each condition must be evaluated to obtain a label. That is, the formation pressure gradient can be *subnormal*, *normal* or *abnormal* according to the established value ranges; for the purposes of this methodology to be pertinent, the formation pressure gradient is required to be "*subnormal*" for an EOR technique to be required for improvement of the well production, in this way, the formation pressure gradient is below normal, which implies that the pressure must be increased with EOR to improve the condition and thus obtain more resources. It should be noted that the well conditions that are selected are not normal, they must be those that need stimulation

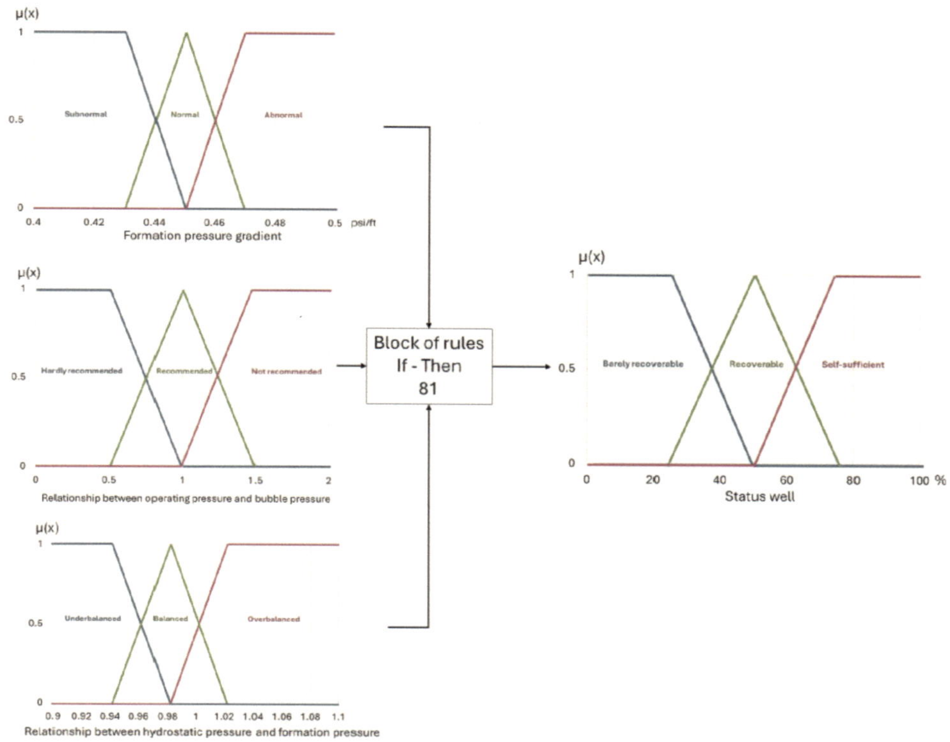

Figure 14.
Fuzzy subsystem structure: status well.

with EOR to be restored to the level required for the well resource to be recovered. The next condition, which is the one that relates the operating pressure and the bubble pressure, can be *hardly recommended*, *recommended* or *not recommended*; the label that must be selected for an EOR to be required is "*hardly recommended*", which means that an external agent is necessary to improve the operation pressure to a value at least equal to the bubble pressure. The last condition that is being considered is the relationship of the hydrostatic pressure with the formation pressure; it can be *underbalanced*, *balanced* or *overbalanced*; In this case, the relationship must be "*underbalanced*" for EOR to be used. The combination of these three labels will allow to categorize the evaluated well, and therefore the well status can be classified as *barely recoverable*, *recoverable* and *self-sufficient*. If the combination of the three conditions are: *subnormal*, *hardly recommended* and *underbalanced*, the status of the evaluated well is *recoverable*, that is, only for this combination DoS =1, which means that is 100% recoverable and EOR must be used to recover the remaining resource (**Figure 15**).

5. Results

To determine the correct performance of the proposed methodology, three values of each of the three defined pressure parameters are used to generate different scenarios that allow for establishing the dependence between the variables and their applicability.

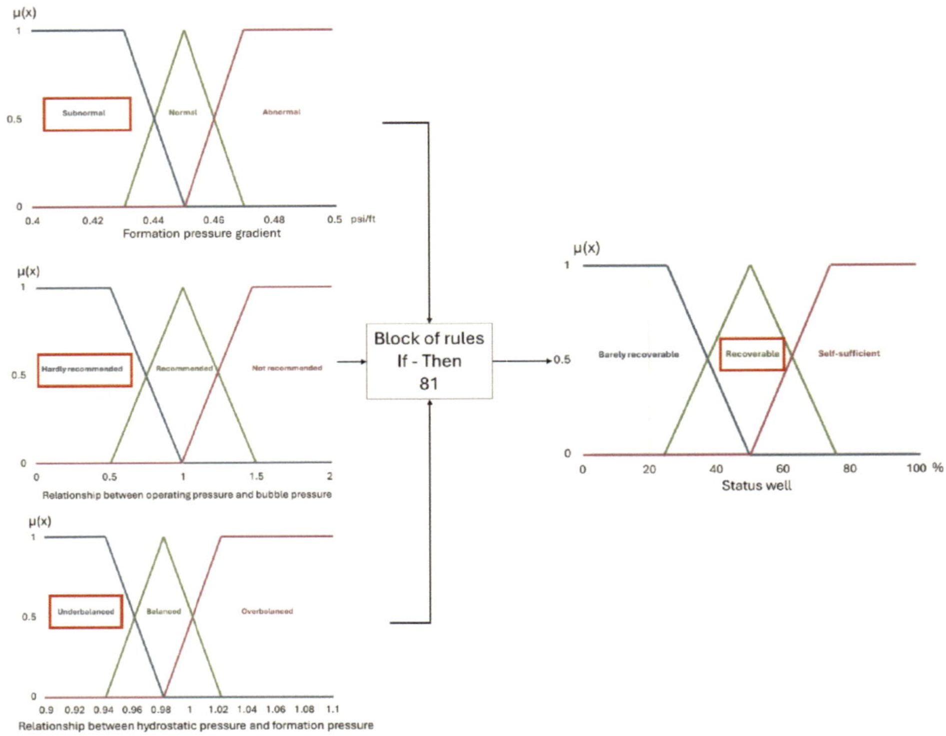

Figure 15.
Qualification of the well to be recovered.

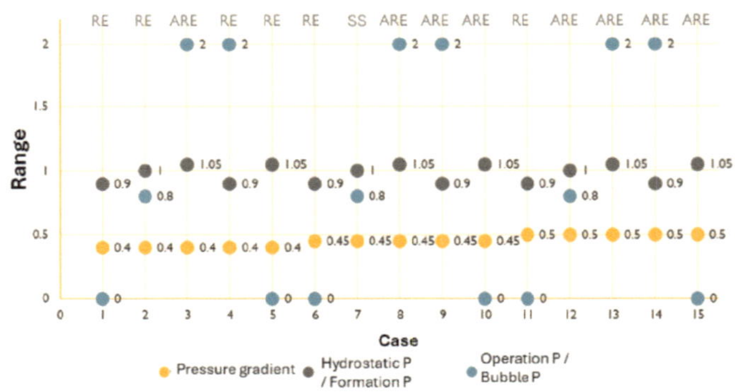

Figure 16.
Pressure gradient effect: variation in hydrostatic pressure/formation pressure and variation in operation pressure/bubble pressure. Their respective well rating is included (capital letters in gray. RE: Recoverable, ARE: Barely recoverable, SS: Self-sufficient).

In the first scenario being analyzed, the pressure gradient is kept constant using three different values (0.4, 0.45, 0.5) while the operating pressure/bubble pressure ratio is varied with three values between 0 and 2 (the typical values being shown in

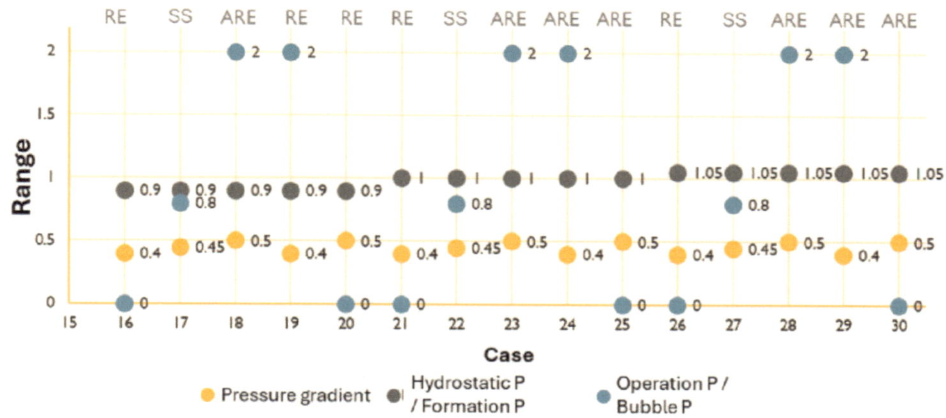

Figure 17.
Effect of hydrostatic pressure and formation pressure ratio: variation in pressure gradient and variation in operation pressure/bubble pressure. Their respective well rating is included (capital letters in gray. RE: Recoverable, ARE: Barely recoverable, SS: Self-sufficient).

Figure 7) and the hydrostatic pressure/formation pressure ratio with three values between 0.9 and 1.05. All the value combinations are shown in Appendixes - B. **Figure 16** illustrates that if there are at least two variables with values close to the lower limit, it leads to the well qualification result being "recoverable" (cases 1, 2, 4, 5, 6, 11); this makes sense, since the block of rules was built under that scheme. Subnormal

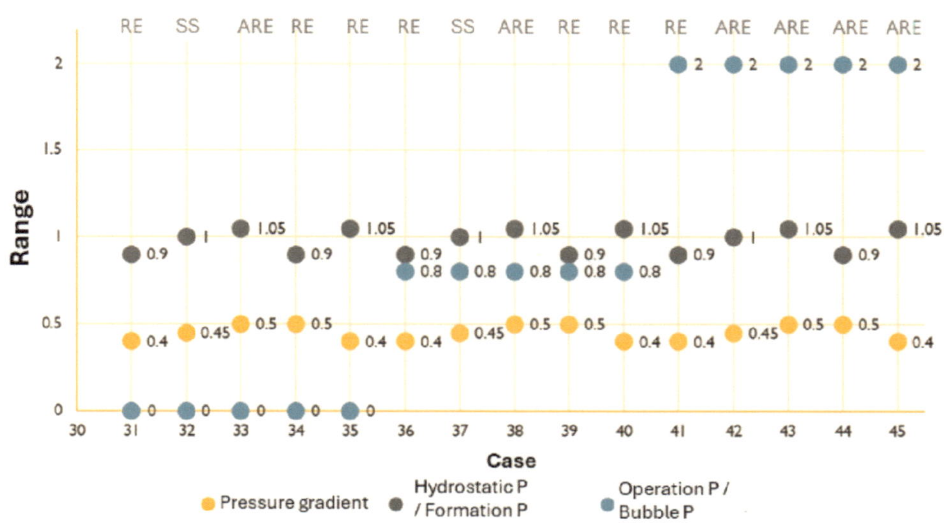

Figure 18.
Effect of operating pressure and bubble pressure ratio: variation in pressure gradient and variation in hydrostatic pressure/formation pressure. Their respective well rating is included (capital letters in gray. RE: Recoverable, ARE: Barely recoverable, SS: Self-sufficient).

pressures, that is, those pressures that are below the normal gradient, can cause lost circulation problems in wells drilled with liquid drilling mud [34]. On the other hand, in case 7, the resulting label is "SS", that is, self-sufficient, this behaves according to the rules that were established in the block of rules, where the values 0.45 for the pressure gradient, 0.8 for the operating pressure/bubble pressure and 1 for hydrostatic pressure with formation pressure are intermediate values in the ranges of each case. It then implies that the conditions are adequate to continue producing; no recovery technique is required. When at least one of the three established values is met, it results in the qualification of barely recoverable as in cases 3, 8, 9, 10 and 12–15. **Figure 17** considers the constant relationship between hydrostatic pressure and formation pressure. It behaves in the same way as the three variables under consideration. That is, if two of them are close to the lower limit, it implies a recoverable well. If two variables are at the middle limit, the well is "self-sufficient" and if two of the three variables tend to the upper limit, the well is classified as "barely recoverable". It should be noted that the fact that the well is "recoverable" means that although it does not have enough energy to continue producing autonomously, it is a candidate well for an EOR technique depending on other factors, such as economic and environmental factors. **Figure 18** shows that with different value ranges of the relationship between operating pressure and constant bubble pressure, it behaves in a similar way. Since pressure variables are being handled, what is sought is for the values to be below the operating limits so that it is possible to classify the well as recoverable.

6. Conclusions

The growing problem of fuel shortages has given rise to the search for alternatives for the best use of existing resources. That is why it is important to categorize oil wells as "subject to recovery" with the intention that by knowing pressure parameters, it can be determined if it is possible to achieve such categorization since currently there are no alternatives that can be easily and quickly qualified. One of the most important facts is that a history of values or databases is not required to generate results. This is a huge advantage of using fuzzy logic for this type of problem.

The methodology was verified with three variables associated with different kinds of pressure, demonstrating that the results obtained are consistent with what is established in the block of rules. As a following stage of this work, other parameters such as saturation, °API, viscosity, compressibility and volume factor will be considered.

It should be noted that this is the first step, and as future work, it is intended that after categorizing the well, the best option for enhanced recovery technique is chosen according to availability, cost and feasibility. Later on, it is intended to promote the use of economic incentives for the incorporation of enhanced recovery techniques. Finally, efforts will also be made to find the best geolocation of the injection wells in the recovery of the wells.

Acknowledgements

This work was supported by UNAM Postdoctoral Program (POSDOC).

Appendices and nomenclature

A. Block of rules

See **Table A1**.

Rule		Pressure gradient		Operating pressure/Bubble pressure		Hydrostatic pressure/ Formation pressure		DoS	Status well
1	If	S	And	AR	And	SU	Then	0	ARE
2	If	S	And	AR	And	SU	Then	1	RE
3	If	S	And	AR	And	SU	Then	0	SS
4	If	S	And	AR	And	B	Then	0	ARE
5	If	S	And	AR	And	B	Then	0.666666667	RE
6	If	S	And	AR	And	B	Then	0.333333333	SS
7	If	S	And	AR	And	SB	Then	0.333333333	ARE
8	If	S	And	AR	And	SB	Then	0.666666667	RE
9	If	S	And	AR	And	SB	Then	0	SS
10	If	S	And	R	And	SU	Then	0	ARE
11	If	S	And	R	And	SU	Then	0.666666667	RE
12	If	S	And	R	And	SU	Then	0.333333333	SS
13	If	S	And	R	And	B	Then	0	ARE
14	If	S	And	R	And	B	Then	0.333333333	RE
15	If	S	And	R	And	B	Then	0.666666667	SS
16	If	S	And	R	And	SB	Then	0.333333333	ARE
17	If	S	And	R	And	SB	Then	0.333333333	RE
18	If	S	And	R	And	SB	Then	0.333333333	SS
19	If	S	And	NR	And	SU	Then	0.333333333	ARE
20	If	S	And	NR	And	SU	Then	0.666666667	RE
21	If	S	And	NR	And	SU	Then	0	SS
22	If	S	And	NR	And	B	Then	0.333333333	ARE
23	If	S	And	NR	And	B	Then	0.333333333	RE
24	If	S	And	NR	And	B	Then	0.333333333	SS
25	If	S	And	NR	And	SB	Then	0.666666667	ARE
26	If	S	And	NR	And	SB	Then	0.333333333	RE
27	If	S	And	NR	And	SB	Then	0	SS
28	If	N	And	AR	And	SU	Then	0	ARE
29	If	N	And	AR	And	SU	Then	0.666666667	RE
30	If	N	And	AR	And	SU	Then	0.333333333	SS
31	If	N	And	AR	And	B	Then	0	ARE

Rule		Pressure gradient		Operating pressure/Bubble pressure		Hydrostatic pressure/ Formation pressure		DoS	Status well
32	If	N	And	AR	And	B	Then	0.333333333	RE
33	If	N	And	AR	And	B	Then	0.666666667	SS
34	If	N	And	AR	And	SB	Then	0.333333333	ARE
35	If	N	And	AR	And	SB	Then	0.333333333	RE
36	If	N	And	AR	And	SB	Then	0.333333333	SS
37	If	N	And	R	And	SU	Then	0	ARE
38	If	N	And	R	And	SU	Then	0.333333333	RE
39	If	N	And	R	And	SU	Then	0.666666667	SS
40	If	N	And	R	And	B	Then	0	ARE
41	If	N	And	R	And	B	Then	0	RE
42	If	N	And	R	And	B	Then	1	SS
43	If	N	And	R	And	SB	Then	0.333333333	ARE
44	If	N	And	R	And	SB	Then	0	RE
45	If	N	And	R	And	SB	Then	0.666666667	SS
46	If	N	And	NR	And	SU	Then	0.333333333	ARE
47	If	N	And	NR	And	SU	Then	0.333333333	RE
48	If	N	And	NR	And	SU	Then	0.333333333	SS
49	If	N	And	NR	And	B	Then	0.333333333	ARE
50	If	N	And	NR	And	B	Then	0	RE
51	If	N	And	NR	And	B	Then	0.666666667	SS
52	If	N	And	NR	And	SB	Then	0.666666667	ARE
53	If	N	And	NR	And	SB	Then	0	RE
54	If	N	And	NR	And	SB	Then	0.333333333	SS
55	If	AB	And	AR	And	SU	Then	0.333333333	ARE
56	If	AB	And	AR	And	SU	Then	0.666666667	RE
57	If	AB	And	AR	And	SU	Then	0	SS
58	If	AB	And	AR	And	B	Then	0.333333333	ARE
59	If	AB	And	AR	And	B	Then	0.333333333	RE
60	If	AB	And	AR	And	B	Then	0.333333333	SS
61	If	AB	And	AR	And	SB	Then	0.666666667	ARE
62	If	AB	And	AR	And	SB	Then	0.333333333	RE
63	If	AB	And	AR	And	SB	Then	0	SS
64	If	AB	And	R	And	SU	Then	0.333333333	ARE
65	If	AB	And	R	And	SU	Then	0.333333333	RE
66	If	AB	And	R	And	SU	Then	0.333333333	SS
67	If	AB	And	R	And	B	Then	0.333333333	ARE

Rule		Pressure gradient		Operating pressure/Bubble pressure		Hydrostatic pressure/ Formation pressure		DoS	Status well
68	If	AB	And	R	And	B	Then	0	RE
69	If	AB	And	R	And	B	Then	0.666666667	SS
70	If	AB	And	R	And	SB	Then	0.666666667	ARE
71	If	AB	And	R	And	SB	Then	0	RE
72	If	AB	And	R	And	SB	Then	0.333333333	SS
73	If	AB	And	NR	And	SU	Then	0.666666667	ARE
74	If	AB	And	NR	And	SU	Then	0.333333333	RE
75	If	AB	And	NR	And	SU	Then	0	SS
76	If	AB	And	NR	And	B	Then	0.666666667	ARE
77	If	AB	And	NR	And	B	Then	0	RE
78	If	AB	And	NR	And	B	Then	0.333333333	SS
79	If	AB	And	NR	And	SB	Then	0.333333333	ARE
80	If	AB	And	NR	And	SB	Then	0	RE
81	If	AB	And	NR	And	SB	Then	0	SS

Table A1.
Complete block of rules where the type of well is qualified, where DoS = 1 is fully confirmed, DoS = 0 is fully ruled out.

B. Variation for results

To test the proposed block of rules, the effect of the change for each variable is shown with the well rating and its degree of belonging to the established label (**Table B1**).

	Pressure gradient	Hydrostatic pressure/formation pressure	Operating pressure/bubble pressure	Grade	Label
1	0.4	0.9	0	50	RE
2	0.4	1	0.8	50	RE
3	0.4	1.05	2	12.5	ARE
4	0.4	0.9	2	50	RE
5	0.4	1.05	0	50	RE
6	0.45	0.9	0	50	RE
7	0.45	1	0.8	87.5	SS
8	0.45	1.05	2	12.5	ARE
9	0.45	0.9	2	12.5	ARE
10	0.45	1.05	0	12.5	ARE
11	0.5	0.9	0	50	RE

	Pressure gradient	Hydrostatic pressure/formation pressure	Operating pressure/bubble pressure	Grade	Label
12	0.5	1	0.8	12.5	ARE
13	0.5	1.05	2	12.5	ARE
14	0.5	0.9	2	12.5	ARE
15	0.5	1.05	0	12.5	ARE
16	0.4	0.9	0	50	RE
17	0.45	0.9	0.8	87.5	SS
18	0.5	0.9	2	12.5	ARE
19	0.4	0.9	2	50	RE
20	0.5	0.9	0	50	RE
21	0.4	1	0	50	RE
22	0.45	1	0.8	87.5	SS
23	0.5	1	2	12.5	ARE
24	0.4	1	2	12.5	ARE
25	0.5	1	0	12.5	ARE
26	0.4	1.05	0	50	RE
27	0.45	1.05	0.8	87.5	SS
28	0.5	1.05	2	12.5	ARE
29	0.4	1.05	2	12.5	ARE
30	0.5	1.05	0	12.5	ARE
31	0.4	0.9	0	50	RE
32	0.45	1	0	87.5	SS
33	0.5	1.05	0	12.5	ARE
34	0.5	0.9	0	50	RE
35	0.4	1.05	0	50	RE
36	0.4	0.9	0.8	50	RE
37	0.45	1	0.8	87.5	SS
38	0.5	1.05	0.8	12.5	ARE
39	0.5	0.9	0.8	50	RE
40	0.4	1.05	0.8	50	RE
41	0.4	0.9	2	50	RE
42	0.45	1	2	12.5	ARE
43	0.5	1.05	2	12.5	ARE
44	0.5	0.9	2	12.5	ARE
45	0.4	1.05	2	12.5	ARE

Table B1.
Combinations of values of each variable and their degree of belonging to the established label.

Author details

Esbeydi Villicaña-García*, Arturo Palacio-Pérez and Silvia Raquel García-Benítez
Institute of Engineering – National Autonomous University of Mexico, Mexico City, Mexico

*Address all correspondence to: esbeydi.villicana@umich.mx

References

[1] Naciones Unidas. El papel de los combustibles fósiles en un sistema energético sostenible. 2015. Available from: https://www.un.org/es/chronicle/article/el-papel-de-los-combustibles-fosiles-en-un-sistema-energetico-sostenible#:~:text=Los%20combustibles%20f%C3%B3siles%20comprenden%20el, emisiones%20globales%20de%20CO2 [Accessed: June 2024]

[2] Our world in data. Fossil fuels. 2024. Available from: https://ourworldindata.org/fossil-fuels [Accessed: June 2024]

[3] National Geographic. Distribution of Fossil Fuels. 2023. Available from: https://education.nationalgeographic.org/resource/distribution-fossil-fuels/ [Accessed: June 2024]

[4] Statista. Volumen de las reservas probadas de petróleo crudo en México de 2005 a 2023. 2024. Available from: https://es.statista.com/estadisticas/606126/volumen-de-las-reservas-probadas-de-petroleo-crudo-mexico/#:~:text=M%C3%A9xico%3A%20volumen%20de%20las%20reservas%20probadas%20de%20petr%C3%B3leo%20crudo%202005%2D2023&text=En%202023%2C%20las%20reservas%20probadas,las%20reservas%20probadas%20en%202022 [Accessed: July 2024]

[5] Comisión Nacional de Hidrocarburos. Lineamientos que regulan el procedimiento de cuantificación y certificación de Reservas de la Nación. 2017. Available from: https://cnh.gob.mx/media/18296/lineamientos-que-regulan-el-procedimiento-de-cuantificacion-y-certificacion-de-reservas-de-la-nacion.pdf [Accessed: July 2024]

[6] El mundo. Los pozos abandonados de petróleo y gas en EEUU siguen emitiendo metano años después de cerrados. 2016. Available from: https://www.elmundo.es/ciencia/2016/11/22/5829bdcb46163fe3668b4577.html [Accessed: July 2024]

[7] Bret-Rouzaut N. Economics of oil and gas production. In: Hafner M, Luciani G, editors. The Palgrave Handbook of International Energy Economics. Cham: Palgrave Macmillan; 2022. DOI: 10.1007/978-3-030-86884-0_1

[8] Eren T, Polat C. Natural gas underground storage and oil recovery with horizontal wells. Journal of Petroleum Science and Engineering. 2020;**187**:106753. DOI: 10.1016/j.petrol.2019.106753

[9] Demirel NC, Demirel T, Deveci M, Vardar G. Location selection for underground natural gas storage using Choquet integral. Journal of Natural Gas Science and Engineering. 2017;**45**:368-379. DOI: 10.1016/j.jngse.2017.05.013

[10] Bhattacharyya S, Vyas A. Application of machine learning in predicting oil rate decline for Bakken shale oil wells. Scientific Reports. 2022;**12**:16154. DOI: 10.1038/s41598-022-20401-6

[11] Aslam N, Khan IU, Alansari A, Alrammah M, Alghwairy A, Alqahtani R, et al. Anomaly detection using explainable random forest for prediction of undesirable events in oil wells. Applied Computational Intelligence and Soft Computing. 2022;**1**:1558381. DOI: 10.1155/2022/1558381

[12] Pan S, Yang B, Wang S, Guo Z, Wang L, Liu J, et al. Oil well production prediction based on CNN-LSTM model

with self-attention mechanism. Energy. 2023;**284**(1):128701. DOI: 10.1016/j.energy.2023.128701

[13] Ghahnavieh MZ, Manesh HH, Sheikhmoradi S. Comparative comparison of fuzzy logic and classical logic. International Journal of Nonlinear Analysis and Applications. 2024;**15**(12): 435-452. DOI: 10.22075/ijnaa.2024.32978.4905

[14] Mazzoleni M, Sarda K, Acernese A, Russo L, Manfredi L, Glielmo L, et al. A fuzzy logic-based approach for fault diagnosis and condition monitoring of industry 4.0 manufacturing processes. Engineering Applications of Artificial Intelligence. 2022;**115**:105317. DOI: 10.1016/j.engappai.2022.105317

[15] Okwu MO, Nwachukwu AN. A review of fuzzy logic applications in petroleum exploration, production and distribution operations. Journal of Petroleum Exploration and Production Technology. 2019;**9**:1555-1568. DOI: 10.1007/s13202-018-0560-2

[16] RTVE. Pozos petrolíferos abandonados: la fantasmagórica bomba de relojería climática. 2023. Available from: https://www.rtve.es/noticias/20230607/pozos-petroliferos-fantasma-bomba-relojeria-climatica/2446514.shtml [Accessed: July 2024]

[17] El financiero. Pozos petroleros abandonados: La 'bronca' que enfrenta el Golfo de México por 30 mil mdd. 2023. Available from: https://www.elfinanciero.com.mx/economia/2023/05/08/pozos-petroleros-abandonados-en-el-golfo-de-mexico-contaminacion-y-altos-costos/ [Accessed: July 2024]

[18] Capitulo 1. Lógica difusa y sistemas de control. 2017. Available from: https://cayetanoguerra.github.io/ia/logicadifusa/Logica%20difusa%20y%20sistemas%20de%20control.pdf [Accessed: June 2024]

[19] Thakur R, Sikka G, Bansal U, Giri J, Mallik S. Deadline – Aware and energy efficient IoT task scheduling using fuzzy logic in fog computing. Multimedia Tools and Applications. 2024. DOI: 10.1007/s11042-024-19509-w

[20] Zadeh LA. Fuzzy logic – a personal perspective. Fuzzy Sets and Systems. 2015;**281**:4-20. DOI: 10.1016/j.fss.2015.05.009

[21] Feltan C, Caballero AL. Principios de lógica difusa. 2016. Available from: https://www.researchgate.net/publication/328052443_Principios_de_logica_difusa [Accessed: June 2024]

[22] Capítulo 2. Conceptos fundamentales de lógica difusa. 2023. Available from: https://www.tdx.cat/bitstream/handle/10803/6887/04Rpp04de11.pdf;sequence=4 [Accessed: January 2024]

[23] Dernoncourt, F. Introduction to Fuzzy Logic. 2013. Available from: https://www.researchgate.net/publication/267041266_Introduction_to_fuzzy_logic [Accessed: June 2024]

[24] Mamdani EH. Application of fuzzy algorithms for control of simple dynamic plant. Proceedings of the Institution of Electrical Engineers. 1974;**121**(12): 1585-1588. DOI: 10.1049/piee.1974.0328

[25] Mamdani EH, Assilian S. An experiment in linguistic synthesis with a fuzzy logic controller. International Journal of Man-Machine Studies. 1975; 7(1):1-13. DOI: 10.1016/S0020-7373(75)80002-2

[26] Chinchilla E, Rojas JD. Diseño e implementación de un controlador difuso que, interviene la inyección

electrónica de combustible en un vehículo, para la utilización de HHO como combustible complementario. 2014. Available from: https://www.researchgate.net/publication/270589487_Diseno_e_implementacion_de_un_controlador_difuso_que_interviene_la_inyeccion_electronica_de_combustible_en_un_vehiculo_para_la_utilizacion_de_HHO_como_combustible_complementario [Accessed: June 2024]

[27] Blej M, Azizi M. Comparison of Mamdani-type and Sugeno-type fuzzy inference systems for fuzzy real time SCHEDULING. International Journal of Applied Engineering Research. 2016; **11**(22):11071-11075

[28] Capítulo 1. Lógica difusa y sistemas de control. 2023. Available from: https://cayetanoguerra.github.io/ia/logicadifusa/Logica%20difusa%20y%20sistemas%20de%20control.pdf [Accessed: December 2023]

[29] Satter A, Iqbal GM. Reservoir Engineering: 4-Reservoir fluid properties2016. pp. 81-105. ISBN: 978-0-12-800219-3. DOI: 10.1016/C2013-0-13485-X

[30] Universidad de Aquino Bolivia. Conceptos básicos de presiones. 2018. Available from: https://s8dcd9b0a8ad1af4e.jimcontent.com/download/version/1526422818/module/5702463066/name/03%20PRESIONES.pdf [Accessed: February 2024]

[31] Instituto Mexicano de Petróleo. Biblioteca Visual del Petróleo. 2015. Available from: https://www.gob.mx/cms/uploads/attachment/file/123944/Biblioteca_Visual_del_Petr_leo.pdf [Accessed: July 2024]

[32] Servicio Geológico Mexicano, Petróleo en México. 2017. Available from: https://www.sgm.gob.mx/Web/MuseoVirtual/Aplicaciones_geologicas/Petroleo-en-Mexico.html [Accessed: July 2024]

[33] Mathworks, Fuzzy Inference Process. Available from: https://la.mathworks.com/help/fuzzy/fuzzy-inference-process.html [Accessed: July 2024]

[34] Birchall T, Senger K, Swarbrick R. Naturally occurring underpressure – A global review. Petroleum Geoscience. 2022;**28**(2):1-22. DOI: 10.1144/petgeo2021-051

Chapter 5

Smart Technologies in Enhanced Oil Recovery: Integrating AI, Nanotechnology, and Sustainable Practices

Nouratan Singh, Poonam Rani and Neeraj Tandan

Abstract

Enhanced oil recovery (EOR) is a critical method for extracting additional oil from mature reservoirs, but it faces increasing pressure to become more efficient and environmentally sustainable. This chapter explores the integration of *smart technologies* such as *artificial intelligence (AI)*, *nanotechnology*, and *sustainable practices* into EOR. AI is revolutionizing EOR operations by optimizing reservoir management, improving real-time monitoring, and reducing operational costs. Nanotechnology enhances oil recovery through the use of *functionalized nanoparticles* and *smart fluids*, which improve oil mobility and reduce chemical consumption. Additionally, sustainable practices, including *CO_2-EOR*, *water-efficient techniques*, and *biodegradable chemicals*, are being adopted to lower the environmental impact of EOR, especially in terms of carbon emissions and water use. While challenges remain—such as the high cost of technology and fluctuating oil prices—the future of EOR holds promise through continuous technological innovation and a growing emphasis on sustainability.

Keywords: enhanced oil recovery (EOR), artificial intelligence (AI), nanotechnology, sustainable practices, carbon capture utilization and storage (CCUS), reservoir management, microbial enhanced oil recovery (MEOR), functionalized nanoparticles, renewable energy integration, water management in EOR, biodegradable polymers, low-carbon EOR, energy transition

1. Introduction

The growing global demand for energy, coupled with the declining availability of easily accessible oil reserves, has necessitated the development of more advanced oil recovery methods. Enhanced oil recovery (EOR) techniques, which aim to extract additional oil beyond what primary and secondary recovery methods can achieve, have played a critical role in increasing oil production. However, conventional EOR methods—including thermal recovery, gas injection, and chemical flooding—are not without limitations. These traditional techniques often face operational inefficiencies, high costs, and environmental concerns, making them less viable as standalone solutions, particularly in challenging reservoir conditions [1].

In response to these challenges, the oil and gas industry is increasingly looking towards *smart technologies* to enhance the effectiveness of EOR operations. These technologies—ranging from *Artificial Intelligence (AI)* and *nanotechnology to sustainability-driven innovations*—have the potential to revolutionize oil recovery by improving efficiency, reducing costs, and addressing environmental concerns [2]. Integrating AI and machine learning with EOR processes allows for data-driven decision-making and predictive modeling, enabling more precise optimization of reservoir management [3]. Nanotechnology, on the other hand, provides new materials and mechanisms for improving oil mobility at the molecular level, significantly boosting recovery rates [4]. Moreover, sustainable methods like *Carbon Capture, Utilization, and Storage (CCUS)* are being implemented to mitigate the environmental impact of enhanced oil recovery (EOR) processes, in accordance with worldwide initiatives to decrease carbon emissions [5].

1.1 The growing need for smart technologies in EOR

The increasing complexity of oil reservoirs—many of which are highly heterogeneous, deep, or involve high-temperature and high-pressure conditions—has exposed the limitations of traditional EOR techniques. For example, thermal recovery methods, while effective in some reservoirs, can be energy-intensive and costly, particularly in deep or offshore reservoirs. Similarly, chemical flooding methods face challenges such as chemical degradation and environmental risks related to wastewater disposal [6]. The integration of AI and machine learning helps overcome some of these limitations by providing real-time data analysis, which enables operators to make informed decisions about how to adjust their recovery strategies based on evolving reservoir conditions. AI models have been successfully used to predict reservoir performance, optimize injection rates, and prevent equipment failures, improving both operational efficiency and production outcomes [7].

At the same time, nanotechnology is providing innovative solutions for EOR. Nanoparticles, due to their small size and large surface area, exhibit unique properties that can enhance fluid flow, reduce interfacial tension, and alter wettability—critical factors in improving oil mobility and recovery rates [8]. For example, nanoparticles are being engineered to improve the efficiency of water flooding, surfactant-polymer flooding, and other chemical EOR methods by increasing the displacement efficiency of oil trapped in the porous rock [9]. Field trials in various locations have demonstrated that nanotechnology can significantly increase oil recovery while reducing the amount of chemicals needed, thus lowering environmental risks [10].

1.2 Sustainability concerns in traditional EOR

The environmental impact of conventional EOR techniques has raised significant concerns in recent years, particularly in the context of climate change and global efforts to reduce greenhouse gas emissions. Conventional methods often involve the use of large quantities of chemicals or high energy inputs, both of which contribute to carbon emissions and environmental degradation [11]. For instance, steam-assisted gravity drainage (SAGD) and other thermal methods generate considerable carbon emissions due to the energy required to produce steam at the surface [12]. Chemical

EOR methods, on the other hand, pose challenges related to chemical toxicity and the safe disposal of produced water.

In light of these challenges, sustainability is becoming a key consideration in the future of EOR. Technologies such as *Carbon Capture, Utilization, and Storage (CCUS)*, which involves capturing CO_2 emissions from industrial processes and injecting them into oil reservoirs to enhance recovery, are emerging as dual-purpose solutions. CCUS not only improves oil production but also mitigates CO_2 emissions, making it a viable solution for reducing the carbon footprint of EOR operations [13]. Projects such as the Weyburn-Midale CCUS project in Canada have already demonstrated the potential of this technology, capturing millions of tons of CO_2 while increasing oil recovery rates [14].

1.3 The convergence of AI, nanotechnology, and sustainability in EOR

The future of enhanced oil recovery lies in the integration of AI, nanotechnology, and sustainable practices. By combining these smart technologies, oil producers can overcome the limitations of traditional methods and maximize recovery efficiency while minimizing environmental impact. AI enables predictive modeling and data-driven optimization, which leads to more efficient operations and reduced downtime [15]. Nanotechnology enhances the physical and chemical processes involved in oil recovery, improving the displacement of trapped oil and increasing the overall recovery factor [16]. Sustainability-driven innovations, such as CCUS, help reduce the carbon footprint of EOR processes, making oil production more environmentally friendly [17].

This convergence of smart technologies offers a promising future for the oil industry, especially as more fields enter maturity and the focus shifts to maximizing the recovery of remaining resources. As these technologies continue to evolve, they will play a pivotal role in the future of energy production, enabling more efficient and sustainable recovery methods that can meet the world's growing energy needs while addressing environmental concerns.

2. Artificial intelligence in enhanced oil recovery

Artificial Intelligence (AI) has emerged as a transformative technology in various industries, and its impact on Enhanced Oil Recovery (EOR) is no exception. In EOR operations, AI's ability to process vast amounts of data, model complex reservoir behaviors, and optimize operational parameters in real-time has led to significant improvements in oil recovery efficiency. Data on AI applications in EOR shown in **Table A1**. By leveraging machine learning algorithms, neural networks, and data analytics, AI has provided the oil industry with a powerful toolkit to address challenges such as declining reservoir performance, high operational costs, and environmental concerns.

2.1 The role of AI in EOR

The integration of AI in EOR focuses primarily on predictive modeling, reservoir characterization, and process optimization. Traditional EOR methods often rely on limited data and are based on simplified reservoir models, which can result in suboptimal decision-making. AI, on the other hand, enables data-driven models that analyze large datasets from reservoirs to provide more accurate predictions of oil recovery potential under different scenarios [18]. These models can simulate reservoir behavior with higher precision, allowing operators to dynamically adjust injection strategies, enhance oil displacement, and maximize recovery rates [19].

One of the key areas where AI is applied in EOR is *reservoir management*. AI-based models can analyze geological, geophysical, and production data to generate highly detailed reservoir simulations. These simulations enable a better understanding of reservoir heterogeneities, fluid flow patterns, and the impact of various recovery methods. For example, AI-powered tools can predict changes in reservoir pressure and fluid saturation during injection processes, providing insights that help optimize the placement of injection wells and the choice of recovery techniques [20].

AI also plays a crucial role in optimizing injection rates and well performance. Machine learning algorithms analyze historical data from wells and reservoirs to determine the optimal rates of injection for water, gas, or chemicals. These models continuously learn and adapt based on real-time data, allowing operators to fine-tune the injection process to maximize oil recovery while minimizing waste and costs [21].

2.2 Applications of machine learning in EOR

Machine learning, a subset of AI, has gained significant traction in EOR operations due to its ability to recognize patterns, handle complex datasets, and make predictions based on past experiences. Various machine learning techniques, such as neural networks, decision trees, and support vector machines, are used to improve decision-making in EOR projects.

The decision-making process in EOR projects is enhanced with the use of several machine learning techniques. These approaches include neural networks, decision trees, and support vector machines.

One prominent application of machine learning in EOR is *reservoir performance forecasting*. Neural networks, in particular, are well suited for analyzing nonlinear relationships between reservoir parameters and production data. These networks can predict future production rates, estimate remaining oil in place, and suggest optimal recovery strategies [22]. By providing accurate forecasts, machine learning models help operators reduce the risk of underperformance and enhance production efficiency.

Another key application of machine learning is *real-time data processing* and *predictive maintenance*. In EOR operations, maintaining equipment reliability is critical to minimizing downtime and avoiding costly failures. Machine learning algorithms are used to monitor the health of equipment such as pumps, compressors, and pipelines. By analyzing sensor data from these machines, AI can detect early warning signs of equipment failure, enabling proactive maintenance and reducing unplanned shutdowns [23]. Predictive maintenance enhances operating efficiency and prolongs equipment longevity, hence decreasing long-term expenses.

2.3 Reinforcement learning for optimizing EOR processes

In addition to machine learning, *reinforcement learning (RL)* has shown promise in optimizing EOR processes. In contrast to conventional machine learning, which emphasizes pattern recognition in data, reinforcement learning entails an agent that acquires knowledge by interaction with an environment, getting feedback in the form of rewards or punishments.

The reinforcement learning (RL) agent learns by interacting with an environment and receiving feedback in the form of rewards or penalties. In the context of EOR, RL can be used to optimize recovery strategies by continuously adjusting injection parameters based on real-time reservoir performance.

For example, in water flooding operations, RL algorithms can be programmed to adjust the water injection rate based on how the reservoir responds to different injection pressures and flow rates. The RL agent learns over time which injection parameters yield the highest oil recovery, leading to more efficient operations [24]. RL is also being explored in the optimization of *gas injection* and *chemical flooding*, where it can help adjust the composition and volume of injected materials to achieve maximum recovery.

2.4 AI for reservoir characterization and monitoring

Reservoir characterization is a critical aspect of EOR, as it provides detailed information about the geological and fluid properties of the reservoir. AI-powered tools enhance reservoir characterization by processing large datasets from seismic surveys, well logs, and core samples. These tools can identify geological features such as faults, fractures, and high-permeability zones that are not easily detectable using traditional methods [25].

AI techniques, particularly deep learning, are increasingly used in the interpretation of *seismic data*. Deep learning models can analyze seismic images and automatically classify reservoir zones, making it easier for geoscientists to identify areas with high oil saturation. This enables more accurate placement of injection and production wells, improving the effectiveness of EOR methods [26].

Moreover, AI is being employed in *reservoir monitoring* through the use of sensor networks and remote sensing technologies. By integrating data from sensors placed throughout the reservoir, AI algorithms can monitor real-time changes in reservoir conditions such as pressure, temperature, and fluid composition. This real-time monitoring allows operators to detect potential issues, such as water breakthrough or gas migration, early and adjust their recovery strategies accordingly [27].

2.5 Case studies of AI in EOR

The application of AI in EOR has led to several successful case studies. For example, *Chevron Corporation* has implemented AI-driven reservoir management systems in its mature fields, leading to significant improvements in oil recovery. By analyzing production data and adjusting injection rates in real time, Chevron reported a 10% increase in recovery rates in certain fields [28].

In another case, *Petrobras*, the Brazilian oil company, utilized machine learning algorithms to optimize the injection of water and gas in its offshore fields. The AI-driven optimization reduced operational costs by 15% and increased oil recovery by over 12% [29]. These examples demonstrate the potential of AI to transform EOR operations by improving efficiency, reducing costs, and enhancing recovery [30].

3. Nanotechnology in enhanced oil recovery

Nanotechnology has emerged as a powerful tool in enhanced oil recovery (EOR), offering new approaches to overcome the limitations of conventional recovery techniques. The application of nanotechnology in EOR primarily focuses on enhancing fluid properties, improving mobility control, and altering reservoir wettability at the nanoscale, thus maximizing oil recovery from mature and challenging reservoirs. Nanoparticles (NPs) exhibit unique physical and chemical properties, such as high surface area-to-volume ratios and tunable functionalities, which make them highly effective for improving oil displacement and sweep efficiency in EOR processes. Properties of functionalized nanoparticles shown in **Table B1**.

3.1 Role of nanoparticles in EOR

Nanoparticles play a crucial role in EOR by altering the interactions between reservoir fluids and rock surfaces. Their small size and high reactivity enable them to penetrate deep into reservoir pores, enhancing oil recovery through mechanisms such as altering wettability, reducing interfacial tension, and improving the viscosity of injection fluids. In particular, nanoparticles can be tailored to work synergistically with existing EOR methods, such as water flooding, chemical flooding, and gas injection, to achieve higher recovery factors.

One of the primary mechanisms through which nanoparticles enhance EOR is *wettability alteration*. In many reservoirs, oil remains trapped in pores due to unfavorable wettability conditions, where the rock surface is oil-wet or mixed-wet, preventing the effective displacement of oil. Nanoparticles can alter the wettability of the reservoir rock to become more water-wet, improving the mobility of the trapped oil and allowing it to be swept more efficiently by the injected fluids [31]. For instance, silica nanoparticles have been widely used for this purpose due to their ability to alter wettability and reduce interfacial tension between oil and water phases [32].

Another important role of nanoparticles is in reducing interfacial tension (IFT). High IFT between oil and water often leads to poor displacement efficiency, with significant amounts of oil remaining in the reservoir. Nanoparticles, particularly those functionalized with surfactants or other surface-active agents, can significantly reduce IFT, enabling better mixing of the injection fluids with the oil phase. This leads to improved mobilization of the residual oil, enhancing the overall recovery factor [33].

3.2 Types of nanoparticles used in EOR

Several types of nanoparticles have been developed and tested for their effectiveness in EOR, each with unique properties that contribute to oil recovery in different ways. Some of the most commonly used nanoparticles in EOR include the following:

Silica nanoparticles (SiO_2): Silica nanoparticles are among the most studied and widely used in EOR due to their ability to alter wettability and reduce interfacial tension. These nanoparticles can be functionalized with various chemical groups to improve their performance in specific reservoir conditions. Silica NPs have shown significant potential in water flooding operations, where they increase oil recovery by improving the displacement efficiency of trapped oil [34].

Metal oxide nanoparticles: Metal oxide nanoparticles, such as titanium dioxide (TiO_2) and zinc oxide (ZnO), are known for their thermal stability and photocatalytic properties, making them suitable for high-temperature reservoirs. These nanoparticles can enhance EOR by altering wettability, reducing interfacial tension, and promoting oil displacement through their unique surface interactions with reservoir fluids [35].

Magnetic nanoparticles: Magnetic nanoparticles, such as *iron oxide (Fe_3O_4)*, have attracted attention for their ability to be manipulated using external magnetic fields. This property allows for more controlled injection and distribution of nanoparticles within the reservoir. Additionally, magnetic nanoparticles can improve oil recovery by enhancing the viscosity of injection fluids and altering the flow characteristics of the reservoir [36].

Carbon-based nanoparticles: Carbon-based materials, such as carbon nanotubes (CNTs) and graphene oxide, offer unique properties, including high surface area, electrical conductivity, and mechanical strength. These materials can improve oil recovery by enhancing fluid properties and altering the wettability of reservoir rocks. CNTs, in particular, have shown potential in thermal EOR applications, where they help improve heat transfer and oil displacement in steam injection processes [37].

3.3 Mechanisms of nanotechnology in EOR

The mechanisms by which nanoparticles enhance oil recovery are complex and involve several interrelated processes. Some of the key mechanisms include the following:

Wettability alteration: As mentioned, altering the wettability of the reservoir rock from oil-wet to water-wet is a critical mechanism in improving oil mobility. Nanoparticles can be engineered to adsorb onto the rock surface, changing its affinity from oil to water and thus facilitating better oil displacement during water flooding or chemical flooding operations [38].

Interfacial tension reduction: Nanoparticles can reduce the interfacial tension between oil and water, improving the ability of injected fluids to mobilize trapped oil. This is particularly important in surfactant-assisted EOR processes, where nanoparticles act as co-surfactants to enhance the emulsification and displacement of oil from the reservoir pores [39].

Viscosity modification: Nanoparticles can modify the viscosity of injection fluids, making them more effective at displacing oil from the reservoir. For example, adding nanoparticles to polymer solutions in polymer flooding can improve the stability and viscosity of the polymer, leading to better sweep efficiency and higher oil recovery [40].

Enhanced thermal conductivity: In thermal EOR methods such as steam injection, nanoparticles can improve the thermal conductivity of the injected fluids, allowing for more efficient heat transfer to the reservoir. This enhances the ability of steam to reduce the viscosity of heavy oil and mobilize it for production [41].

3.4 Field applications and case studies

Numerous field trials and laboratory studies have demonstrated the effectiveness of nanoparticles in enhancing oil recovery. For instance, Petroleum Development Oman (PDO) conducted a field trial using silica nanoparticles in a mature reservoir to improve water flooding efficiency. The trial resulted in a 10% increase in oil recovery, highlighting the potential of nanoparticles to enhance conventional EOR methods [42].

Similarly, *Chevron Corporation* has explored the use of metal oxide nanoparticles in their steam injection operations. By incorporating titanium dioxide nanoparticles into the steam, Chevron was able to improve heat transfer and reduce the viscosity of heavy oil, leading to a 15% increase in recovery rates from high-temperature reservoirs [43].

In another case, Saudi Aramco conducted a study on the use of magnetic nanoparticles in chemical flooding operations. The magnetic properties of the nanoparticles allowed for better control of fluid injection and distribution, resulting in more efficient sweep and higher recovery rates [44]. These examples illustrate the versatility and effectiveness of nanotechnology in enhancing oil recovery across a variety of reservoir conditions.

3.5 Challenges and future outlook

Despite the promising potential of nanotechnology in EOR, several challenges remain. The cost of producing and scaling up nanoparticle-based EOR solutions can be high, particularly for more complex nanoparticles such as carbon nanotubes and magnetic nanoparticles. Additionally, there are concerns about the environmental impact of nanoparticles, particularly in terms of their long-term behavior in the subsurface environment and potential toxicity to ecosystems [45].

However, ongoing research and development are addressing these challenges by focusing on the development of more cost-effective and environmentally friendly nanoparticles. As the technology matures, it is expected that nanotechnology will play an increasingly important role in the future of EOR, particularly in the context of complex and mature reservoirs where conventional methods are less effective.

4. Sustainable practices in enhanced oil recovery

Enhanced oil recovery (EOR) has traditionally been viewed as a means of extending the life of declining reservoirs and maximizing oil recovery. However, as global concerns over environmental sustainability and climate change intensify, the oil and gas industry has come under increased scrutiny for its environmental impact. Sustainable practices in EOR aim to balance the demand for increased oil production with the need to reduce greenhouse gas emissions, minimize environmental degradation, and ensure the responsible use of resources. This shift towards sustainability has led to the adoption of greener technologies, carbon management strategies, and resource-efficient methods that reduce the overall environmental footprint of EOR operations [46].

4.1 The importance of sustainability in EOR

EOR processes are energy-intensive and often involve the use of chemicals, steam, or gases that can have detrimental environmental effects if not managed properly. For instance, steam injection, a common thermal EOR technique, requires significant amounts of water and energy, leading to increased CO_2 emissions. Similarly, chemical flooding can introduce potentially harmful substances into the environment. In response to these concerns, there is a growing emphasis on incorporating sustainability into EOR projects to ensure that they not only improve oil recovery but also reduce environmental harm.

Sustainable EOR practices are being developed to address the following key areas:

Carbon emissions reduction: As global efforts to mitigate climate change intensify, reducing carbon emissions in EOR operations has become a critical focus. This is being achieved through carbon capture and storage (CCS), as well as carbon capture, utilization, and storage (CCUS) techniques [13].

Water management: The efficient use of water in EOR processes is essential for reducing the environmental impact of operations, especially in regions where water scarcity is a concern [47].

Chemical minimization: Sustainable EOR practices aim to reduce the reliance on harmful chemicals and develop biodegradable or less toxic alternatives [48].

Energy efficiency: Sustainable EOR also focuses on increasing the energy efficiency of operations, utilizing renewable energy sources, and reducing the overall energy consumption of EOR methods [49].

4.2 Carbon management in EOR: CO_2-EOR

One of the most prominent sustainable practices in EOR is CO_2-EOR, which simultaneously enhances oil recovery and mitigates carbon emissions. CO_2-EOR involves the injection of carbon dioxide into the reservoir to improve oil displacement by reducing oil viscosity and increasing its mobility. This technique enhances oil recovery while facilitating the long-term sequestration of CO_2 in the reservoir, hence diminishing the volume of CO_2 emitted into the environment.

This dual benefit of CO_2-EOR makes it one of the most promising sustainable EOR techniques.

In CO_2-EOR, collected CO_2 from industrial sources, such power plants or refineries, is compressed and pumped into the reservoir. Carbon dioxide amalgamates with the oil, diminishing its viscosity and facilitating its flow into producing wells. After the injection process, a portion of the CO_2 remains trapped in the reservoir, providing a form of carbon sequestration that helps reduce greenhouse gas emissions [50]. This method has been successfully implemented in numerous fields worldwide, particularly in North America, where it has been used to enhance recovery from mature fields while simultaneously reducing emissions.

The main environmental advantage of CO_2-EOR is its ability to permanently store CO_2 in geological formations, effectively removing it from the atmosphere. Studies estimate that CO_2-EOR can store up to 0.6 metric tons of CO_2 for every barrel of oil produced, making it a key technology for achieving climate goals while continuing oil production [51]. However, the long-term effectiveness of CO_2 storage, potential leakage risks, and monitoring challenges remain critical areas of research and development [52].

4.3 Water management in EOR

Water is a critical resource in many EOR methods, particularly in water flooding, steam injection, and polymer flooding. However, the large volumes of water required for these processes can lead to water scarcity issues, especially in arid regions. Sustainable water management practices in EOR focus on reducing freshwater consumption, reusing produced water, and minimizing the environmental impact of water disposal.

One approach to sustainable water management is the use of produced water—the water that is co-produced with oil and gas during extraction. Instead of disposing of produced water, which can be expensive and environmentally harmful, many EOR operations are increasingly reusing it for injection purposes. Water recycling technologies are being developed to treat produced water so that it meets the quality standards required for reinjection into the reservoir [53]. This not only reduces the demand for freshwater but also minimizes the environmental impact of wastewater disposal.

In steam-assisted gravity drainage (SAGD), a thermal EOR method, efficient water management is crucial due to the large volumes of steam required for heavy oil production [54].

Closed-loop water systems are being developed to recycle the steam condensate, reducing freshwater consumption and improving the overall sustainability of SAGD operations. Moreover, innovations in desalination and water treatment technologies are enabling the use of saline or brackish water in EOR processes, further decreasing the reliance on freshwater resources [55].

4.4 Minimizing chemical use in EOR

Chemical flooding, particularly surfactant flooding and polymer flooding, involves the injection of chemicals into the reservoir to improve oil displacement. However, the use of large quantities of synthetic chemicals poses environmental risks, such as soil and water contamination if they are not properly managed. To address these concerns, sustainable EOR practices aim to develop biodegradable or environmentally friendly chemicals that reduce environmental harm while maintaining the effectiveness of EOR processes.

Research is being conducted on biosurfactants and biopolymers, which are derived from natural, renewable resources and are less harmful to the environment than traditional chemicals. For instance, rhamnolipids, a type of biosurfactant produced by bacteria, have shown potential in reducing interfacial tension and enhancing oil recovery without the environmental risks associated with synthetic surfactants [56]. Similarly, biodegradable polymers, such as xanthan gum, are being tested as alternatives to synthetic polymers for improving sweep efficiency in polymer flooding [57].

Moreover, nanotechnology is playing an increasingly important role in reducing the chemical use in EOR. Nanoparticles can be used to enhance the performance of surfactants and polymers, allowing for the same or better oil recovery with lower chemical concentrations. This reduces the overall chemical load on the environment while maintaining the effectiveness of the EOR process [45].

Enhanced oil recovery (EOR) is a critical technology in the petroleum industry, aimed at maximizing the extraction of hydrocarbons from reservoirs [58]. Among the various EOR methods, chemical EOR—including surfactant flooding and polymer flooding—plays a significant role [59]. However, the use of chemicals in these processes poses environmental and economic challenges, necessitating strategies to minimize chemical usage while maintaining efficiency.

4.4.1 Important chemicals used in surfactant flooding and chemical flooding

Surfactant flooding: Surfactant flooding is one of the most widely used methods in enhanced oil recovery (EOR), which improves oil recovery by reducing the interfacial tension between oil and water. This process involves the injection of surfactants to mobilize oil from the reservoir [60]. The key chemicals involved in surfactant flooding include:

Anionic surfactants: Anionic surfactants are negatively charged and are highly effective in reducing interfacial tension. These surfactants are used extensively in surfactant flooding due to their ability to improve oil displacement. Sodium dodecyl sulfate (SDS), a commonly used anionic surfactant, is known for its strong efficiency in reducing interfacial tension [61].

Cationic surfactants: Cationic surfactants carry a positive charge and are used in reservoirs where the rock surface is negatively charged. These surfactants are often employed to enhance oil recovery in certain types of formations. Cetyltrimethylammonium bromide (CTAB) has been studied for its effectiveness in oil recovery in cationic surfactant systems [62].

Nonionic surfactants: Nonionic surfactants do not carry any charge, making them less sensitive to the salinity of the reservoir water. These surfactants are particularly useful in high-salinity environments, where ionic surfactants tend to lose their effectiveness. Polyethylene glycol (PEG) derivatives are commonly used in nonionic surfactant systems due to their stability and low sensitivity to salinity [63].

Amphoteric surfactants: Amphoteric surfactants possess both positive and negative charges, allowing them to work efficiently in various pH and salinity conditions. These surfactants provide flexibility in different reservoir conditions. Lauryl betaine has been used in laboratory studies for its ability to perform well across a range of pH and ionic conditions [64].

Co-surfactants and solvents: Co-surfactants and solvents are often used to enhance the solubility and efficiency of primary surfactants. These compounds work by altering the physicochemical properties of the surfactant solution. Alcohols like butanol and isopropyl alcohol (IPA) are common co-surfactants that help improve the stability of surfactants in oil [65].

Chemical flooding: Chemical flooding involves the injection of chemicals that alter the properties of reservoir fluids, improving oil recovery by increasing the viscosity of water or breaking down oil components [66, 67]. The chemicals commonly used in chemical flooding include the following:

Polymers: Polymers are often injected to increase the viscosity of the water, which helps improve the sweep efficiency by reducing the mobility of water in comparison to oil. Partially hydrolyzed polyacrylamide (HPAM) is the most commonly used polymer for chemical flooding due to its high viscosity and cost-effectiveness [68].

Alkaline agents: Alkaline agents are injected to raise the pH of the reservoir, promoting the formation of soap-like compounds that can emulsify crude oil, thus aiding in oil recovery. Sodium hydroxide (NaOH) and potassium hydroxide (KOH) are commonly used as alkaline agents in chemical flooding processes [69].

Polymer-surfactant mixtures: The combination of surfactants and polymers is often used to improve both displacement and sweep efficiencies. The polymer increases the viscosity, while the surfactant reduces interfacial tension. A mixture of polyacrylamide with surfactants like SDS or Triton X-100 has been demonstrated to improve the overall recovery factor in chemical flooding operations [70].

Viscosifiers: Viscosifiers are chemicals used to increase the viscosity of injected water or fluids, making them more effective in displacing oil. Xanthan gum is widely used as a viscosifier in chemical flooding because of its excellent stability and high viscosity at low concentrations [71].

Enzymes: In some chemical flooding processes, enzymes are used to break down high-viscosity oil components, making it easier to recover the oil from the reservoir. Cellulases and lipases are examples of enzymes that have been tested for their ability to degrade oil components [72].

4.4.2 Strategies to minimize chemical usage

Optimized formulation: Developing tailored chemical formulations for specific reservoir conditions can reduce the quantity of chemicals required [73].

Nanotechnology integration: Incorporating nanoparticles into surfactants and polymers can improve their efficiency, thereby reducing the amount needed [74].

Smart water injection: Modifying the ionic composition of the injected water can enhance oil recovery, potentially reducing the reliance on chemicals [75].

Advanced monitoring techniques: Real-time monitoring and modeling of EOR processes can optimize chemical injection and minimize wastage [76].

Biodegradable and eco-friendly alternatives: Developing and using biodegradable chemicals can reduce environmental impact and enhance sustainability [77].

4.5 Energy efficiency and renewable energy integration in EOR

EOR processes, particularly thermal methods like steam injection, are highly energy-intensive, contributing to increased carbon emissions and operational costs. To improve sustainability, there is a growing focus on energy efficiency and the integration of renewable energy sources into EOR operations.

One promising approach is the use of solar energy to generate steam for thermal EOR methods. In regions with abundant sunlight, solar thermal plants can be used to produce the steam required for steam injection, reducing the reliance on fossil fuels and lowering the carbon footprint of the operation [78].

Geothermal energy is another renewable energy source being explored for EOR. Geothermal heat can be harnessed to generate the steam or hot water needed for thermal EOR processes, providing a sustainable and cost-effective energy solution [79].

In addition to renewable energy integration, energy recovery systems are being developed to capture waste heat from EOR operations and reuse it within the process. By improving energy efficiency and reducing waste, these systems help lower the overall environmental impact of EOR operations [80].

4.6 Challenges and future directions in sustainable EOR

While sustainable practices in EOR have shown significant potential, several challenges remain. The cost of implementing sustainable technologies, such as carbon capture and renewable energy systems, can be prohibitively high for some operators. Additionally, the long-term environmental impacts of certain sustainable methods, such as nanoparticle use and CO_2 storage, are still being studied. There is also the challenge of regulatory and policy frameworks, which vary by region and can influence the adoption of sustainable EOR practices [81].

Looking forward, continued innovation and investment in sustainable EOR technologies will be critical to reducing the environmental impact of oil production. The development of low-carbon technologies, renewable energy integration, and resource-efficient methods will help ensure that EOR can continue to contribute to global energy needs while aligning with climate and sustainability goals [82].

5. Future prospects and challenges in enhanced oil recovery (EOR)

As the global energy landscape evolves, the future of enhanced oil recovery (EOR) lies at the intersection of technological advancements, environmental sustainability, and economic viability. While EOR has proven to be an effective strategy for increasing oil recovery from mature fields, significant challenges remain. These challenges include the cost and complexity of EOR technologies, environmental concerns, and fluctuating oil prices. However, continued innovation and research are paving the way for more efficient, environmentally friendly, and cost-effective EOR techniques.

5.1 Prospects for technological advancements in EOR

EOR technologies are expected to evolve rapidly in the coming decades, driven by advancements in artificial intelligence (AI), nanotechnology, and sustainable practices. The integration of these technologies promises to improve the efficiency and success rates of EOR operations while reducing their environmental impact.

AI-driven optimization: The use of AI in EOR is expected to grow, with advanced algorithms helping to optimize reservoir management, predict production rates, and reduce operational costs. AI models can also analyze vast datasets to recommend the most effective EOR methods for specific reservoirs. AI can improve real-time monitoring, automate decision-making processes, and optimize the allocation of resources, all of which enhance recovery rates and reduce environmental risks [83]. In the future, AI systems could further integrate data from multiple sources, enabling even more precise control over EOR operations.

Nanotechnology: The future of nanotechnology in EOR holds immense potential for revolutionizing the industry. Nanoparticles can improve the mobility of trapped oil, enhance chemical reactions, and reduce the amount of chemicals required in EOR processes. As research progresses, functionalized nanoparticles—those designed for specific reservoir conditions—will likely become more prevalent. These particles can be customized to alter their properties under different temperature, pressure, and salinity conditions, making them highly versatile for a range of EOR applications [84]. Furthermore, the use of smart fluids with nanomaterials may provide a more efficient and sustainable method for oil recovery, reducing the environmental impact of traditional chemical flooding techniques.

Biotechnology: Microbial enhanced oil recovery (MEOR) is another emerging technology with potential future applications. By utilizing bacteria and other microorganisms, MEOR can enhance oil recovery by altering the reservoir's fluid properties, reducing oil viscosity, and creating natural surfactants. While MEOR is still in its infancy compared to other EOR techniques, advances in biotechnology and microbiology are expected to improve its efficiency and adaptability in different reservoir environments [85].

5.2 Environmental and sustainability challenges

Despite the advancements in EOR technologies, the industry still faces significant challenges related to environmental sustainability. The extraction and recovery of hydrocarbons contribute to greenhouse gas (GHG) emissions and environmental degradation. As global efforts to combat climate change intensify, EOR projects must align with sustainability goals to remain viable.

Carbon emissions: One of the primary environmental concerns in EOR operations is the generation of CO_2 and other GHGs, particularly in thermal EOR processes that require large amounts of energy. While CO_2-EOR offers a solution by injecting captured carbon into reservoirs, its long-term success depends on the ability to securely store CO_2 without leakage [86]. Regulatory frameworks and advanced monitoring techniques are essential to ensuring that CO_2 injected into reservoirs remains permanently sequestered [87].

Water usage: Water-intensive methods, such as steam injection and water flooding, continue to pose sustainability challenges, especially in regions facing water scarcity. Future EOR operations will need to adopt more water-efficient techniques, such as closed-loop systems that recycle water or the use of alternative fluids like CO_2 or nitrogen for injection [77]. Additionally, advanced water treatment technologies will be necessary to safely reuse produced water, reducing the environmental impact of wastewater disposal. Water-efficient EOR technologies shown in **Table C1**.

Sustainable chemical use: While chemical flooding has proven effective in many reservoirs, it involves the use of synthetic chemicals that can harm the environment if not properly managed. To address this, future EOR methods must prioritize the use of environmentally friendly chemicals and biodegradable materials that reduce the potential for contamination and minimize the environmental footprint [88]. Continued research into biosurfactants and biopolymers will likely play a significant role in achieving this goal.

5.3 Economic challenges and future viability

The economic viability of EOR is heavily influenced by the fluctuating prices of oil and the costs associated with EOR technologies. As oil prices remain volatile, the high capital and operational costs of many EOR methods, such as steam injection and CO_2-EOR, can deter investment in EOR projects, especially in low-margin fields.

Cost of advanced technologies: While technologies like AI, nanotechnology, and renewable energy integration promise to improve EOR efficiency, their upfront costs remain a barrier for widespread adoption. CO_2 capture, utilization, and storage (CCUS), for example, is a critical technology for reducing emissions, but its cost is often prohibitively high without financial incentives or regulatory frameworks that encourage investment [89]. Overcoming these economic barriers will require both technological innovations that reduce costs and supportive policies that promote the adoption of low-carbon technologies.

Uncertain oil demand: Another economic challenge facing EOR is the uncertain future demand for oil, as the world transitions towards renewable energy sources. While oil will likely continue to play a significant role in the global energy mix for decades, the pace of the energy transition could reduce the need for EOR in certain regions. However, EOR projects that prioritize sustainability and minimize environmental impact may be better positioned to succeed in a low-carbon future [90].

5.4 Future directions and policy considerations

Looking ahead, the future of EOR will be shaped not only by technological advancements but also by regulatory and policy frameworks designed to address climate change and environmental sustainability. Governments and industry stakeholders will need to work together to create policies that incentivize the adoption of sustainable EOR technologies and reduce the overall environmental impact of oil production.

Carbon pricing and incentives: Implementing carbon pricing mechanisms, such as carbon taxes or cap-and-trade systems, could provide the necessary financial incentives for operators to adopt low-carbon EOR technologies. Policies that support investment in CCUS and other carbon management technologies will be crucial for the future viability of CO_2-EOR and other sustainable EOR methods [91].

Regulations on water use and chemical management: Future policies will likely impose stricter regulations on water use and chemical management in EOR operations.

Governments may encourage the use of recycled water and environmentally friendly chemicals through subsidies or other incentives. In addition, monitoring and reporting standards will need to be strengthened to ensure that EOR projects comply with environmental regulations and minimize their impact on local ecosystems [92].

Collaboration and innovation: The future of EOR will also depend on continued collaboration between industry, academia, and government agencies. Research partnerships will be essential for advancing the next generation of EOR technologies, from AI and nanotechnology to sustainable chemical solutions. By fostering a culture of innovation, the oil and gas industry can ensure that EOR remains a viable and environmentally responsible method for meeting global energy demands.

6. The ability to minimize environmental impact

The sustainable EOR practices aim to develop biodegradable or environmentally friendly chemicals that reduce environmental harm while maintaining the effectiveness of EOR processes. Recent advancements in this area have led to the development of several environmentally friendly chemicals that are being tested for their ability to minimize environmental impact as shown in **Tables 1** and **2**. Some examples include the following.

6.1 Biodegradable surfactants

These surfactants are designed to break down into non-toxic components after use, reducing long-term environmental impact. For example, surfactants derived from renewable resources, such as fatty acids or sugar-based surfactants, have been shown to exhibit biodegradability while maintaining oil recovery efficiency. In one study, it was found that biodegradable surfactants could reduce the toxicity of effluent by up to 30% compared to traditional surfactants like sodium dodecyl sulfate (SDS) [93]. While the cost of biodegradable surfactants may be higher (about 10–20% more expensive than conventional ones), their environmental benefits and regulatory acceptance make them a viable option for sustainable EOR.

Parameter	Conventional EOR practices	Sustainable EOR practices
Cost of surfactants (per kg)	\$2–\$4 (for synthetic surfactants like SDS)	\$2–\$3 (for biodegradable surfactants or biosurfactants)
Environmental toxicity	High (toxic effluents in wastewater)	25–30% reduction in toxicity (biodegradable options)
Cost of polymers (per kg)	\$1.5–\$2 (for synthetic HPAM)	\$2–\$2.5 (for biopolymers like xanthan gum)
Wastewater contamination	High, requires expensive treatment	30–35% reduction in contamination (biopolymers, enzymes)
Oil recovery efficiency	40–60% (depending on reservoir conditions)	40–60% (comparable with improved sustainability)
CO_2 emission reduction (in CO_2-EOR)	Moderate	50% reduction in emissions (natural alkali agents)

Table 1.
Statistical comparison of old vs. new practices.

Category	Traditional EOR	AI-based EOR	Nanotechnology EOR	Sustainable EOR
Techniques used	Water, gas, and chemical flooding	AI-driven reservoir modeling, predictive analytics	Nanoparticles, nanofluids	CO_2 sequestration, biosurfactants, renewable energy
Efficiency	Moderate	High (optimized through AI insights)	High (enhanced fluid properties)	Moderate to high (depends on approach)
Cost	High (due to chemical and energy use)	Lower (efficiency reduces operational costs)	Moderate to high (depends on nanoparticle cost)	Moderate (cost offset by sustainable benefits)
Environmental impact	High (hazardous chemicals, water use)	Moderate (depends on implementation)	Low to moderate (depends on nanoparticle recycling)	Low (focus on minimizing emissions)
Scalability	Established, widely implemented	High (scalable with existing data)	Moderate (requires advanced materials)	Moderate to high (depends on policy support)
Technological barriers	Minimal (well-understood processes)	Requires expertise in AI and data science	Advanced material development and handling	Integration of sustainable energy and processes
Case studies	Extensive	Emerging (e.g., predictive maintenance)	Field applications still growing	Proven in CO_2-EOR and bio-based EOR
Key challenges	High chemical and water use, emissions	Data availability and computational resources	Cost of nanoparticles, field deployment	Policy and upfront costs

Table 2.
A comparison showcasing the features, advantages, and challenges of different enhanced oil recovery (EOR) technologies.

6.2 Biosurfactants

Biosurfactants, produced by microorganisms, are an environmentally friendly alternative to traditional synthetic surfactants. They are biodegradable, non-toxic, and effective in reducing interfacial tension in EOR. Rhamnolipids, produced by *Pseudomonas aeruginosa*, have been shown to achieve oil recovery rates similar to conventional synthetic surfactants. Studies show that rhamnolipids can be produced at a cost of approximately $2–$3 per kilogram, which is comparable to some synthetic surfactants used in the industry [62]. Additionally, biosurfactants exhibit a lower environmental impact due to their natural origins and biodegradability. A study comparing biosurfactants to synthetic surfactants reported a 25% reduction in environmental toxicity in wastewater treatment processes.

6.3 Polymer-based flooding using biopolymers

Biopolymers, such as xanthan gum, which are naturally derived, can be used in chemical flooding. These biopolymers offer an eco-friendly alternative to synthetic polymers like partially hydrolyzed polyacrylamide (HPAM). A field study comparing the cost-effectiveness of xanthan gum to HPAM showed that the latter was about 15–20% less expensive; however, the biodegradability of xanthan gum resulted in a more sustainable long-term solution, with a 30% reduction in the environmental footprint [94].

6.4 Enzyme-based EOR

The use of enzymes, such as lipases and cellulases, for breaking down high-viscosity oil components provides an environmentally sustainable approach. These enzymes are biodegradable and less harmful compared to chemical additives. A comparative study of enzyme-based EOR and conventional chemical flooding found that enzyme-based methods led to a 20% reduction in chemical usage, while maintaining similar oil recovery rates [72, 95]. The overall environmental impact was also reduced by 35%, as enzymes do not contribute to the same level of contamination or toxicity in wastewater compared to synthetic chemicals.

6.5 Environmentally friendly chemical additives for CO_2-EOR

Chemical additives designed to improve CO_2-EOR processes while being environmentally safe have also been developed. For instance, using natural alkali agents like sodium bicarbonate instead of conventional chemicals such as potassium hydroxide leads to a 10% reduction in operational costs while offering a 50% reduction in environmental contamination associated with wastewater disposal [73]. These natural agents are less toxic, easily recyclable, and more environmentally friendly than traditional alkaline agents used in CO_2-EOR.

By shifting towards these environmentally friendly chemicals, the oil and gas industry can move towards more sustainable EOR practices that reduce environmental degradation while ensuring the continued effectiveness of oil recovery methods. The combination of lower toxicity, biodegradability, and a reduced environmental footprint makes these newer technologies a better choice for the future of EOR.

A bar chart comparing the relative costs of traditional EOR methods versus those employing advanced technologies like AI, nanotechnology, and sustainability-focused approaches. This visualization highlights the cost efficiency of modern practices in enhanced oil recovery as shown in **Figure 1**.

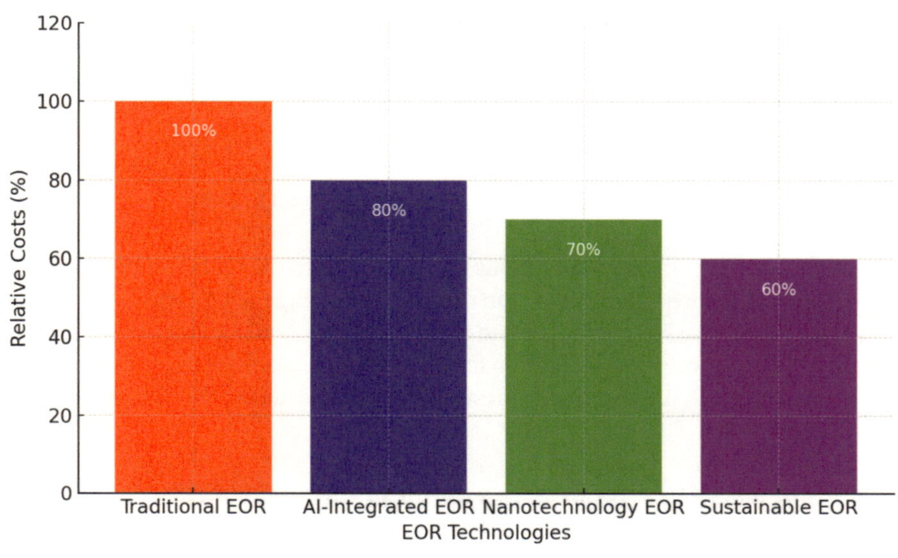

Figure 1.
Cost comparison: traditional vs. advanced EOR technology.

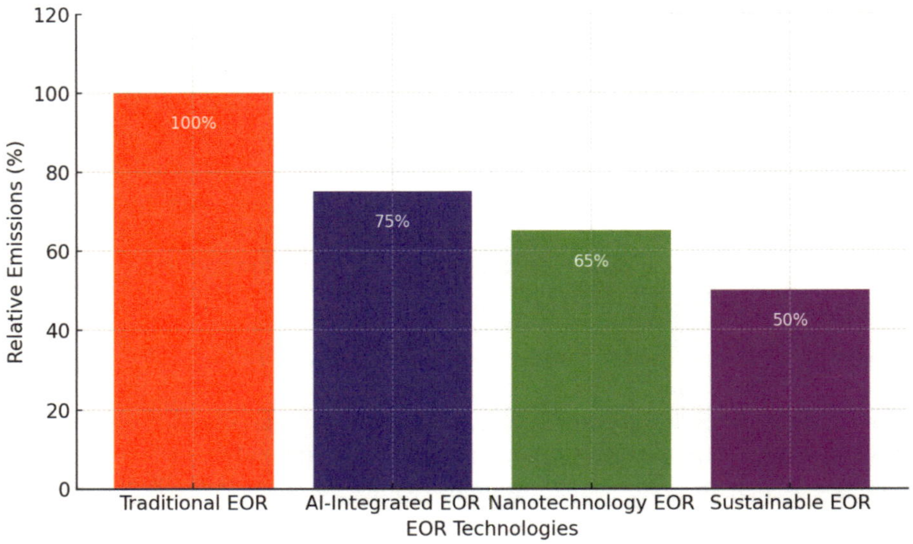

Figure 2.
Emission reduction: traditional vs. advanced EOR technology.

A bar chart visualizing the reduction in hazardous emissions achieved through sustainable enhanced oil recovery (EOR) practices compared to traditional methods. It demonstrates the significant environmental benefits of integrating AI, nanotechnology, and sustainability into EOR processes as shown in **Figure 2**.

7. Recent technologies in enhanced oil recovery (EOR) to reduce costs and minimize hazardous emissions

Enhanced oil recovery (EOR) methods have been pivotal in maximizing oil production from mature reservoirs. However, traditional EOR techniques can be expensive and environmentally challenging due to high chemical usage, energy consumption, and greenhouse gas emissions. Recent technological advancements aim to address these issues by improving cost-effectiveness, reducing chemical and energy consumption, and minimizing environmental impacts. Below are key innovations that are reshaping EOR:

7.1 Carbon capture and storage (CCS) integrated with CO_2-EOR

One of the most significant innovations in reducing both the costs and environmental impact of EOR is the integration of carbon capture and storage (CCS) with CO_2-EOR. In this process, carbon dioxide is injected into the reservoir, not only aiding in oil recovery but also acting as a long-term storage solution for CO_2. This approach helps mitigate climate change by trapping CO_2 underground, preventing it from entering the atmosphere [96].

Benefits:

- *Cost reduction:* CO_2-EOR can extend the life of existing oil fields, reducing the need for new drilling and the associated costs.

- *Environmental impact:* By sequestering CO_2, this method significantly reduces greenhouse gas emissions, contributing to carbon neutrality goals.

7.2 Nanotechnology for environmentally friendly EOR

Nanotechnology has the potential to enhance EOR efficiency while reducing environmental hazards. *Nanofluids*, which consist of nanoparticles suspended in fluids, are used in EOR to improve sweep efficiency and reduce oil viscosity. Moreover, *nanomaterials* can be engineered to minimize the use of harmful chemicals and optimize the recovery process [16, 97].

Benefits:

- *Cost efficiency*: Nanoparticles can be used in smaller quantities, reducing the need for large chemical volumes in the recovery process.

- *Reduced environmental impact*: The use of environmentally friendly nanoparticles (e.g., silica nanoparticles) minimizes the toxicity and environmental risks associated with traditional EOR chemicals.

7.3 Biotechnology and microbial enhanced oil recovery (MEOR)

Microbial enhanced oil recovery (MEOR) utilizes bacteria and microbial metabolism to improve oil recovery. This biotechnology-based approach involves injecting specific microorganisms or nutrients into the reservoir to either alter the viscosity of the oil, reduce surface tension, or produce gases that help displace the oil [85, 98].

Benefits:

- *Cost reduction*: Microbial agents are often less expensive than chemical additives and can be produced in situ, reducing transportation and handling costs.

- *Environmental sustainability*: MEOR is an eco-friendly approach as it involves the use of natural organisms, which can be less toxic and more biodegradable than conventional chemicals.

7.4 Advanced polymer flooding with bio-based polymers

Polymer flooding has been a well-established technique in EOR, where polymers like *HPAM* (partially hydrolyzed polyacrylamide) are injected to increase the viscosity of the water, improving the sweep efficiency of injected fluids. However, recent innovations involve the use of *bio-based polymers*, which are derived from renewable sources and are biodegradable, reducing the environmental footprint of the process [94, 99].

Benefits:

- *Cost reduction*: Bio-based polymers are often cheaper to produce than synthetic polymers.

- *Environmental sustainability*: They are more eco-friendly, biodegradable, and less harmful to reservoir ecosystems.

7.5 Renewable energy-powered EOR systems

Incorporating *renewable energy* sources such as *solar* and *wind energy* to power EOR operations can significantly reduce the carbon footprint associated with conventional fossil fuel-based energy sources. Renewable energy-powered *injection systems* can reduce the need for traditional energy sources, lowering operational costs and emissions [80, 100].

Benefits:

- *Cost efficiency*: Solar and wind energy can offer long-term savings by reducing reliance on expensive grid power.
- *Sustainability*: These systems can substantially decrease the carbon emissions associated with EOR operations, aligning with the global push towards sustainable energy practices.

8. Conclusion

Enhanced oil recovery (EOR) continues to play a pivotal role in meeting global energy demands by maximizing oil production from mature and declining reservoirs. However, as environmental sustainability and economic efficiency become increasingly critical, the integration of advanced technologies and greener practices in EOR has emerged as the key to its future viability. This chapter has explored the profound impact of *artificial intelligence (AI)*, *nanotechnology*, and *sustainable practices* on transforming EOR into a more efficient, eco-friendly, and economically sustainable method.

Artificial intelligence has proven instrumental in optimizing EOR operations, allowing for data-driven decision-making, real-time monitoring, and process automation, thereby enhancing recovery rates while reducing costs. The application of nanotechnology offers innovative solutions for improving fluid mobility, reducing chemical usage, and tailoring recovery methods to specific reservoir conditions. *Sustainable practices*, including carbon management through CO_2-EOR, efficient water usage, the minimization of harmful chemicals, and the integration of renewable energy sources, have underscored the industry's commitment to reducing its environmental footprint.

Despite these promising advancements, significant challenges remain. The high costs associated with implementing advanced technologies, coupled with concerns about carbon emissions, water consumption, and long-term sustainability, require continued research, investment, and regulatory support. However, with global efforts to mitigate climate change and transition to a low-carbon future, EOR's role in ensuring energy security while aligning with environmental goals has become more important than ever.

Recent advancements in technology offer promising solutions to reduce the costs and environmental impact of enhanced oil recovery (EOR). By integrating carbon capture, nanotechnology, biotechnology, bio-based polymers, and renewable energy, the oil and gas industry can enhance production efficiency while adhering to sustainable practices. These innovations are pivotal in ensuring the continued viability of EOR while minimizing its ecological footprint.

The future of EOR will be shaped by ongoing innovations in AI, nanotechnology, and sustainable practices, as well as the industry's ability to navigate the economic and environmental challenges it faces. By fostering collaboration between industry, government, and academia, and by embracing forward-thinking policies that support sustainability and technological advancement, EOR has the potential to remain a vital and responsible contributor to global energy supply. As EOR technologies evolve and become more sustainable, they will ensure that oil recovery is not only more efficient but also more aligned with the growing demand for environmental stewardship.

Acknowledgements

The completion of this chapter, *"Smart Technologies in Enhanced Oil Recovery: Integrating AI, Nanotechnology, and Sustainable Practices,"* would not have been possible without the contributions and support of many individuals and organizations. I am deeply grateful to the editorial team for providing an opportunity to contribute to this book and for their invaluable guidance throughout the writing process. Their constructive feedback has significantly enhanced the quality of this work. Special thanks to my senior Mr. Neeraj Singh, and colleagues Mr. S. K. Swami, Mr. R. K. Meena, Mr. Brajesh Yadav and Mr. Premlal. Thanks to special guidance of Prof. (Dr.) S. P. Singh (Medical Superintendent), Dr. Aaditya Shivhare, Dr. Shweta Chaudhary, Dr. Jyoti Kala Bharti, Dr. Yatendra Mohan, Dr. Arvind Kumar Singh, faculties, whose insightful discussions and collaborative spirit have inspired innovative perspectives in this field. I also acknowledge the contributions of researchers and practitioners whose pioneering work laid the foundation for advancements in enhanced oil recovery. I would like to extend my heartfelt thanks to the research community and industry experts whose ground breaking contributions have significantly shaped the field of enhanced oil recovery (EOR). Their dedication and innovation continue to inspire the development of advanced and sustainable practices in the energy sector. I am also thankful for the collaborative spirit of professionals and institutions that encourage interdisciplinary approaches, fostering the integration of Artificial Intelligence, nanotechnology, and sustainability in EOR. This chapter is dedicated to the pursuit of technological advancements that address global energy challenges while promoting environmental stewardship and resource efficiency. Finally, I extend my heartfelt appreciation to my family and friends for their unwavering support and encouragement during the preparation of this chapter. This work is dedicated to the advancement of sustainable and intelligent technologies that drive the future of energy innovation. We acknowledge the use of *Grammarly AI* for the language polishing of the manuscript.

Funding

No external funding was received for the preparation of this chapter.

Conflict of interest

The author declares no conflict of interest regarding the publication of this chapter. All efforts have been made to ensure the objectivity and integrity of the information presented. There are no financial, professional, or personal relationships that could have influenced the content or conclusions of this work.

Ethical approval

This work does not involve any studies with human participants or animals performed by the author.

Data availability statement

No datasets were generated or analyzed during the preparation of this chapter. All information are based on publicly available literature and scientific research.

Nomenclature

EOR	Enhanced Oil Recovery – Techniques used to increase oil extraction from mature fields
AI	Artificial Intelligence – Technology that simulates human intelligence in decision-making processes
Nanotechnology	The use of nanoparticles to improve oil recovery by altering fluid properties or enhancing chemical reactions
CO_2-EOR	A method of EOR using carbon dioxide to increase oil displacement and storage in reservoirs
CCUS	Carbon Capture, Utilization, and Storage – Technology for capturing and storing carbon emissions from industrial processes
Smart fluids	Fluids that incorporate advanced materials like nanoparticles to improve flow and oil recovery efficiency
Functionalized nanoparticles	Nanoparticles that are engineered to perform specific functions in EOR, such as reducing oil viscosity or improving mobility
MEOR	Microbial Enhanced Oil Recovery – A technique using microorganisms to improve oil recovery by altering reservoir properties
Water-flooding	An EOR method where water is injected into the reservoir to displace oil and increase recovery
Sustainable EOR	Environmentally conscious approaches to EOR that focus on reducing resource use and emissions

Acronyms and abbreviations

EOR	enhanced oil recovery
AI	artificial intelligence
CO_2-EOR	carbon dioxide enhanced oil recovery
CCUS	carbon capture, utilization, and storage
MEOR	microbial enhanced oil recovery
NPs	nanoparticles
SNPs	surface-functionalized nanoparticles
IoT	Internet of things
ML	machine learning
NN	neural networks
RF	reservoir fluid
ANN	artificial neural networks
PVT	pressure-volume-temperature
NMR	nuclear magnetic resonance
SLN	smart liquid nanoparticles
FTIR	Fourier transform infrared spectroscopy
SDS	sodium dodecyl sulfate
SWCNT	Single-walled carbon nanotubes

LCA	life cycle assessment
TDS	total dissolved solids
FPS	field performance simulation

Appendices

Appendix A: Data on AI applications in EOR

Application	Description	Impact on EOR
Predictive analytics	AI algorithms predict production rates and optimize recovery methods	Increases accuracy in decision-making, reduces operational risks
Real-time monitoring	Sensors and AI analyze data from reservoirs and wells	Enhances real-time adjustments, improving recovery efficiency
Automation in drilling	AI-driven automation of drilling and reservoir management	Reduces costs, increases precision, and minimizes human error

Table A1.
Data on AI applications in enhanced oil recovery (EOR).

Appendix B: Properties of functionalized nanoparticles

Nanoparticle type	Function in EOR	Key benefits
Silica nanoparticles	Reduce oil viscosity, improve mobility	Enhanced oil displacement, reduced chemical use
Metal-oxide nanoparticles	Improve thermal properties for thermal EOR techniques	Improved heat transfer and fluid behavior
Carbon nanotubes	Adsorption of oil and contaminants	Increased recovery efficiency, less environmental contamination

Table B1.
Properties of functionalized nanoparticles used in EOR.

Appendix C: Water-efficient EOR technologies

Technology	Description	Impact on sustainability
Closed-loop systems	Reuse and recycle water within EOR operations	Minimizes freshwater usage, reduces wastewater production
CO_2 and nitrogen injection	Alternative to water-based flooding methods	Reduces dependence on water, lowers environmental impact
Advanced water treatment	Technologies to treat and reuse produced water	Enables the safe reuse of water, reduces disposal challenges

Table C1.
Water-efficient technologies applied in enhanced oil recovery (EOR).

Author details

Nouratan Singh[1*], Poonam Rani[2] and Neeraj Tandan[3]

1 U.P. University of Medical Sciences, Saifai, Etawah, India

2 Hansraj College, University of Delhi, New Delhi, India

3 New Delhi, India Chapter of The American University, USA & UNUGP, USA, India

*Address all correspondence to: nouratansingh@gmail.com

References

[1] Gibrata MA, Ameish G, Eldali M, Huang Q, Rouis L, Dernika M, et al. An integrated core, reservoir characterization and PVT for hybrid enhanced oil recovery in unconsolidated sandstone reservoir. In: Paper Presented at the Mediterranean Offshore Conference, Alexandria, Egypt. Oct 2024. p. D011S006R004. DOI: 10.2118/223163-MS

[2] Manikandan S et al. Artificial intelligence-driven sustainability: Enhancing carbon capture for sustainable development goals–A review. Sustainable Development. 2024. DOI: 10.1002/sd.3222

[3] Cheraghi Y, Kord S, Mashayekhizadeh V. Application of machine learning techniques for selecting the most suitable enhanced oil recovery method; challenges and opportunities. Journal of Petroleum Science and Engineering. 2021;**205**:108761. DOI: 10.1016/j.petrol.2021.108761

[4] Rizvi SM. Nanotechnology applications in enhanced oil recovery (EOR). International Journal of Scientific Research and Management. 4 Jun 2024;**12**(06):135-143. DOI: 10.18535/ijsrm/v12i06.c03

[5] Ampomah W et al. CCUS perspectives: Assessing historical contexts, current realities, and future prospects. Energies. 2024;**17**(17):4248. DOI: 10.3390/en17174248

[6] Ahmadi MA, Pournik M. A predictive model of chemical flooding for enhanced oil recovery purposes: Application of least square support vector machine. Petroleum. 2016;**2**(2):177-182. DOI: 10.1016/j.petlm.2015.10.002

[7] Huerta V, Meza K. Application of machine learning techniques for EOR screening in heavy oilfields in Peru. In: Fourth HGS/EAGE Conference on Latin America. Vol. 2022. No. 1. European Association of Geoscientists & Engineers; 2022. pp. 1-4. DOI: 10.3997/2214-4609.202282027

[8] Rostami P, Sharifi M, Aminshahidy B, Fahimpour J. Enhanced oil recovery using silica nanoparticles in the presence of salts for wettability alteration. Journal of Dispersion Science and Technology. 2020;**41**(3):402-413. DOI: 10.1080/01932691.2019.1583575

[9] Pandey A et al. Advanced multi-wall carbon nanotube-optimized surfactant-polymer flooding for enhanced oil recovery. Fuel. 2024;**355**:129463. DOI: 10.1016/j.fuel.2023.129463

[10] Afolabi RO. Enhanced oil recovery for emergent energy demand: Challenges and prospects for a nanotechnology paradigm shift. International Nano Letters. 2019;**9**(1):1-5. DOI: 10.1007/s40089-018-0248-0

[11] Millemann RE, Haynes RJ, Boggs TA, Hildebrand SG. Enhanced oil recovery: Environmental issues and state regulatory programs. Environment International. 1982;**7**(3):165-177. DOI: 10.1016/0160-4120(82)90103-9

[12] Prada L et al. Nanoparticles technology for improving steam-assisted gravity drainage process performance: A review. Industrial & Engineering Chemistry Research. 2024;**63**(30):13047-13077. DOI: 10.1021/acs.iecr.4c01377

[13] Hanson E, Nwakile C, Hammed VO. Carbon capture, utilization, and storage (CCUS) technologies: Evaluating the

effectiveness of advanced CCUS solutions for reducing CO_2 emissions. Results in Surfaces and Interfaces. 2024;**18**:100381. DOI: 10.1016/j.rsurfi.2024.100381

[14] Zaluski W et al. Monitoring technology ranking methodology for CO_2-EOR sites using the Weyburn-Midale field as a case study. International Journal of Greenhouse Gas Control. 2016;**54**:466-478. DOI: 10.1016/j.ijggc.2016.06.012

[15] Jambol DD, Sofoluwe OO, Ukato A, Ochulor OJ. Transforming equipment management in oil and gas with AI-driven predictive maintenance. Computer Science & IT Research Journal. 2024;**5**(5):1090-1112. DOI: 10.51594/csitrj.v5i5.1117

[16] Negin C, Ali S, Xie Q. Application of nanotechnology for enhancing oil recovery–A review. Petroleum. 2016;**2**(4):324-333. DOI: 10.1016/j.petlm.2016.10.002

[17] International Energy Agency (IEA). CCUS in clean energy transitions. Energy Technology Perspectives. 2020. Available from: https://www.iea.org/reports/ccus-in-clean-energy-transitions

[18] Su S, Zhang N, et al. Investigation and optimization of EOR screening by implementing machine learning algorithms. Applied Sciences. 2023;**13**(22):12267. DOI: 10.3390/app132212267

[19] Vaziri P, Ahmadi S, Daneshfar F, Sedaee B, Alimohammadi H, Rasaei MR. Machine learning techniques in enhanced oil recovery screening using semisupervised label propagation. SPE Journal. 11 Sep 2024;**29**(09):4557-4578. DOI: 10.2118/221475-PA

[20] Arinze CA, Izionworu VO, Isong D, Daudu CD, Adefemi A. Predictive maintenance in oil and gas facilities, leveraging AI for asset integrity management. International Journal of Frontiers in Engineering and Technology Research. 2024;**6**(1):16-26. DOI: 10.53294/ijfetr.2024.6.1.0026

[21] Madhumaya A, Vyas A. Application of machine learning in screening the optimal enhanced oil recovery technique. In: Paper Presented at the Mediterranean Offshore Conference, Alexandria, Egypt. Oct 2024. p. D011S004R004. DOI: 10.2118/223150-MS

[22] Tariq Z et al. A systematic review of data science and machine learning applications to the oil and gas industry. Journal of Petroleum Exploration and Production Technology. 2021;**11**(12):1-36. DOI: 10.1007/s13202-021-01302-2

[23] Aderamo AT, Olisakwe HC, Adebayo YA, Esiri AE. AI-enabled predictive safeguards for offshore oil facilities: Enhancing safety and operational efficiency. Comprehensive Research and Reviews in Engineering and Technology. 2024;**2**(1):23-43. DOI: 10.57219/crret.2024.2.1.0060

[24] Hourfar F, Bidgoly HJ, Moshiri B, Salahshoor K, Elkamel A. A reinforcement learning approach for waterflooding optimization in petroleum reservoirs. Engineering Applications of Artificial Intelligence. 2019;**77**:98-116. DOI: 10.1016/j.engappai.2018.09.019

[25] Hamilton L, Rauch M. Artificial intelligence integration for optimal reservoir data analysis and pattern recognition. In: Unconventional Resources Technology Conference (URTeC). Houston, Texas, USA, Richardson, Texas, USA: Society of Petroleum Engineers (SPE); 17-19 June 2024. pp. 2968-2988. DOI: 10.15530/urtec-2024-4054687

[26] Sircar A, Yadav K, Rayavarapu K, Bist N, Oza H. Application of machine

learning and artificial intelligence in oil and gas industry. Petroleum Research. 2021;**6**(4):379-391. DOI: 10.1016/j.ptlrs.2021.05.009

[27] Lake LW, Pope GA, Schechter RS. Reservoir Characterization and Enhanced Oil Recovery Research. Austin, TX (United States): Texas Univ., Center for Petroleum and Geosystems Engineering; 1992. Available from: https://digital.library.unt.edu/ark:/67531/metadc1089063/

[28] Sheng JJ, editor. Enhanced Oil Recovery Field Case Studies. Netherlands: Gulf Professional Publishing; 2013. DOI: 10.1016/B978-0-12-803734-8.00016-3

[29] Mezzomo RF. Program advances oil recovery technology in Brazil. Oil & Gas Journal. 1999;**97**(40):56-60. Available from: https://www.ogj.com/home/article/17230959/program-advances-oil-recovery-technology-in-brazil

[30] Monigha A. AI-driven cost optimization in oil and gas projects. International Journal of Petroleum and Gas Engineering Research. 2024;**7**(2):17-32. Available from: https://tudr.org/id/eprint/3017/

[31] Afekare D, Garno J, Rao D. Enhancing oil recovery using silica nanoparticles: Nanoscale wettability alteration effects and implications for shale oil recovery. Journal of Petroleum Science and Engineering. 2021;**203**:108897. DOI: 10.1016/j.petrol.2021.108897

[32] Ali JA, Kolo K, Manshad AK, Mohammadi AH. Recent advances in application of nanotechnology in chemical enhanced oil recovery: Effects of nanoparticles on wettability alteration, interfacial tension reduction, and flooding. Egyptian Journal of Petroleum. 2018;**27**(4):1371-1383. DOI: 10.1016/j.ejpe.2018.09.006

[33] Cheraghian G, Hendraningrat L. A review on applications of nanotechnology in the enhanced oil recovery part A: Effects of nanoparticles on interfacial tension. International Nano Letters. 2016;**6**:129-138. DOI: 10.1007/s40089-015-0173-4

[34] Mohajeri M, Rasaei MR, Hekmatzadeh M. Experimental study on using SiO_2 nanoparticles along with surfactant in an EOR process in micromodel. Petroleum Research. 2019;**4**(1):59-70. DOI: 10.1016/j.ptlrs.2018.09.001

[35] Esfandyari Bayat A et al. Impact of metal oxide nanoparticles on enhanced oil recovery from limestone media at several temperatures. Energy & Fuels. 2014;**28**(10):6255-6266. DOI: 10.1021/ef5013616

[36] Hasani M, Jafari A. Electromagnetic field's effect on enhanced oil recovery using magnetic nanoparticles: Microfluidic experimental approach. Fuel. 2022;**307**:121718. DOI: 10.1016/j.fuel.2021.121718

[37] Shen M, Zhang C, Yan X, Wang L, Wu Y, Jin X. Research progress and prospects of utilizing carbon-based nanomaterials in enhanced oil recovery. Advances in Geo-Energy Research. 2024;**14**(3):201-214. DOI: 10.46690/ager.2024.12.05

[38] Agi A, Junin R, Gbadamosi A. Mechanism governing nanoparticle flow behaviour in porous media: Insight for enhanced oil recovery applications. International Nano Letters. 2018;**8**:49-77. DOI: 10.1007/s40089-018-0237-3

[39] Liu X, Kang Y, Yan L, Tian J, Li J, You L. Implication of interfacial tension

reduction and wettability alteration by surfactant on enhanced oil recovery in tight oil reservoirs. Energy Reports. 2022;**8**:13672-13681. DOI: 10.1016/j.egyr.2022.10.052

[40] Gbadamosi A, Patil S, Kamal MS, Adewunmi AA, Yusuff AS, Agi A, et al. Application of polymers for chemical enhanced oil recovery: A review. Polymers. 2022;**14**(7):1433. DOI: 10.3390/polym14071433

[41] Mustafin R, Manasrah AD, Vitale G, Askari R, Nassar NN. Enhanced thermal conductivity and reduced viscosity of aegirine-based VR/VGO nanofluids for enhanced thermal oil recovery application. Journal of Petroleum Science and Engineering. 2020;**185**:106569. DOI: 10.1016/j.petrol.2019.106569

[42] Sun X, Zhang Y, Chen G, Gai Z. Application of nanoparticles in enhanced oil recovery: A critical review of recent progress. Energies. 2017;**10**(3):345. DOI: 10.3390/en10030345

[43] Megayanti R et al. Effect of titanium dioxide nanoparticles on surfactants and their impact on the interfacial properties of the oil–water–rock system. ACS Omega. 2023;**8**(41):38539-38545. DOI: 10.1021/acsomega.3c05365

[44] Bera A, Kumar S. Applications of magnetic nanoparticles in thermal enhanced oil recovery. In: Fundamentals and Industrial Applications of Magnetic Nanoparticles. Cambridge, UK: Woodhead Publishing; 2022. pp. 527-553. DOI: 10.1016/B978-0-12-822819-7.00005-3

[45] Hosny R et al. Nanotechnology impact on chemical-enhanced oil recovery: A review and bibliometric analysis of recent developments. ACS Omega. 2023;**8**(49):46325-46345. DOI: 10.1021/acsomega.3c06206

[46] Seidy-Esfahlan M, Tabatabaei-Nezhad SA, Khodapanah E. Comprehensive review of enhanced oil recovery strategies for heavy oil and bitumen reservoirs in various countries: Global perspectives, challenges, and solutions. Heliyon. 11 Sep 2024;**10**:1-20. DOI: 10.1016/j.heliyon.2024.e37826

[47] Nasiri M, Jafari I, Parniankhoy B. Oil and gas produced water management: A review of treatment technologies, challenges, and opportunities. Chemical Engineering Communications. 2017;**204**(8):990-1005. DOI: 10.1080/00986445.2017.1330747

[48] Massarweh O, Abushaikha AS. Towards environmentally sustainable oil recovery: The role of sustainable materials. Energy Reports. 2024;**12**:95-119. DOI: 10.1016/j.egyr.2024.06.013

[49] Farajzadeh R et al. Improved oil recovery techniques and their role in energy efficiency and reducing CO_2 footprint of oil production. Journal of Cleaner Production. 2022;**369**:133308. DOI: 10.1016/j.jclepro.2022.133308

[50] You J et al. Machine learning based co-optimization of carbon dioxide sequestration and oil recovery in CO_2-EOR project. Journal of Cleaner Production. 2020;**260**:120866. DOI: 10.1016/j.jclepro.2020.120866

[51] Yang W, Peng B, Liu Q, Wang S, Dong Y, Lai Y. Evaluation of CO_2 enhanced oil recovery and CO_2 storage potential in oil reservoirs of Bohai Bay basin, China. International Journal of Greenhouse Gas Control. 2017;**65**:86-98. DOI: 10.1016/j.ijggc.2017.08.012

[52] Wang PT, Wu X, Ge G, Wang X, Xu M, Wang F, et al. Evaluation of CO_2 enhanced oil recovery and CO_2 storage potential in oil reservoirs of petroliferous sedimentary basin, China. Science and

Technology for Energy Transition. 2023;**78**:3. DOI: 10.2516/stet/2022022

[53] Al-Kindi S, Al-Bahry S, Al-Wahaibi Y, Taura U, Joshi S. Partially hydrolyzed polyacrylamide: Enhanced oil recovery applications, oil-field produced water pollution, and possible solutions. Environmental Monitoring and Assessment. 2022;**194**(12):875. DOI: 10.1007/s10661-022-10569-9

[54] Canbolat S, Serhat A, Kovscek AR. A study of steam-assisted gravity drainage performance in the presence of noncondensable gases. In: Paper Presented at the SPE/DOE Improved Oil Recovery Symposium, Tulsa, Oklahoma. Apr 2002. p. SPE-75130. DOI: 10.1021/acsomega.2c01939

[55] Mielke E, Anadon LD, Narayanamurti V. Water Consumption of Energy Resource Extraction, Processing, and Conversion. Cambridge, MA: Belfer Center for Science and International Affairs; 2010. Available from: https://www.solunesco.com/wp-content/uploads/2018/02/harvard-study.pdf

[56] Karlapudi AP et al. Role of biosurfactants in bioremediation of oil pollution - A review. Petroleum. 2018;**4**(3):241-249. DOI: 10.1016/j.petlm.2018.03.007

[57] Muhammed NS, Haq MB, Al-Shehri D, Rahaman MM, Keshavarz A, Hossain SZ. Comparative study of green and synthetic polymers for enhanced oil recovery. Polymers. 2020;**12**(10):2429. DOI: 10.3390/polym12102429

[58] Karimov D, Toktarbay Z. Enhanced oil recovery: Techniques, strategies, and advances. ES Materials & Manufacturing. 2023;**23**(2):1005. Available from: https://www.espublisher.com/journals/articledetails/1005

[59] Jain S, Pachisia H, Sharma A, Patel S, Patel S, Ragunathan B. A systematic review–chemical EOR using surfactants and polymers. Materials Today: Proceedings. 2022;**62**:7220-7223. DOI: 10.1016/j.matpr.2022.03.602

[60] Anto R, Bhui UK. Surfactant flooding for enhanced oil recovery–insights on oil solubilization through optical spectroscopic approach. Journal of Petroleum Science and Engineering. 2022;**208**:109803. DOI: 10.1016/j.petrol.2021.109803

[61] Pan F, Zhang Z, Zhang X, Davarpanah A. Impact of anionic and cationic surfactants interfacial tension on the oil recovery enhancement. Powder Technology. 2020;**373**:93-98. DOI: 10.1016/j.powtec.2020.06.033

[62] Xu L, Han M, Cao D, Fuseni A. Catanionic surfactants for improving oil production in carbonate reservoirs. SPE Reservoir Evaluation & Engineering. 2023;**26**(01):88-98. DOI: 10.2118/200182-PA

[63] Belhaj AF et al. The effect of surfactant concentration, salinity, temperature, and pH on surfactant adsorption for chemical enhanced oil recovery: A review. Journal of Petroleum Exploration and Production Technology. 2020;**10**:125-137. DOI: 10.1007/s13202-019-0685-y

[64] Sarkar R, Pal A, Rakshit A, Saha B. Properties and applications of amphoteric surfactant: A concise review. Journal of Surfactants and Detergents. 2021;**24**(5):709-730. DOI: 10.1002/jsde.12542

[65] Rawal SU, Patel MM. Lipid nanoparticulate systems: Modern versatile drug carriers. In: Lipid Nanocarriers for Drug Targeting. Norwich, NY: William Andrew Publishing;

2018. pp. 49-138. DOI: 10.1016/B978-0-12-813687-4.00002-5

[66] Terry RE. Enhanced oil recovery. Encyclopedia of Physical Science and Technology. 2001;**18**(1):503-518. DOI: 10.1016/B0-12-227410-5/00868-1

[67] Mandal A. Chemical flood enhanced oil recovery: A review. International Journal of Oil, Gas and Coal Technology. 2015;**9**(3):241-264. DOI: 10.1504/IJOGCT.2015.069001

[68] Seright RS, Wang D. Polymer flooding: Current status and future directions. Petroleum Science. 2023;**20**(2):910-921. DOI: 10.1016/j.petsci.2023.02.002

[69] Dai C, You Q, Zhao M, Zhao G, Zhao F. Alkaline Flooding. In: Principles of Enhanced Oil Recovery. Singapore: Springer Nature Singapore; 2023. pp. 133-145. DOI: 10.1007/978-981-99-0193-7_6

[70] Kamari A, Gharagheizi F, Shokrollahi A, Arabloo M, Mohammadi AH. Integrating a robust model for predicting surfactant–polymer flooding performance. Journal of Petroleum Science and Engineering. 2016;**137**:87-96. DOI: 10.1016/j.petrol.2015.10.034

[71] Mazhit B, Abbas AH, Pourafshary P. Evaluation of flaxseed hydrocolloid's potential in improving oil recovery. Journal of King Saud University-Engineering Sciences. 2024;**36**(8, Part A):611-618. DOI: 10.1016/j.jksues.2023.10.003

[72] Khusainova A. Enhanced Oil Recovery with Application of Enzymes. Kgs. Lyngby: Technical University of Denmark; 2016. 114 p. Available from: https://orbit.dtu.dk/en/publications/enhanced-oil-recovery-with-application-of-enzymes

[73] Gbadamosi AO, Junin R, Manan MA, Agi A, Yusuff AS. An overview of chemical enhanced oil recovery: Recent advances and prospects. International Nano Letters. 2019;**9**:171-202. DOI: 10.1007/s40089-019-0272-8

[74] Tavakkoli O, Kamyab H, Shariati M, Mohamed AM, Junin R. Effect of nanoparticles on the performance of polymer/surfactant flooding for enhanced oil recovery: A review. Fuel. 2022;**312**:122867. DOI: 10.1016/j.fuel.2021.122867

[75] Aghaeifar Z, Strand S, Puntervold T, Austad T, Sajjad FM. Smart water injection strategies for optimized EOR in a high temperature offshore oil reservoir. Journal of Petroleum Science and Engineering. 2018;**165**:743-751. DOI: 10.1016/j.petrol.2018.02.021

[76] Saputelli L et al. Success cases and lessons learned after 20 years of oilfield digitalization efforts. In: SPE Annual Technical Conference and Exhibition. Richardson, Texas: Society of Petroleum Engineers; 2024. p. D021S012R005. DOI: 10.2118/220932-MS

[77] Virginia EO, Kosi S, Junior J, Obada OR, Peter Y. Enhanced oil recovery methods using biodegradable materials in different reservoirs. Global Scientific Journal. 2024;**12**(9):648-657. Available from: https://www.researchgate.net/publication/384604504

[78] Temizel C, Aydin H, Hosgor FB, Yegin C, Kabir CS. Green energy sources reduce carbon footprint of oil and gas industry processes: A review. Journal of Energy and Power Technology. 2023;**5**(1):1-25. DOI: 10.21926/jept.2301004

[79] Wang K, Yuan B, Ji G, Wu X. A comprehensive review of geothermal energy extraction and utilization in

oilfields. Journal of Petroleum Science and Engineering. 2018;**168**:465-477. DOI: 10.1016/j.petrol.2018.05.012

[80] Ericson SJ, Engel-Cox J, Arent DJ. Approaches for Integrating Renewable Energy Technologies in Oil and Gas Operations. Golden, CO (United States): National Renewable Energy Lab.(NREL); 2019. DOI: 10.2172/1491378

[81] Ahmed AM. Geological and environmental implications of enhanced oil recovery techniques: Balancing efficiency with sustainability. EarthArXiv. 31 Oct 2024. DOI: 10.31223/X5M69G [Preprint]

[82] Muggeridge A et al. Recovery rates, enhanced oil recovery and technological limits. Philosophical Transactions of the Royal Society A: Mathematical, Physical and Engineering Sciences. 2006;**2014**(372):20120320. DOI: 10.1098/rsta.2012.0320

[83] Arinze CA, Jacks BS. A comprehensive review on AI-driven optimization techniques enhancing sustainability in oil and gas production processes. Engineering Science & Technology Journal. 2024;**5**(3):962-973. DOI: 10.51594/estj.v5i3.950

[84] Ahmadi Y, Hemmati M, Vaferi B, Gandomkar A. Applications of nanoparticles during chemical enhanced oil recovery: A review of mechanisms and technical challenges. Journal of Molecular Liquids. 2024;**415**:126287. DOI: 10.1016/j.molliq.2024.126287

[85] Sen R. Biotechnology in petroleum recovery: The microbial EOR. Progress in Energy and Combustion Science. 2008;**34**(6):714-724. DOI: 10.1016/j.pecs.2008.05.001

[86] Roefs P, Moretti M, Welkenhuysen K, Piessens K, Compernolle T. CO_2-enhanced oil recovery and CO_2 capture and storage: An environmental economic trade-off analysis. Journal of Environmental Management. 2019;**239**:167-177. DOI: 10.1016/j.jenvman.2019.03.007

[87] Melzer LS. Carbon Dioxide Enhanced Oil Recovery (CO_2 EOR): Factors Involved in Adding Carbon Capture, Utilization and Storage (CCUS) to Enhanced Oil Recovery. Arlington, VA, USA: Center for Climate and Energy Solutions; 2012. pp. 1-7. Available from: https://carboncapturecoalition.org/wp-content/uploads/2018/01/Melzer_CO2EOR_CCUS_Feb2012.pdf

[88] Hou S, Gu D, Yang D, Yang S, Zhao M. Integrated optimization of hybrid steam-solvent injection in post-CHOPS reservoirs with consideration of wormhole networks and foamy oil behavior. SPE Journal. 2024;**29**(08):4319-4339. DOI: 10.2118/212145-PA

[89] Gowd SC, Ganeshan P, Vigneswaran VS, Hossain MS, Kumar D, Rajendran K, et al. Economic perspectives and policy insights on carbon capture, storage, and utilization for sustainable development. Science of the Total Environment. 2023;**883**:163656. DOI: 10.1016/j.scitotenv.2023.163656

[90] Wang Z, Li S, Jin Z, Li Z, Liu Q, Zhang K. Oil and gas pathway to net-zero: Review and outlook. Energy Strategy Reviews. 2023;**45**:101048. DOI: 10.1016/j.esr.2022.101048

[91] Cabrera G, Dickson A, Nimubona AD, Quigley J. Carbon capture, utilisation and storage: Incentives, effects and policy. International Journal of Greenhouse Gas Control. 2022;**120**:103756. DOI: 10.1016/j.ijggc.2022.103756

[92] Cordes EE et al. Environmental impacts of the deep-water oil and gas industry: A review to guide management

strategies. Frontiers in Environmental Science. 2016;**4**:58. DOI: 10.3389/fenvs.2016.00058

[93] Shaikhah D et al. New trends in biosurfactants: From renewable origin to green enhanced oil recovery applications. Molecules. 2024;**29**(2):301. DOI: 10.3390/molecules29020301

[94] Petri DF. Xanthan gum: A versatile biopolymer for biomedical and technological applications. Journal of Applied Polymer Science. 2015;**132**(23):42035, 1-13. DOI: 10.1002/app.42035

[95] Elemuo NG, Ikiensikimama SS, Wachikwu-Elechi VU. The use of biological catalyst (enzyme) for enhanced oil recovery in Niger Delta. Heliyon. 2024;**10**(4):e25294. DOI: 10.1016/j.heliyon.2024.e25294

[96] Mota J, Brander M, Herrera D. Carbon accounting methods for the system-wide evaluation of carbon capture, utilisation and storage: A case study in Mexico's southeast region. Journal of Cleaner Production. 2024;**453**:142159, 1-12. Available from: https://hdl.handle.net/10419/280120

[97] Davoodi S, Al-Shargabi M, Wood DA, Rukavishnikov VS, Minaev KM. Experimental and field applications of nanotechnology for enhanced oil recovery purposes: A review. Fuel. 2022;**324**:124669. DOI: 10.1016/j.fuel.2022.124669

[98] Ke CY et al. Microbial enhanced oil recovery (MEOR): Recent development and future perspectives. Critical Reviews in Biotechnology. 2024;**44**(6):1183-1202. DOI: 10.1080/07388551.2023.2270578

[99] Rostami A, Kalantari-Meybodi M, Karimi M, Tatar A, Mohammadi AH. Efficient estimation of hydrolyzed polyacrylamide (HPAM) solution viscosity for enhanced oil recovery process by polymer flooding. Oil & Gas Sciences and Technology–Revue d'IFP Energies Nouvelles. 2018;**73**:22. DOI: 10.2516/ogst/2018006

[100] Hosseinpour M, Soltani M, Nathwani J. Renewable energy integration with hot compressed water in heavy oil upgrading: A practice toward sustainability. Journal of Cleaner Production. 2022;**334**:130268. DOI: 10.1016/j.jclepro.2021.130268

Section 3

Nanotechnology, Chemical Methods and Case Studies

Chapter 6

Nanoparticles-Assisted Polymer Flooding: A New Approach toward Enhanced Oil Recovery

Khalaf Gad Salem, Adel Salem, Sayed Gomaa and Abdelaziz El-hoshoudy

Abstract

Polymer flooding is widely regarded as the most effective method for chemical-enhanced oil recovery due to its ability to reduce water mobility and improve sweep efficiency. However, traditional polymer flooding has some limitations, such as the degradation of polymer structure in harsh reservoir conditions. Recently, experimental investigations have shown that the synergy of nanoparticles (NPs) and polymer can enhance flooding performance and increase oil recovery compared to traditional polymer flooding. The focus of this chapter is on nanoparticle-assisted polymer flooding and its use to enhance oil recovery. Therefore, the purpose of this chapter is to provide a brief but thorough overview of key aspects necessary to understand this technology as a new enhanced oil recovery process. This chapter covers various aspects, including preparations, enhanced oil recovery (EOR) mechanisms, displacement tests, numerical simulations, and economic assessments of nanoparticle-assisted polymer flooding. Additionally, the chapter identifies the challenges and future directions for polymeric nanofluids. Overall, the polymeric nanofluids significantly enhance rheological properties, reduce interfacial tension (IFT), and alter wettability for EOR purposes. Moreover, polymeric nanofluids demonstrate higher incremental oil recovery than conventional polymer flooding.

Keywords: enhanced oil recovery, polymer flooding, nanoparticles, experimental investigation, polymeric nanofluid, IFT reduction, polymer viscosity

1. Introduction

Typically, nearly half of the oil and gas initially present in a reservoir remains unrecovered after primary and secondary recovery processes. This is primarily due to capillary forces and an unfavorable mobility ratio between the aqueous and oleic phases, which cause the oil to either be bypassed or become trapped within the reservoir [1, 2]. To maximize the amount of recoverable oil, enhanced oil recovery (EOR) techniques must be carefully planned and implemented [3]. **Figure 1** exhibits the classification of various oil-recovery terminology: primary, secondary

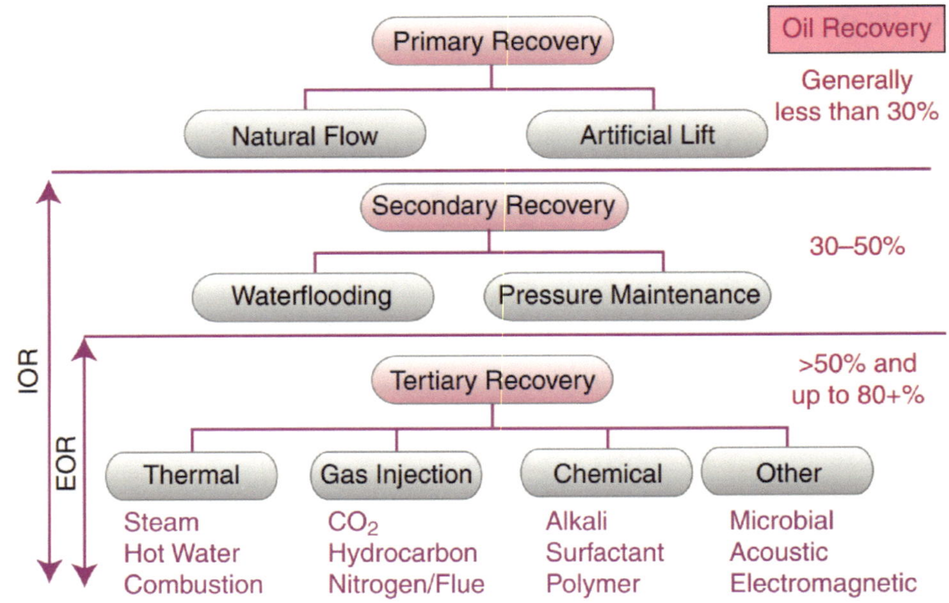

Figure 1.
Classification of oil recovery [4].

and tertiary recovery. The primary recovery typically involves extracting oil using the reservoir's natural energy or with the help of artificial lift systems. This stage usually recovers less than 30% of the initial oil in place. Once the natural energy is depleted, leading to a decline in reservoir pressure, or when it becomes uneconomical to continue production, additional energy must be introduced to maintain pressure and sustain oil production. This is achieved through secondary recovery methods, such as injecting water or gas to support reservoir pressure. The secondary recovery phase can increase the total recovery to 30–50% of the oil initially in place. Next, the enhanced oil-recovery stage is applied to increase the oil recovered beyond the range that conventional primary and secondary recovery can reach. EOR methods may be categorized into four classes: chemical [2, 5], thermal [6], miscible gas [7], and others [8]. Polymer flooding stands out as a well-established and effective method within the realm of chemically enhanced oil recovery (EOR) [9]. This technique works by lowering the mobility ratio between the water and oil phases, thereby enhancing the efficiency of the sweep process. This improvement is illustrated by the following equation [10–13]:

$$M = \frac{\text{Mobility of oil phase}}{\text{Mobility of water phase}} = \frac{\mu o \,/\, ko}{\mu w \,/\, kw} \tag{1}$$

where M is mobility ratio; μo is oil viscosity; ko is oil permeability; μw is water viscosity; kw is water permeability.

Despite its effectiveness, traditional polymer flooding faces several practical challenges, especially in the high-temperature and high-salinity environments typical of many reservoirs [14–16]. Consequently, researchers are actively working to enhance the performance of conventional polymers to better suit these demanding conditions for EOR applications.

2. Applications of nanotechnology in enhanced oil recovery

The groundbreaking advancements in nanotechnology have significantly transformed researchers' perspectives on the application of nanomaterials in enhanced oil recovery (EOR). This shift in perception underscores the potential of nanotechnology to revolutionize EOR techniques, making them more efficient and effective [17–21]. A nanofluid is characterized as a base fluid containing nanoparticles (NPs) with an average size of less than 100 nanometers in a colloidal suspension [20]. Thanks to their diminutive size, these nanoparticles exhibit an exceptionally high surface-to-volume ratio, as illustrated in **Figure 2**. This extensive surface area significantly boosts the proportion of atoms present on the surface of the nanoparticles. In enhanced oil recovery (EOR) applications, nanofluids can be categorized into four distinct groups based on the formulation of nanoparticles (NPs) used in the flooding process: (a) only NPs in the injected water [21–23], (b) a mixture of NPs and polymer [24–26], (c) a mixture of NPs and surfactant [25, 27, 28], and (d) a mixture of NPs, polymer, and surfactant [27–29]. Recently, nanoparticle-induced chemical-enhanced oil recovery (EOR) has emerged as a promising alternative to traditional chemical EOR techniques [20, 30, 31]. The integration of polymers and nanoparticles (NPs) is seen as a viable replacement for conventional polymer flooding in EOR [16, 32]. The primary reason for combining nanoparticles with polymer solutions is to enhance the properties of the polymers, particularly their thermal, chemical, and mechanical stability, thereby achieving higher oil-recovery rates than traditional polymer injection [11, 32]. Furthermore, nanoparticle-induced polymer flooding has been shown to reduce interfacial tension (IFT) [24, 33], alter wettability [26, 34], and decrease polymer adsorption in porous media [35–37], all of which contribute to improved oil recovery performance. This innovative class of EOR techniques is referred to as polymeric nanofluid, nanoparticle-assisted polymer flooding, or nanocomposite polymer flooding [11, 38–40].

3. Preparation of nanoparticles-assisted polymer flooding

According to the literature, two primary methods are employed to prepare polymeric nanofluids: the polymer nano-grafting method and the polymer

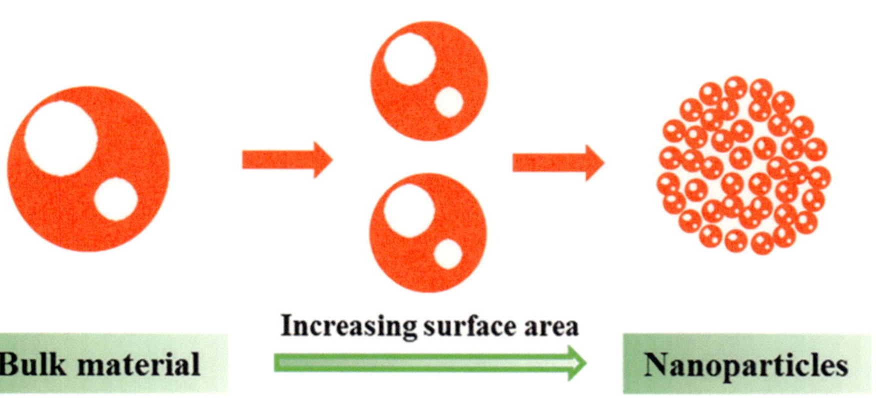

Figure 2.
Schematic diagram of nanoparticles with high surface-to-volume ratio [20].

nanosuspension method. Before formulating polymeric nanofluids, it is essential to treat the nanoparticle (NP) surfaces either physically or chemically to enhance dispersion compatibility [16]. Physical treatments involve coating the NP surfaces with polymers or surfactants. In contrast, chemical treatments may involve modifications using coupling agents like silane and zirconate or grafting polymers onto the NP surfaces.

In polymer nano-grafting (PNG), polymers are chemically bonded to nanomaterials, enhancing stability and reducing particle agglomeration by filling the interstitial spaces of nanoparticles (NPs) [41]. This method offers significant advantages, including improved stability and reduced agglomeration. There are two primary techniques for chemically grafting polymers onto NP surfaces, as illustrated in **Figure 3**.

1. *"Grafting to"*: This involves the covalent attachment of polymer ends to the nanomaterial surface.

2. *"Grafting from"*: This method initiates polymerization directly from the nanoparticle surface.

"Grafting from" is typically executed using free radical polymerization, atom-transfer radical polymerization, nitroxide-mediated radical polymerization, ring-opening polymerization, and reversible addition-fragmentation chain transfer. Additionally, *"grafting in"* refers to the attachment of a pre-synthesized polymer chain to the surface of a nanoparticle [41]. This process is facilitated by a coupling agent that interacts with both the nanoparticle surface and the polymer chain. For instance, silica NP surfaces can be functionalized with silanes containing reactive functional groups, allowing the polymer to attach *via* a chemical reaction. On the other hand, *"grafting from"* involves initiating polymerization from the nanoparticle surface using a suitable initiator [42]. After the initiator is attached, a monomer can be polymerized from the surface, leading to the development of a polymer chain that is covalently bonded to the nanoparticle, as illustrated in **Figure 3**.

In the second method, known as polymer nanosuspension (PNS), nanoparticles are incorporated into the polymer matrix through methods like sonication and high-pressure homogenization [32]. Optimizing the dispersion process is crucial to

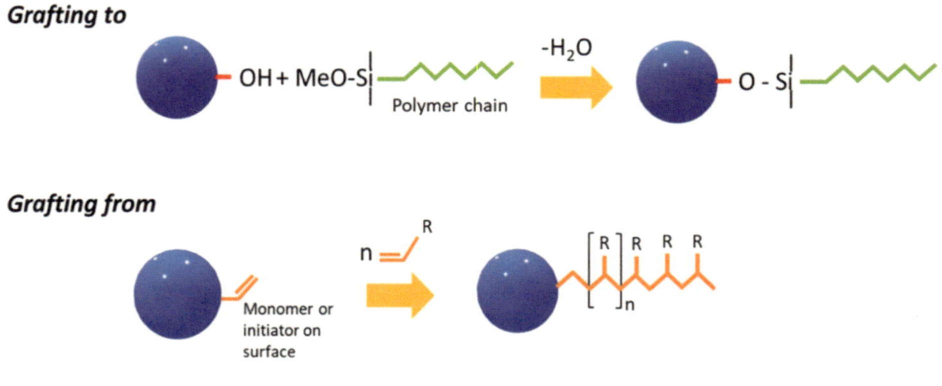

Figure 3.
The types of grafting techniques [16].

Polymer		NPs			Method	Remarks	Author
Type	Concentration	Type	Size, nm	Concentration			
HPAM	2000 ppm	Al_2O_3 SiO_2	7	2000 ppm	Polymer nanosuspension	The polymeric nanofluid demonstrated excellent performance in terms of polymer viscosity, interfacial tension (IFT), wettability alteration, and enhanced oil recovery.	[43]
HPAM	1000–4000 ppm	SiO_2 and Al_2O_3			Polymer nano-grafting	The nanocomposite polymer exhibited excellent performance in terms of polymer rheology, interfacial tension (IFT), and oil recovery.	[24]
HPAM	2000 ppm	Al_2O_3 SiO_2	15–20	1000 ppm	Polymer nanosuspension	The polymeric nanofluid demonstrated excellent performance related to polymer rheology, wettability alteration, and oil recovery.	[26]
Xanthan gum	2000 ppm	ZnO/SiO_2	under 100	2000 ppm	Polymer nanosuspension	The polymeric nanofluid demonstrated excellent performance in terms of oil recovery, interfacial tension (IFT), emulsion stability, and wettability alteration.	[39]
Acrylamide		SiO_2	20–30	2.5 wt.%	Polymer nano-grafting	Excellent outcomes in oil recovery, interfacial tension (IFT), and wettability alteration.	[34]
Xanthan gum	2000 ppm	TiO_2	20	0.01–1 wt%	Polymer nanosuspension	Excellent outcomes in polymer rheology, interfacial tension (IFT), wettability alteration, and oil recovery.	[44]
Xanthan gum	0.5 wt%	SiO_2, TiO_2, and Al_2O_3	15–20 nm	0.5 wt.%	Polymer nanosuspension	Excellent outcomes in polymer rheology, interfacial tension (IFT), and oil recovery.	[33]
Gum Arabic	0.05 g/L	Graphene	190–1100 nm	0.05 mg/mL	Polymer nano-grafting and polymer nanosuspension	Excellent outcomes in polymer rheology, interfacial tension (IFT), wettability alteration, and oil recovery.	[45]
HAPAM	0.15 wt%	SiO_2	10 nm	0–3% w/w	Polymer nano-grafting	The nanohybrid polymer improved oil recovery by 7.82%, compared to just 2.53% achieved by the conventional polymer solution.	[46]

Table 1.
Examples of polymeric nanofluids Formulation.

ensure a stable and uniform distribution of nanoparticles. This method is favored by researchers due to its simplicity and ease of formulation. However, it does have some drawbacks, such as the challenge of achieving a homogeneous distribution of nanoparticles within the polymer mixture. This difficulty arises from the nanoparticles' strong tendency to aggregate and form agglomerates. Additional examples of polymeric nanofluid preparation are illustrated in **Table 1**. Researchers typically utilized two main techniques: grafting and polymer nanosuspension [45]. Among these, polymer nanosuspension emerged as the more commonly adopted method, as evidenced in **Table 1**.

4. Characterization and stability evaluation

It is important to emphasize that nanofluid characterization is crucial for evaluating their performance. This involves measuring the physical and chemical properties of nanoparticles (NPs) using various instruments such as scanning electron microscopy (SEM), transmission electron microscopy (TEM), and dynamic light scattering (DLS). Additionally, Fourier transform infrared (FTIR) spectroscopy is commonly used for characterizing polymeric nanofluids. SEM and TEM are typically employed to analyze the surface morphology of nanoparticles. SEM provides detailed information on the morphology and size of NPs, while TEM offers insights into their appearance, size, defects, crystal structure, and electronic state. DLS is used to determine NP size and to analyze their Brownian motion in suspension, which is related to the velocity and size of the nanoparticles.

The stability of nanofluids is a significant challenge in nanofluid flooding. The stability of nanoparticles (NPs) in various fluids is determined by the balance between attractive and repulsive forces on NP surfaces [47]. When repulsive forces dominate, the suspension remains stable, preventing nanoparticle aggregation. Therefore, it is essential to evaluate the stability of nanoparticle-enhanced chemical-enhanced oil recovery (EOR) before injecting them into reservoirs.

Several techniques are employed to assess nanofluid stability, including dynamic light scattering (DLS), zeta potential measurement, spectral absorption, and structural analysis. Zeta potential measurement is a widely used method to determine nanofluid stability by balancing van der Waals forces (attractive) and electrostatic forces (repulsive) [47, 48]. An increase in the absolute value of the zeta potential indicates improved nanofluid stability due to stronger electrostatic repulsion around the particles. **Table 2** illustrates the relationship between zeta potential values (in millivolts) and the degree of stability.

Absolute value of Z potential (mv)	Level of stability
0	No stability
15	Fair
30	Moderate
45	Good
60	Very good

Table 2.
The zeta potential and the stability level of nanofluids [47].

5. Mechanisms of nanoparticles-assisted polymer flooding

Through experimental studies, the introduction of nanomaterials into polymer solutions has revealed several mechanisms by which polymeric nanofluids improve oil recovery. Researchers have identified these mechanisms as the enhancement of polymer rheological properties, alteration of wettability, reduction of interfacial tension, and decreased polymer retention.

5.1 Enhancement of polymer viscosity

A key aspect of polymer flooding is the use of a mobility control agent, which reduces the mobility ratio and thereby enhances sweep efficiency. Achieving a favorable mobility ratio is possible when the viscosity of the injectant exceeds that of the displaced fluid (oil), resulting in a piston-like displacement with a more uniform front [2, 10]. In traditional polymer flooding, the viscosity of the polymer can decrease under harsh reservoir conditions such as high temperatures and salinity. Polymers, especially synthetic ones, also degrade due to high shear rates. However, incorporating nanoparticles (NPs) into the polymer enhances viscosity because NPs form a network structure through hydrogen bonding, which also influences shear stress. Consequently, polymeric nanofluids exhibit superior viscosity performance and surpass pure polymer solutions [24, 49].

Many studies on nanoparticle-enhanced polymers have focused on changes in the viscosity of polymer solutions. Liu et al. [46] used in situ polymerization to synthesize a silica nanoparticle-polymer, which they then compared to conventional polymers. Their findings revealed that nanohybrid polymers exhibited superior thickening capabilities at equivalent polymer concentrations. Moreover, the silica nanoparticles were uniformly incorporated into the polymer matrix, significantly boosting its viscosity. They also studied the impact of high temperatures on polymer behavior, concluding that the viscosity of the nanohybrid polymer, which was double that of conventional polymers, exhibited the best temperature resistance. In one study, Cao et al. [50] explored how the degree of hydrolysis affects the rheological properties of both pure synthetic and nanohybrid polymers. Their analysis revealed that polymers with low degrees of hydrolysis did not enhance viscosity. In contrast, nanohybrid polymers with high hydrolysis levels showed a marked increase in viscosity, especially when the nanoparticle concentration was kept below 0.25 weight percent. Bera et al. [40] observed that the viscosity of guar gum increases with the addition of silica nanoparticles (NPs) across various temperatures and NP concentrations. They also found that higher NP concentrations enhance the viscosity at elevated temperatures, enabling guar gum to maintain its functionality under such conditions. El-hoshoudy et al. [24] examined how reservoir salinity, temperature, and shear rate affect the viscosity of polyacrylamide and the nanocomposite polymer (HAPAM–SiO_2). Their findings indicated that nanocomposite polymers perform better than conventional polymers in terms of salt tolerance, temperature stability, and shear resistance.

5.2 IFT reduction

It is important to note that reducing oil–water interfacial tension (IFT) is a beneficial mechanism for enhanced oil recovery (EOR). Typically, high IFT prevents oil droplets from passing through pore throats. However, lowering IFT enables oil droplets to pass through, mobilizing trapped oil in reservoir pores, as shown in **Figure 4**.

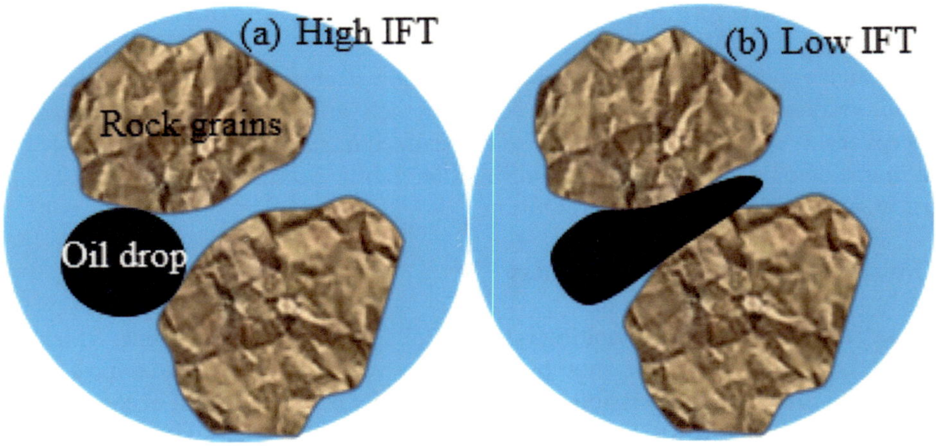

Figure 4.
Schematic of IFT phenomena in porous media (a) trapped oil due to high IFT and (b) mobilization of oil due to low IFT [31].

Notably, reducing IFT is not a mechanism for enhancing oil recovery through conventional polymer flooding. However, it has been proven effective in EOR when using nanohybrid polymer injection due to the presence of nanoparticles in the polymer structure. Polymeric nanofluids reduce IFT by forming micelle-like assemblies, which decrease the oil-water interfacial tension. This reduction in interfacial tension increases the capillary number, thereby reducing residual oil and enhancing oil recovery:

$$N_{ca} = \frac{\mu_w v}{\sigma_{ow}} \tag{2}$$

where N_{ca} represents the capillary number; μ_w denotes the viscosity of displacing fluid; v is the velocity of the injection fluid, and σ_{ow} signifies the interfacial tension (IFT) between the displacing and displaced fluids.

El-hoshoudy et al. [24] engineered the HPAM-SiO_2 nanocomposite polymer by integrating silica into its backbone structure. Their findings demonstrated a substantial reduction in the interfacial tension (IFT) between the nanocomposite polymer and brine solution. This reduction in IFT is attributed to the presence of amphiphilic structures that promote intermolecular associations. Additionally, the inclusion of hydrophobic groups in the polymer's backbone helps form micelles, which decreases the polymer's affinity for water and further lowers the IFT. Jie Cao et al. [51] created a nano-silica polymer and investigated its impact on reducing interfacial tension (IFT). They found that adding nanoparticles significantly lowered the IFT between crude oil and the polymer solution. However, their experiments also showed that once the nanoparticle concentration surpassed a certain threshold, further reduction in IFT was not possible. Salem et al. [52] formulated three nanocomposite polymers: HPAM-SiO2, HPAM-Al_2O_3, and HPAM-ZrO_2. These nanocomposites have also been found to be a remarkable agent for reducing interfacial tension.

5.3 Wettability alteration

A reservoir's wettability refers to the propensity of a fluid to either spread across or adhere to a solid surface when other immiscible liquids are present. This property

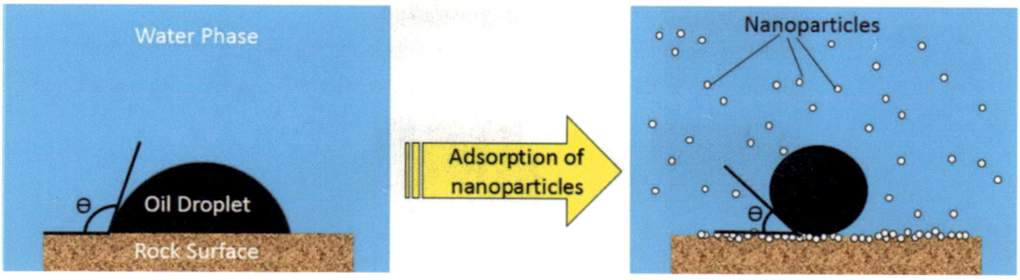

Figure 5.
Schematic of wettability changes in porous media due to NPs adhering to the surface of rocks [53].

is crucial for controlling displacement efficiency and oil recovery. It should be highlighted that changing the wettability of rock from strongly oil-wet to water-wet is often seen as an effective strategy for enhancing oil recovery. Recent studies have shown that incorporating nanoparticles into polymer flooding techniques can significantly alter the wettability of porous media. This modification enhances oil recovery by integrating nanoparticles into the polymer matrix. The change in wettability is primarily due to the adsorption of nanoparticles onto the rock surface, which forms a new surface layer and substantially decreases the water phase's contact angle, as illustrated in **Figure 5**. In their research, Fan et al. [54] explored the effects of incorporating silica dioxide nanoparticles (NPs) into polymers. Their findings revealed that at low concentrations of silica, there was a notable reduction in the contact angle. They identified that the optimal concentration of NPs was 3000 ppm, as increasing the concentration beyond this point did not result in any further decrease in the contact angle. Liu et al. [46] synthesized a nanohybrid polymer through in situ polymerization and examined its contact angle properties. They noted that the contact angle increased from 54.8° to 72.8° as the concentration of silica nanoparticles in the polymer rose to 3 wt.%. Insertion of NP in the polymer was advantageous for modifying rock wettability, they also impacted the polymer's rheological characteristics. Gbadamosi et al. [55] studied the wettability alteration of HPAM with nanoparticles. They found that oil-wet cores immersed in a polymeric nano solution at a concentration of 1000 ppm exhibited a reduction in contact angles across various temperatures. At 27°C, the contact angles for SiO_2-polymer and Al_2O_3-polymer were 91.6° and 71.9°, respectively. At 90°C, the contact angles decreased to 78.6° for SiO_2-polymer and 60.6° for Al_2O_3-polymer. Al_2O_3 nanoparticles demonstrated a greater potential for altering wettability on sandstone surfaces compared to SiO_2 due to their stronger electrostatic affinity and better thermal stability.

5.4 Polymer adsorption reduction

During flow in porous media, interactions between polymer molecules and the rock matrix cause polymers to be retained within the medium, as depicted in **Figure 6**. Polymer adsorption in porous media typically leads to a loss of viscosity and hinders its propagation. This phenomenon is measured in laboratory using static and dynamic methods. Some researchers have explored how nanomaterials affect polymer adsorption by conducting both types of tests. They found that nanoparticle-enhanced polymer solutions exhibit lower adsorption and retention compared to pure polymer solutions. Adding nanomaterials to polymer formulations not only enhances polymer performance in terms of rheological properties and interactions with fluids and rocks

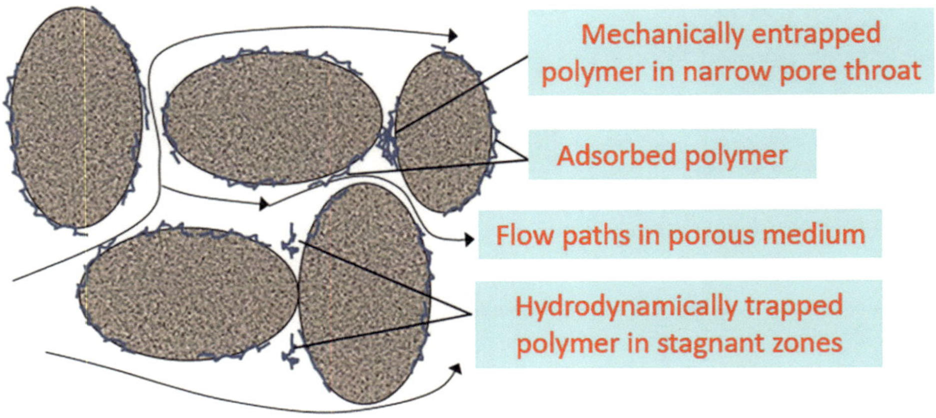

Figure 6.
Schematic of polymer retention in porous media [56].

but also reduces polymer loss to formation rock due to adsorption. Khalilinezhad et al. [57] performed static adsorption tests on silica NP-HPAM and pure HPAM solutions in sandstone rock. They found that polymer adsorption was lower for silica NP-polymer solutions than for conventional HPAM solutions, suggesting that reduced polymer adsorption is a key mechanism in the enhanced oil recovery (EOR) of NP-assisted polymer flooding. Cheraghian et al. [35] conducted an adsorption test to examine how clay and silicon oxide nanoparticles (NPs) affect polymer adsorption in carbonate and sandstone reservoirs. Their static adsorption tests revealed that NP-assisted polymer solutions exhibited lower adsorption and higher oil recovery than conventional polymers. Additionally, the adsorption of nano-polymers was greater in carbonate rocks than in sandstone. Giraldo et al. [58] also used adsorption tests to demonstrate the reduced adsorption of polymer nanosuspensions made from silicon oxide NP and HPAM. These findings suggest that the need for polymer dosages could be minimized, thereby reducing operational costs in enhanced oil recovery (EOR) applications.

5.5 Structural disjoining pressure

Disjoining pressure has been identified as a crucial mechanism in nanofluid-enhanced oil recovery (EOR). This pressure arises from the surface energy effects caused by the accumulation of millions of nanoparticles (NPs) at the interface [59]. Nanofluids play a crucial role in modifying the disjoining pressure between two immiscible fluids. In three-phase systems, nanoparticles (NPs) are dispersed in the liquid form wedge-shaped structures that exert forces toward the oil-solid contact area. This action enhances the spreading and wetting properties of the nanofluid on the solid surface. Wasan et al. [60] showed that nanoparticles can reduce oil adsorption on rock surfaces by increasing the structural disjoining pressure between the oil and the rock. This is achieved by establishing a wedge-film structure on rock surfaces, which helps release oil droplets and thus improves oil recovery, as schematically illustrated in **Figure 7**. Additionally, structural disjoining pressure is a key mechanism for altering the wettability of the rock-oil-water system. The components of structural disjoining pressure include van der Waals forces and a combination of electrostatic and van der Waals forces [61, 62].

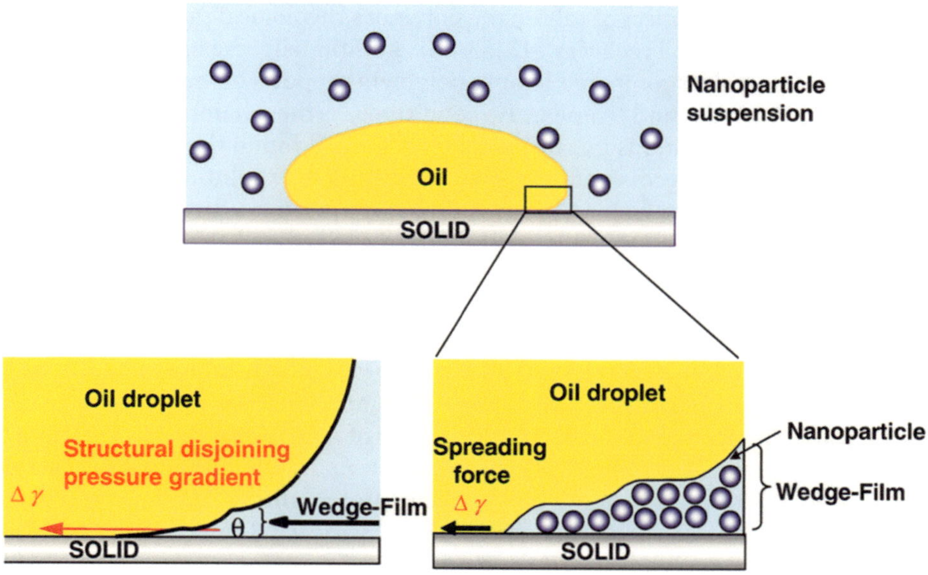

Figure 7.
The disjoining pressure mechanism [60].

6. Core flooding experiments

The effectiveness of enhanced oil recovery (EOR) methods is typically assessed by the incremental recovery factor (RF). This factor represents the extra oil recovered through the EOR process relative to the initial oil in place [2]. Silica nanoparticles (NPs) have been the most extensively studied in core flooding experiments and have shown increased oil recovery. Additionally, core flooding experiments with aluminum, titanium, iron, zinc oxides, and graphene, especially when combined with polymers, have yielded improved results. Gbadamosi et al. [55] conducted oil displacement tests on sandstone cores using two polymeric nanofluids: SiO_2-polymer and Al_2O_3-polymer. They found that Al_2O_3-polymer achieved higher oil recovery than SiO_2 polymer, with cumulative oil recoveries of 65.3% and 60.81% of the original oil in place (OOIP), respectively. Additionally, Bera et al. [40] assessed oil recovery through core displacement tests using a mixture of silica nanoparticles and guar gum. They reported an incremental oil recovery ranging from 10.7% to 27.3% of OOIP. The enhanced oil recovery was attributed to the enhancement of polymer viscosity and the modification of core wettability from intermediate wet to water wet.

6.1 Factors affecting the performance of nanoparticles-assisted polymer flooding

Nanoparticles-assisted polymer flooding is a promising enhanced oil recovery (EOR) technique. Understanding and optimizing these factors can significantly enhance the performance of nanoparticle-assisted polymer flooding, making it a more effective EOR technique. Several factors influence its performance:

6.1.1 Effect of the salinity

Typically, water salinity can adversely affect polymer performance in reservoirs. However, the combined effect of polymers and nanoparticles in nanohybrid polymers mitigates the negative impact of salt on the polymer's rheological properties, thereby enhancing incremental oil recovery [45, 63–65]. Additionally, the improved viscosity of nanohybrid polymers compared to pure polymers supports higher incremental oil recovery in high salinity and temperature conditions, as they maintain their viscosity better under harsh conditions [52, 66]. Maghzi et al. [67] found that under similar salinity conditions, the recovery factor with a nano-polymer solution is about 10% higher than with a pure polymer. When saline is introduced to a pure polymer, the polymer's ionic charge (–CO-O–) is neutralized by the cationic charge, reducing electrostatic repulsions. In contrast, when mixed with nanoparticles (NPs), an additional positive charge from the NPs enhances electrostatic repulsion among the polymer chains. Haruna et al. [63] performed flooding tests to examine the impact of adding NPs to polymer flooding on additional oil recovery under high-salinity conditions. They utilized two distinct displacing fluids: API brine, which has a salinity of around 100,000 ppm, and formation brine, with a salinity of approximately 138,000 ppm. In the first set of experiments with API brine, additional oil recovery was 4.30% for pure polymer, 7.61% for SiO_2-polymer, and 11.28% for modified SiO_2-polymer. In the second set with formation brine, additional oil recovery was 3.56% for pure polymer, 5.23% for SiO_2-polymer, and 10.79% for modified SiO_2-polymer. These results suggest that nanohybrid polymers perform better under harsh salinity conditions.

6.1.2 Effect of the temperature

The effectiveness of polymeric nanofluids was examined at various reservoir temperatures in the literature. Helmi et al. [68] explored how temperature affects the performance of nanocomposite polymer flooding. They tested flooding at temperatures of 40, 60, 80, 90, and 100°C, keeping slug size, nanocomposite concentration, and salinity constant at 40% PV, 2000 ppm, and 30,000 ppm, respectively. Their findings indicated that higher flooding temperatures reduced the cumulative oil recovery with nanocomposite polymer flooding. They also noted that the nanocomposite polymer could withstand temperatures up to 90°C, beyond which no additional oil recovery was observed due to thermal degradation. Additionally, Agi et al. [69] conducted core flooding tests to simulate polymeric nanofluid flooding under extreme temperature conditions of 120°C. They recorded incremental oil recoveries of 13%, 17%, and 23% for SiO_2 polymeric nanofluid, Al_2O_3 polymeric nanofluid, and crystalline starch nanofluid, respectively. Their results suggest that polymeric nanofluids had long-term stability and could withstand high-temperature degradation, suggesting that they are a viable alternative to traditional polymer flooding.

6.1.3 Effect of polymer concentration

Generally, a high polymer concentration is needed to achieve the desired viscosity of the displacing fluid [40]. Increasing the molecular weight or polymer concentration can enhance the polymer's viscosity [70]. Salem et al. [14] found that higher polymer concentrations led to greater incremental oil recovery, positively impacting overall oil recovery. Both synthetic and natural polymers were used in core flood tests of polymeric nanofluids, with the required polymer concentration varying by type. For biopolymers, higher concentrations are used compared to synthetic polymers like polyacrylamide, due to their typically lower molecular weight. Since polymer viscosity increases with molecular weight, higher concentrations are necessary for low molecular weight polymers to achieve the desired viscosity. In medium and high permeability conditions, polymers that require lower concentrations but still provide high displacement efficiency, such as the hyperbranched polymer synthesized by Liu et al., are more advantageous [71, 72].

6.1.4 Effect of the NPs concentration and NPs type

The concentration and type of NPs in polymeric nanofluids are crucial factors that directly influence oil recovery. In their study, Salem et al. [14] indicated the most commonly reported concentrations range from 0.05 to 3 wt.%. Researchers have focused on determining the optimal NP concentration and its benefits for reducing interfacial tension (IFT) and modifying wettability [43]. The reported results indicate that oil recovery improves with increasing NP concentration up to an optimal level. However, NP concentrations above 3 wt.% may reduce reservoir permeability [73]. Maghzi et al. [67] found that increasing NP concentration enhances the viscosity of polymeric nanofluids due to ion-dipole interactions between cations and NPs. Core flooding tests by Gbadamosi et al. [33] showed that Al_2O_3-xanthan gum, TiO_2-xanthan gum, and SiO_2-xanthan gum nanofluids achieved up to 28.4% OOIP incremental oil recovery after water flooding. They also noted that metal oxide polymer nanofluids (TiO_2-XG and Al_2O_3-XG) had higher recoveries than SiO_2-XG. Conversely, Hu et al. [43] reported that SiO_2-polymer flooding resulted in better oil recovery than Al_2O_3-polymer. For SiO_2 nanoparticles, increasing NP concentration reduced the interfacial tension (IFT) for oil-water and oil-polymer but beyond 0.2 wt.%, there was no significant change in contact angle. Therefore, laboratory tests are crucial for each reservoir to determine the optimal NP concentration before field-scale application. Overall, incremental oil recovery improves significantly with increasing NP concentration [14]. Understanding and optimizing these factors can significantly enhance the performance of nanoparticle-assisted polymer flooding, making it a more effective EOR technique.

7. Numerical simulation

Numerical simulation plays a crucial role in forecasting the oil recovery efficiency of polymer flooding. The first step in enhanced oil recovery (EOR) modeling is to select a suitable simulator that can accurately represent the behavior of the proposed EOR method [74]. Polymer flooding and other EOR techniques can be simulated using various commercial reservoir simulators such as IMEX, STARS, and GEM (by CMG company); T-Navigator (by RFD company); ECLIPSE (by Schlumberger company); and UTCHEM (by the University of Texas at Austin) [75, 76]. While these simulators are generally similar in performing numerical simulations, they differ slightly in how they model polymer viscosity. Therefore, it is essential to input accurate polymer rheology properties into the simulator to create an accurate model for polymer transport through porous media. Key factors derived from experimental studies that are valuable for modeling polymer injection include resistance factors (RF), polymer adsorption, and residual resistance factors [77].

Numerical simulation studies have been conducted to evaluate the effectiveness of nanoparticle-enhanced polymer flooding. Loaiza et al. [78] performed numerical studies to simulate polymeric nanofluid injection for EOR. They included polymer adsorption and NP retention in their models, validating their results with commercial software and core flood tests.

Khalilinezhad et al. [37] explored the effects of nanoparticles (NPs) on polymer injection performance through both experimental assessments and numerical modeling. They validated their numerical model with laboratory experiments to ensure its reliability. Their results indicated that the presence of nanoparticles reduced polymer adsorption and increased polymer viscosity. Additionally, they found that nanoparticle-assisted polymer flooding resulted in less degradation of polymer solutions compared to conventional polymer flooding. They also developed a 3D simulation model using the UTCHEM simulator for field trials of nanoparticle-assisted polymer flooding, which demonstrated significant incremental oil recovery for both clay NP-assisted and silica NP-assisted polymer flooding compared to standard polymer flooding.

In another study, Pal and Mandal [27] used the CMG STARS simulator to model surfactant/polymer/nanoparticle flooding and incorporated experimental data to construct their model. Their simulations showed incremental oil recoveries of 15.06%, 17.42%, and 18.49% of OOIP for surfactant, surfactant-polymer, and surfactant-polymer-nano-silica flooding, respectively. Oil saturation decreased from 80% to 29.30% following surfactant-polymer-nano-silica flooding. Maurya et al. [34] simulated polymer-grafted nanoparticles using the CMG STARS simulator, creating a one-dimensional model to replicate sand pack experiments.

In another study, Badiri et al. [79] conducted numerical simulations to compare clay nanoparticle (NP)-assisted polymers with conventional polymers. They created a 3D synthetic model using the UTCHEM commercial simulator. Their simulations examined the effects of permeability heterogeneity, well patterns, and salinity on the performance of clay NP-assisted polymer-enhanced oil recovery (EOR). The inverted two to five-spot well pattern yielded the highest oil recovery compared to other patterns. Their results also showed that NP-assisted polymer flooding outperformed traditional polymer flooding. However, permeability heterogeneity reduced oil recovery in both secondary and tertiary recovery stages.

Khalilinezhad et al. [57] used UTCHEM to simulate the impact of silica NPs on polymer adsorption in porous media. Their simulations indicated that silica NPs could reduce polymer adsorption and enhance polyacrylamide performance. Druetta and Picchioni [80] developed an in-house simulator based on a mathematical model from a previous study, replacing surfactants with NPs to simulate nano-polymer injection. This model, which considered two dimensions, two phases (aqueous and oleic), and five constituents (oil, water, polymer, NPs, and salt), accounted for NP retention in porous media and the interaction between NPs and polymer particles in calculating aqueous phase viscosity. They also modeled core-scale nanoparticle-induced polymer injection, finding that mobility ratio modification significantly influenced oil recovery.

8. Economics of polymeric nanofluid

The main goals of enhanced oil recovery (EOR) are to maximize oil extraction and, consequently, increase the overall revenue of the oilfield. When implementing polymer flooding EOR, three main cost categories are typically considered: (a) costs for experimental research and simulation, (b) expenses for project design, construction, installation, and startup of the injection process, and (c) chemical costs. Notably, the majority of polymer flooding expenses, about 75%, are attributed

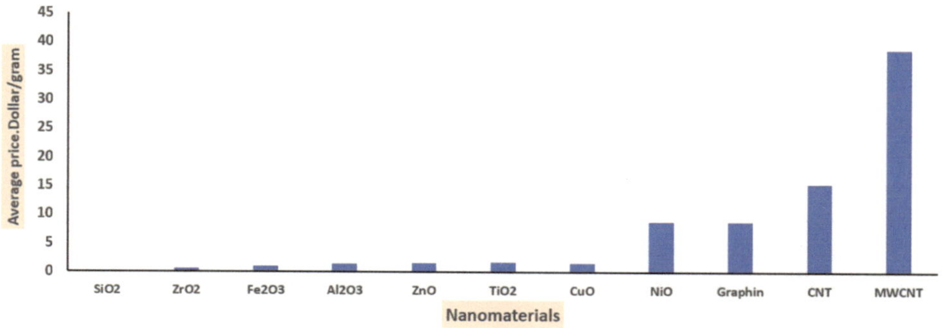

Figure 8.
Costs of nanomaterials [66, 81].

to chemical injection [77]. For polymer nanofluids, the cost of nanomaterials is an additional expense. The feasibility of these processes can be improved by reducing the amount of polymer needed with the help of nanoparticles. The cost of nanomaterials used in EOR depends on various factors, including the type and quality of nanoparticles, the quantity required, and the synthesis or fabrication method. The costs of nanomaterials for EOR are illustrated in **Figure 8**. Silica oxide and metal oxide nanoparticles typically cost between $0.1 and $0.5 per gram. On the other hand, manufacturing costs for carbon-based nanoparticles are generally high. The cost of carbon-based nanoparticles, like graphene oxide and carbon nanotubes, ranges from $8 to $39 per gram.

For polymeric nanofluids to be viable on a field scale, they require low-cost nanomaterials that are also high-quality and can be produced on a large scale. The synthesis costs of nanoparticles used as additives in the oil industry are shown in **Figure 8**. Silica (SiO_2), zirconium (ZrO_2), iron (Fe_2O_3), aluminum (Al_2O_3), zinc (ZnO), and titanium (TiO_2) oxides are relatively inexpensive. In contrast, nickel oxide (NiO), graphene, carbon nanotubes (CNT), and multi-walled carbon nanotubes (MWCNT) are more costly. Due to their high preparation costs, graphene, CNT, and MWCNT are often not economically feasible. To evaluate the economic feasibility of polymeric nanofluid enhanced oil recovery (EOR), it is essential to compare the total costs of the method with the incremental oil recovery it achieves. The expenses associated with polymeric nanofluid EOR must also be weighed against those of conventional polymer flooding. Additionally, factors such as reservoir characteristics, oil prices, and the regulatory environment can influence the economic viability of polymeric nanofluid EOR. Decision-makers in upstream companies aim to implement cost-effective strategies with minimal nanoparticle (NP) usage, such as NP-assisted EOR. The economic success of NP-assisted polymer flooding largely depends on its ability to enhance oil recovery and the prevailing oil prices [5].

9. Challenges and future perspective of NP-assisted polymer flooding (EOR)

Polymeric nanofluids have shown great promise in enhancing oil recovery (EOR) due to their ability to improve the viscosity, shear stability, heat resistance, and salt tolerance of polymers. However, several gaps and challenges need to be addressed for their widespread application:

9.1 Gaps and challenges

- Despite the promising results of experimental investigations, such a polymeric nanofluid-EOR process has not been implemented. To advance and initiate nano-EOR processes under field-scale conditions, it is essential to conduct pilot tests of polymeric nanofluids. This requires fostering effective collaborations among researchers, upstream operators, and industry professionals in the EOR sector. Such partnerships are crucial for translating laboratory findings into practical applications in oil fields.

- The outcomes of core flooding experiments with polymeric nanofluids have demonstrated their potential to enhance oil recovery. However, significant gaps remain in the literature concerning the theoretical understanding of nanoparticle interactions with reservoir media and the adsorption processes and chemical reactions occurring within the reservoir.

- Due to their inherent physical instability and high energy content, nanomaterials are prone to clustering and agglomeration within polymer solutions. This issue can lead to the formation of inhomogeneous solutions, thereby adversely affecting the functionality of polymeric nanofluids. Consequently, ensuring the stability of polymeric nanofluid dispersions remains a significant challenge. Stability is a critical factor in the performance of polymeric nanofluids, particularly in the flooding process in porous media. Therefore, it is essential to rigorously assess the stability of these nanofluids and explore innovative methods to enhance their suspension stability.

9.2 Future perspective

- Despite the encouraging results observed in laboratory settings, the application of polymeric nanofluids for enhanced oil recovery (EOR) at the field scale remains relatively limited. There is considerable potential for wider adoption of these nanofluids as additional field trials are conducted and their efficacy is confirmed. Employing cost-effective and compatible nanomaterials, such as silica oxide, alongside polymers can significantly enhance the scalability of nanoparticle-induced polymer flooding for broader applications in the oil industry. With ongoing research and development focused on improving their efficiency and sustainability, polymeric nanofluids hold promise as a viable EOR agent.

- Incorporating nanoparticles (NPs) into polymer solutions can reduce the amount of polymer needed for enhanced oil recovery (EOR). This is because NPs enhance the viscosity and stability of the injected polymer, allowing for more efficient and effective use of the polymer in the recovery process.

- Comprehensive economic evaluation is necessary for research studies on nanoparticle-assisted polymer flooding. A thorough economic analysis will resolve the ongoing debate about the cost-effectiveness of this method, providing clearer insights into its financial viability.

10. Conclusions

This chapter explored the current experimental status and emerging trends in the development of nanoparticle-assisted polymer flooding. It covered the preparation, characterization, and stability assessment of polymeric nanofluids and the mechanisms that enhance their effectiveness as EOR agents. The chapter also reviewed numerical simulations, economic evaluations, and field applications of these nanofluids. Additionally, it summarized the challenges faced and provided an outlook on future advancements in this innovative field.

Nanoparticle-assisted polymer flooding has demonstrated superior oil recovery compared to conventional polymer floods, owing to its unique properties such as altered interfacial tension (IFT) and wettability, improved rheological behavior, and reduced polymer adsorption. Displacement tests of polymeric nanofluids have shown excellent oil recovery performance not only under conventional reservoir conditions but also in challenging reservoir conditions.

Abbreviations

Al_2O_3	aluminum oxide
CMG	computer modeling group
CNT	carbon nanotubes
CuO	copper oxide
DLS	dynamic light scattering
EOR	enhanced oil recovery
Fe_2O_3	iron oxide
FTIR	fourier transform infrared
HPAM	hydrolyzed polyacrylamide
IFT	interfacial tension
IMEX	implicit-explicit black-oil simulator
MWCNT	multi-wall carbon nanotubes
NiO	nickle oxide
NP	nanoparticle
NPs	nanoparticles
OOIP	original oil in place
PNS	polymer nanosuspension
ppm	part per million
PV	pore volume
RF	recovery factor
SEM	scanning electron microscopy
SiO_2	silicon dioxide
STARS	thermal and advanced process simulator
TiO_2	titanium dioxide
TEM	transmission electron microscope
UTCHEM	the university of Texas chemical flood
wt. %	weight percent
XG	xanthan gum
ZnO	zinc oxide
ZrO_2	zirconium oxide

Author details

Khalaf Gad Salem[1*], Adel Salem[2,4], Sayed Gomaa[3,4] and Abdelaziz El-hoshoudy[5,6]

1 Department of Reservoir Engineering, South Valley Egyptian Petroleum Holding Company (GANOPE), Cairo, Egypt

2 Department of Petroleum Engineering, Faculty of Petroleum and Mining Engineering, Suez University, Suez, Egypt

3 Mining and Petroleum Engineering Department, Faculty of Engineering, Al-Azhar University, Cairo, Egypt

4 Department of Petroleum Engineering, Faculty of Engineering and Technology, Future University in Egypt (FUE), Cairo, Egypt

5 PVT Lab, Production Department, Egyptian Petroleum Research Institute, Cairo, Egypt

6 PVT Service Center, Egyptian Petroleum Research Institute, Cairo, Egypt

*Address all correspondence to: khalaf.saib@pme.suezuni.edu.eg

References

[1] Lake LW. Eaglewood Cliffs, New Jersey: Enhanced Oil Recovery; 1989

[2] Sheng JJ. Modern Chemical Enhanced Oil Recovery: Theory and Practice. Houston, Texas: Gulf Professional Publishing; 2010

[3] Muggeridge A, Cockin A, Webb K, Frampton H, Collins I, Moulds T, et al. Recovery rates, enhanced oil recovery and technological limits. Philosophical Transactions of the Royal Society A: Mathematical, Physical and Engineering Sciences. 2006;**2014**(372):20120320

[4] Denney D. EOR potential in the middle east: Current and future trends. Journal of Petroleum Technology. 2012;**64**(01):70-73

[5] Gbadamosi AO, Junin R, Manan MA, Agi A, Yusuff AS. An overview of chemical enhanced oil recovery: Recent advances and prospects. International Nano Letters. 2019;**9**(3):171-202

[6] Gomaa S, Salem KG, El-hoshoudy A. Enhanced heavy and extra heavy oil recovery: Current status and new trends. Petroleum. 2024;**10**(3):399-410

[7] Kumar N, Sampaio MA, Ojha K, Hoteit H, Mandal A. Fundamental aspects, mechanisms and emerging possibilities of CO_2 miscible flooding in enhanced oil recovery: A review. Fuel. 2022;**330**:125633

[8] Niu J, Liu Q, Lv J, Peng B. Review on microbial enhanced oil recovery: Mechanisms, modeling and field trials. Journal of Petroleum Science and Engineering. 2020;**192**:107350

[9] El-hoshoudy A, Salem KG. Investigation of sulfonated polyethylene glycol-polyacrylate copolymer for enhanced oil recovery through core and field-scale simulation. Journal of Molecular Liquids. 2025;**420**:126820

[10] Sheng JJ, Leonhardt B, Azri N. Status of polymer-flooding technology. Journal of Canadian Petroleum Technology. 2015;**54**(02):116-126

[11] El-hoshoudy A. Experimental and theoretical investigation for synthetic polymers, biopolymers and polymeric nanocomposites application in enhanced oil recovery operations. Arabian Journal for Science and Engineering. 2022;**47**(9):10887-10915

[12] Firozjaii AM, Saghafi HR. Review on chemical enhanced oil recovery using polymer flooding: Fundamentals, experimental and numerical simulation. Petroleum. 2020;**6**(2):115-122

[13] Abd-Elaal AA, Tawfik SM, Abd-Elhamid A, Salem KG, El-hoshoudy A. Experimental and theoretical investigation of cationic-based fluorescent-tagged polyacrylate copolymers for improving oil recovery. Scientific Reports. 2024;**14**(1):27689

[14] Salem KG, Tantawy MA, Gawish AA, Gomaa S, El-hoshoudy A. Nanoparticles assisted polymer flooding: Comprehensive assessment and empirical correlation. Geoenergy Science and Engineering. 2023;**226**:211753

[15] Standnes DC, Skjevrak I. Literature review of implemented polymer field projects. Journal of Petroleum Science and Engineering. 2014;**122**:761-775

[16] Corredor LM, Husein MM, Maini BB. A review of polymer nanohybrids for

oil recovery. Advances in Colloid and Interface Science. 2019;**272**:102018

[17] Agista MN, Guo K, Yu Z. A state-of-the-art review of nanoparticles application in petroleum with a focus on enhanced oil recovery. Applied Sciences. 2018;**8**(6):871

[18] Panchal H, Patel H, Patel J, Shah M. A systematic review on nanotechnology in enhanced oil recovery. Petroleum Research. 2021;**6**(3):204-212

[19] Sircar A, Rayavarapu K, Bist N, Yadav K, Singh S. Applications of nanoparticles in enhanced oil recovery. Petroleum Research. 2022;**7**(1):77-90

[20] Sun X, Zhang Y, Chen G, Gai Z. Application of nanoparticles in enhanced oil recovery: A critical review of recent progress. Energies. 2017;**10**(3):345

[21] Hendraningrat L, Li S, Torsæter O. A coreflood investigation of nanofluid enhanced oil recovery. Journal of Petroleum Science and Engineering. 2013;**111**:128-138

[22] Youssif MI, El-Maghraby RM, Saleh SM, Elgibaly A. Silica nanofluid flooding for enhanced oil recovery in sandstone rocks. Egyptian Journal of Petroleum. 2018;**27**(1):105-110

[23] Salem, Ragab AM, Hannora AE. A Comparative investigation of nano particle effects for improved oil recovery – experimental work. Paper presented at the SPE Kuwait Oil and Gas Show and Conference, Mishref, Kuwait. October 2015

[24] El-Hoshoudy A, Desouky S, Betiha M, Alsabagh A. Use of 1-vinyl imidazole based surfmers for preparation of polyacrylamide–SiO_2 nanocomposite through aza-Michael addition copolymerization reaction for rock wettability alteration. Fuel. 2016;**170**:161-175

[25] Sharma T, Iglauer S, Sangwai JS. Silica nanofluids in an oilfield polymer polyacrylamide: Interfacial properties, wettability alteration, and applications for chemical enhanced oil recovery. Industrial & Engineering Chemistry Research. 2016;**55**(48):12387-12397

[26] Gbadamosi AO, Junin R, Manan MA, Agi A, Oseh JO, Usman J. Synergistic application of aluminium oxide nanoparticles and oilfield polyacrylamide for enhanced oil recovery. Journal of Petroleum Science and Engineering. 2019;**182**:106345

[27] Pal N, Mandal A. Numerical simulation of enhanced oil recovery studies for aqueous Gemini surfactant-polymer-nanoparticle systems. AICHE Journal. 2020;**66**(11):e17020

[28] Joshi D, Maurya NK, Kumar N, Mandal A. Experimental investigation of silica nanoparticle assisted surfactant and polymer systems for enhanced oil recovery. Journal of Petroleum Science and Engineering. 2022;**216**:110791

[29] Pal N, Mandal A. Enhanced oil recovery performance of Gemini surfactant-stabilized nanoemulsions functionalized with partially hydrolyzed polymer/silica nanoparticles. Chemical Engineering Science. 2020;**226**:115887

[30] Kumar D, Lashari N, Ganat T, Ayoub MA, Soomro AA, Chandio TA. A review on application of nanoparticles in cEOR: Performance, mechanisms, and influencing parameters. Journal of Molecular Liquids. 2022;**353**:118821

[31] Foroozesh J, Kumar S. Nanoparticles behaviors in porous media: Application to enhanced oil recovery. Journal of Molecular Liquids. 2020;**316**:113876

[32] Gbadamosi AO, Junin R, Manan MA, Yekeen N, Agi A, Oseh JO. Recent advances and prospects in polymeric nanofluids application for enhanced oil recovery. Journal of Industrial and Engineering Chemistry. 2018;**66**:1-19

[33] Gbadamosi A, Yusuff A, Agi A, Muruga P, Junin R, Jeffrey O. Mechanistic study of nanoparticles-assisted xanthan gum polymer flooding for enhanced oil recovery: A comparative study. Journal of Petroleum Exploration and Production Technology. 2022;**12**(1):207-213

[34] Maurya NK, Kushwaha P, Mandal A. Studies on interfacial and rheological properties of water soluble polymer grafted nanoparticle for application in enhanced oil recovery. Journal of the Taiwan Institute of Chemical Engineers. 2017;**70**:319-330

[35] Cheraghian G, Khalili Nezhad SS, Kamari M, Hemmati M, Masihi M, Bazgir S. Adsorption polymer on reservoir rock and role of the nanoparticles, clay and SiO_2. International Nano Letters. 2014;**4**(3):1-8

[36] Khalili Nezhad SS, Cheraghian G. Mechanisms behind injecting the combination of nano-clay particles and polymer solution for enhanced oil recovery. Applied Nanoscience. 2016;**6**(6):923-931

[37] Khalilinezhad SS, Cheraghian G, Karambeigi MS, Tabatabaee H, Roayaei E. Characterizing the role of clay and silica nanoparticles in enhanced heavy oil recovery during polymer flooding. Arabian Journal for Science and Engineering. 2016;**41**:2731-2750

[38] Agi A, Junin R, Abdullah MO, Jaafar MZ, Arsad A, Sulaiman WRW, et al. Application of polymeric nanofluid in enhancing oil recovery at reservoir condition. Journal of Petroleum Science and Engineering. 2020;**194**:107476

[39] Ali JA, Kolo K, Manshad AK, Stephen KD. Potential application of low-salinity polymeric-nanofluid in carbonate oil reservoirs: IFT reduction, wettability alteration, rheology and emulsification characteristics. Journal of Molecular Liquids. 2019;**284**:735-747

[40] Bera A, Shah S, Shah M, Agarwal J, Vij RK. Mechanistic study on silica nanoparticles-assisted guar gum polymer flooding for enhanced oil recovery in sandstone reservoirs. Colloids and Surfaces A: Physicochemical and Engineering Aspects. 2020;**598**:124833

[41] Tsubokawa N. Surface grafting of polymers onto nanoparticles in a solvent-free dry-system and applications of polymer-grafted nanoparticles as novel functional hybrid materials. Polymer Journal. 2007;**39**(10):983-1000

[42] Asai M, Zhao D, Kumar SK. Role of grafting mechanism on the polymer coverage and self-assembly of hairy nanoparticles. ACS Nano. 2017;**11**(7):7028-7035

[43] Hu Y, Zhao Z, Dong H, Vladimirovna Mikhailova M, Davarpanah A. Hybrid application of nanoparticles and polymer in enhanced oil recovery processes. Polymers. 2021;**13**(9):1414

[44] Keykhosravi A, Vanani MB, Aghayari C. TiO_2 nanoparticle-induced xanthan gum polymer for EOR: Assessing the underlying mechanisms in oil-wet carbonates. Journal of Petroleum Science and Engineering. 2021;**204**:108756

[45] Hamdi SS, Al-Kayiem HH, Alsabah MS, Muhsan AS. A comparative study of dispersed and grafted nanofluids of graphene nanoplatelets

with natural polymer in high salinity brine for enhanced oil recovery. Journal of Petroleum Science and Engineering. 2022;**210**:110004

[46] Liu J, Wang S, He L, Hu P, Gao Y. Preparation and properties of nano-silica hybrid hydrophobic associated polyacrylamide for polymer flooding. Journal of Petroleum Science and Engineering. 2022;**208**:109434

[47] Ghadimi A, Saidur R, Metselaar H. A review of nanofluid stability properties and characterization in stationary conditions. International Journal of Heat and Mass Transfer. 2011;**54**(17-18):4051-4068

[48] Rezk MY, Allam NK. Impact of nanotechnology on enhanced oil recovery: A mini-review. Industrial & Engineering Chemistry Research. 2019;**58**(36):16287-16295

[49] Elhaei R, Kharrat R, Madani M. Stability, flocculation, and rheological behavior of silica suspension-augmented polyacrylamide and the possibility to improve polymer flooding functionality. Journal of Molecular Liquids. 2021;**322**:114572

[50] Cao J, Song T, Wang X, Zhu Y, Wang S, Zhao M, et al. Studies on the rheological properties of amphiphilic nanosilica and a partially hydrolyzed polyacrylamide hybrid for enhanced oil recovery. Chemical Engineering Science. 2019;**206**:146-155

[51] Cao J, Song T, Zhu Y, Wang X, Wang S, Yu J, et al. Aqueous hybrids of amino-functionalized nanosilica and acrylamide-based polymer for enhanced oil recovery. RSC Advances. 2018;**8**(66):38056-38064

[52] Salem KG, Salem AM, Tantawy MA, Gawish AA, Gomaa S, El-hoshoudy A. A comprehensive investigation of nanocomposite polymer flooding at reservoir conditions: New insights into enhanced oil recovery. Journal of Polymers and the Environment. 2024;**32**:5915-5935

[53] Iravani M, Khalilnezhad Z, Khalilnezhad A. A review on application of nanoparticles for EOR purposes: History and current challenges. Journal of Petroleum Exploration and Production Technology. 2023;**13**(4):959-994

[54] Fan G, Li M, Chen X, Palyanitsina A, Timoshin A. Polymer-nanosilica-assisted to evaluate oil recovery performances in sandstone reservoirs. Energy Reports. 2021;**7**:2588-2593

[55] Gbadamosi AO, Junin R, Manan MA, Agi A, Oseh JO, Usman J. Effect of aluminium oxide nanoparticles on oilfield polyacrylamide: Rheology, interfacial tension, wettability and oil displacement studies. Journal of Molecular Liquids. 2019;**296**:111863

[56] Sorbie K, Sorbie K. Polymer retention in porous media. Polymer-Improved Oil Recovery. 1991:126-164. DOI: 10.1007/978-94-011-3044-8_5

[57] Khalilinezhad SS, Cheraghian G, Roayaei E, Tabatabaee H, Karambeigi MS. Improving heavy oil recovery in the polymer flooding process by utilizing hydrophilic silica nanoparticles. Energy Sources, Part A: Recovery, Utilization, and Environmental Effects. 2017;**12**:1-10

[58] Giraldo LJ, Giraldo MA, Llanos S, Maya G, Zabala RD, Nassar NN, et al. The effects of SiO_2 nanoparticles on the thermal stability and rheological behavior of hydrolyzed polyacrylamide based polymeric solutions. Journal of Petroleum Science and Engineering. 2017;**159**:841-852

[59] Birkeland MÅ. Investigation of nanoparticle effect on wettability and interfacial tension: An experimental study of a two-phase system of heavy oil and di water. Master's thesis, Institutt for petroleumsteknologi og anvendt geofysikk. 2013

[60] Wasan D, Nikolov A, Kondiparty K. The wetting and spreading of nanofluids on solids: Role of the structural disjoining pressure. Current Opinion in Colloid & Interface Science. 2011;**16**(4):344-349

[61] Zhang H, Nikolov A, Wasan D. Enhanced oil recovery (EOR) using nanoparticle dispersions: Underlying mechanism and imbibition experiments. Energy & Fuels. 2014;**28**(5):3002-3009

[62] Chengara A, Nikolov AD, Wasan DT, Trokhymchuk A, Henderson D. Spreading of nanofluids driven by the structural disjoining pressure gradient. Journal of Colloid and Interface Science. 2004;**280**(1):192-201

[63] Haruna MA, Gardy J, Yao G, Hu Z, Hondow N, Wen D. Nanoparticle modified polyacrylamide for enhanced oil recovery at harsh conditions. Fuel. 2020;**268**:117186

[64] Zhu D, Wei L, Wang B, Feng Y. Aqueous hybrids of silica nanoparticles and hydrophobically associating hydrolyzed polyacrylamide used for EOR in high-temperature and high-salinity reservoirs. Energies. 2014;**7**(6):3858-3871

[65] Izadi N, Koochi MM, Amrollahi A, Pourkhalil M. Investigation of functionalized polyelectrolyte polymer-coated Fe_3O_4 nanoparticles stabilized in high salinity brine at high temperatures as an EOR agent. Journal of Petroleum Science and Engineering. 2019;**178**:1079-1091

[66] Salem KG, Tantawy MA, Gawish AA, Salem AM, Gomaa S, El-hoshoudy A. Key aspects of polymeric nanofluids as a new enhanced oil recovery approach: A comprehensive review. Fuel. 2024;**368**:131515

[67] Maghzi A, Kharrat R, Mohebbi A, Ghazanfari MH. The impact of silica nanoparticles on the performance of polymer solution in presence of salts in polymer flooding for heavy oil recovery. Fuel. 2014;**123**:123-132

[68] Helmi M, Abu El Ela M, Desouky S, Sayyouh M. Effects of nanocomposite polymer flooding on Egyptian crude oil recovery. Journal of Petroleum Exploration and Production Technology. 2020;**10**(8):3937-3945

[69] Agi A, Junin R, Gbadamosi A, Manan M, Jaafar MZ, Abdullah MO, et al. Comparing natural and synthetic polymeric nanofluids in a mid-permeability sandstone reservoir condition. Journal of Molecular Liquids. 2020;**317**:113947

[70] Saboorian-Jooybari H, Dejam M, Chen Z. Heavy oil polymer flooding from laboratory core floods to pilot tests and field applications: Half-century studies. Journal of Petroleum Science and Engineering. 2016;**142**:85-100

[71] Liu R, Pu W, Sheng JJ, Du D. Star-like hydrophobically associative polyacrylamide for enhanced oil recovery: Comprehensive properties in harsh reservoir conditions. Journal of the Taiwan Institute of Chemical Engineers. 2017;**80**:639-649

[72] Liu R, Pu W-F, Du D-J. Synthesis and characterization of core–shell associative polymer that prepared by oilfield formation water for chemical flooding. Journal of Industrial and Engineering Chemistry. 2017;**46**:80-90

[73] Eltoum H, Yang Y-L, Hou J-R. The effect of nanoparticles on reservoir

wettability alteration: A critical review. Petroleum Science. 2021;**18**(1):136-153

[74] Pal N, Mandal A. Compositional simulation model and history-matching analysis of surfactant-polymer-nanoparticle (SPN) nanoemulsion assisted enhanced oil recovery. Journal of the Taiwan Institute of Chemical Engineers. 2021;**122**:1-13

[75] Goudarzi A, Delshad M, Sepehrnoori K. A chemical EOR benchmark study of different reservoir simulators. Computers & Geosciences. 2016;**94**:96-109

[76] Goudarzi A, Delshad M, Sepehrnoori K. A critical assessment of several reservoir simulators for modeling chemical enhanced oil recovery processes. Paper presented at the SPE Reservoir Simulation Symposium, The Woodlands, Texas, USA. 2013

[77] Thomas A. Polymer flooding. In: Chemical Enhanced Oil Recovery (cEOR)-a Practical Overview. London, United Kingdom: IntechOpen; 2016. pp. 55-99

[78] Loaiza CS, Patiño JF, Mejía JM. Numerical evaluation of a combined chemical enhanced oil recovery process with polymer and nanoparticles based on experimental observations. Journal of Petroleum Science and Engineering. 2020;**191**:107166

[79] Badiri S, Mahmoodi O, Zakavi M, Anousha A. Oil recovery improvement in heterogeneous layered reservoirs using polymer nanohybrids: A numerical approach. Arabian Journal for Science and Engineering. 2022;**47**(9):11481-503

[80] Druetta P, Picchioni F. Polymer and nanoparticles flooding as a new method for enhanced oil recovery. Journal of Petroleum Science and Engineering. 2019;**177**:479-495

[81] Davoodi S, Al-Shargabi M, Wood DA, Rukavishnikov VS, Minaev KM. Experimental and field applications of nanotechnology for enhanced oil recovery purposes: A review. Fuel. 2022;**324**:124669

Chapter 7

Perspective Chapter: Additives for Viscosity Reduction of Heavy Oils

Juan Carlos García-Mayorga and Vladimir Escobar-Barrios

Abstract

Nowadays, most oil reserves contain heavy and extra-heavy oils, which need to be added with light compounds to facilitate their transportation, and then the cost of oil barrels is increased. The use of additives that are capable of integrating with the different oil fractions reduces the viscosity of such oils by one order of magnitude, at least. In this chapter, bibliometric analysis regarding the use of such additives is presented. In addition, additives based on alternative chemistry to those currently used open the possibility of facilitating oil transportation, reducing the energy demand for pumping, and even stimulating production. In this chapter, the use of additives based on imidazoline derivatives is described for reducing the viscosity of heavy oil, even at quantities as low as 75 ppm.

Keywords: enhanced oil recovery (EOR), chemical additives, heavy oils, bibliometric analysis, viscosity reduction

1. Introduction

The constant growth of the global population necessitates an increasing supply of goods and services to enhance societal well-being. In contemporary society, the reliance on products derived from petroleum is profoundly pervasive, encompassing diverse applications from textiles to aerospace technology. This extensive utilization of polymers and other chemicals and synthetic materials derived from petrochemical sources illustrates the substantial dependence of modern civilization on these resources.

Consequently, petroleum extraction and refinement remain critical to obtain components for daily life. However, it is important to note that the current landscape of petroleum reservoirs reveals a predominance of heavy and extra-heavy crude oils, while reserves of light and medium oils are dwindling.

Various technological approaches exist to manage heavy and extra-heavy oils, particularly in terms of viscosity reduction, which is vital for facilitating their transport and subsequent refining processes.

One prevalent method for modifying the viscosity of crude oil is thermal treatment, which involves the application of heat to decrease viscosity. This technology proves particularly effective in both offshore and onshore drilling environments, although it entails significant economic and energy expenditures.

An alternative method to thermal treatment involves the use of diluents composed of lighter fractions, such as naphtha. While effective, this strategy may require the addition of diluents in quantities of up to 30% by weight, thereby escalating the overall cost per barrel of oil.

On the other hand, the use of chemical additives presents an economically viable and effective solution for viscosity reduction. Various additives are employed, including nanoparticles, plasticizers, and more recent advancements involving chemicals derived from amines, as well as ionic liquids.

Each alternative for the viscosity reduction of heavy and extra-heavy oils carries its own set of advantages and disadvantages, encompassing both technical and economic considerations. Nevertheless, the implementation of such technologies remains essential for optimizing oil extraction and transportation processes, ensuring that refineries operate under safer and more continuous conditions.

2. Bibliometric analysis

In this section, a bibliometric analysis is based on the keywords and connectors as follows: "EOR OR enhanced oil recovery AND additives AND viscosity reduction AND heavy oils." Using the Scopus database and the VOS viewer free software, it was possible to analyze the interconnection between such keywords as well as the research trend over time.

Figure 1 shows the correlation between the different used keywords. It is clear that there are four clusters.

The red cluster is related to viscosity reduction and heavy oil production, mainly. Such a relationship is clear and natural since the heavy oil has high viscosity due to the chemical oil composition, and the heavy fraction predominates along with medium to high sulfur content [1].

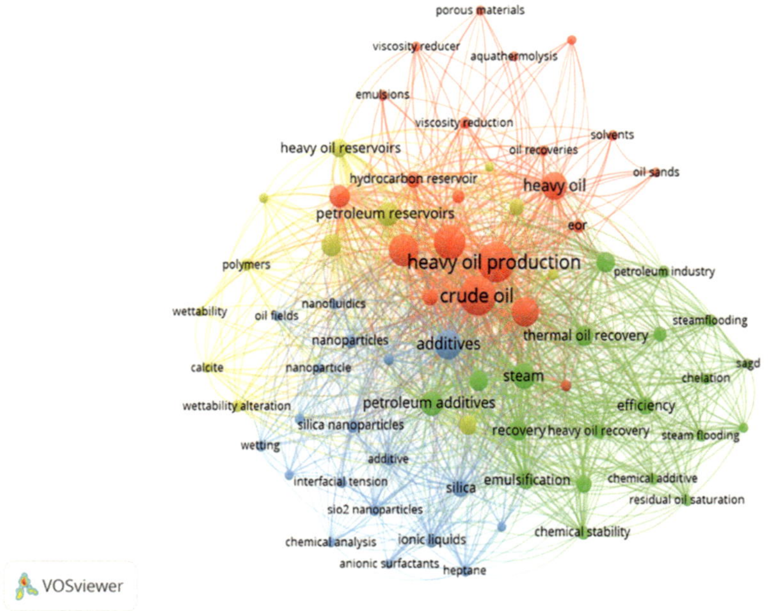

Figure 1.
Network visualization.

Following the analysis, the green cluster relates different methods used for heavy oil production, involving the viscosity reduction as thermal oil recovery, steam flooding, and chemical additives [2].

The blue and lime clusters are very close and related to different additives used for viscosity reduction and, consequently, enhanced oil recovery. Thus, silica particles and nanoparticles, surfactants, nanofluids, and ionic liquids are the main reported additives. These additives are used for modifying the interfacial tension between the liquid portion (oil or water) and the solids (rock) to facilitate the oil/water separation and consequent extraction.

Also, the viscosity reduction can be achieved by using some of those additives. For instance, recently, some researchers reported the use of ionic liquids that are able to modify substantially both the interfacial tension and viscosity, depending on the solvent used as a carrier [3]. In addition, the viscosity reduction using nanofluids based on silica or zirconium nanoparticles has been reported during the last 10 years [4–6]. It is important to mention that permanent viscosity reduction can be achieved by catalytic methods like aquathermolysis and aquaprocessing, which implies the water vapor molecules break down to generate hydrogen *in situ* and are used for hydrogenation of heavy fractions [7–9].

Figure 2 depicts the research trend regarding the viscosity reduction in heavy oil production.

From **Figure 2**, the light green-yellow terms are the most recent reported documents. Thus, the viscosity reduction, nanoparticles, chemical additive, EOR, and emulsification are the terms on which the researchers are focused on. As it was previously mentioned, one key factor is the viscosity reduction, which can be achieved in permanent or transient mode. From the financial point of view, it seems that the use of chemical additives has a better cost-benefit balance since they are not just used during the well stimulation but also during the oil transportation, which is

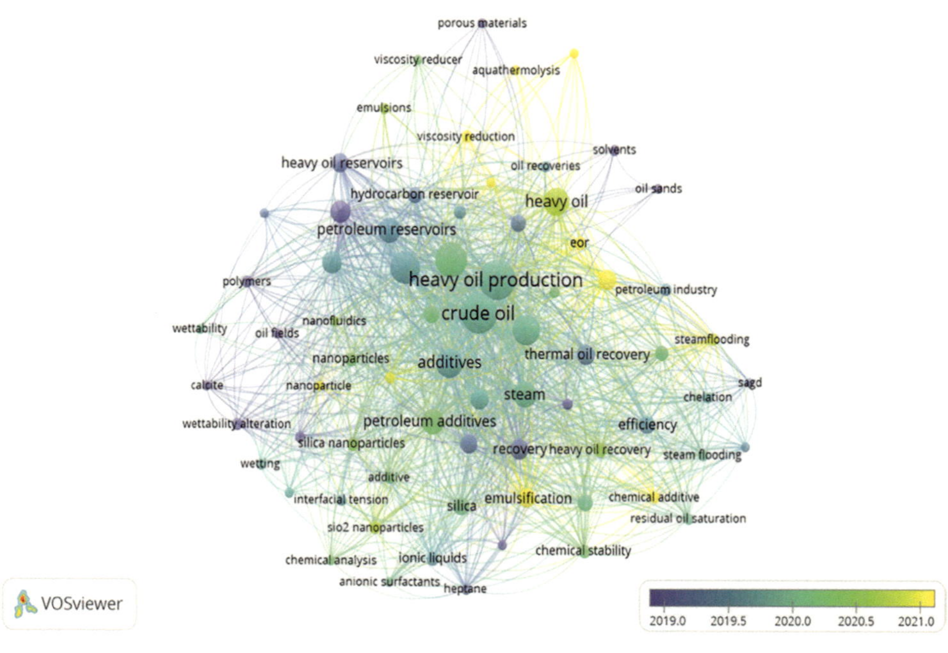

Figure 2.
Research trends over time.

very important for such balance. In addition, the required infrastructure for additive injection is considerably less complex than that required for catalytic processing of the oil. These aspects explain why the research trend is focused on the development and use of chemical additives. Although the potential of mixed methodologies could be considered to improve the viscosity reduction for enhanced oil recovery.

3. Thermal technologies for EOR

The application of thermal technologies in enhanced oil recovery (EOR) typically yields a high oil recovery factor, primarily due to the viscosity reduction that results from the heating of crude oil. The conventional thermal methods employed in this context include:

a. Hot fluid injection: This approach aims to reduce oil viscosity, though it is generally regarded as the least effective among thermal technologies due to its limited impact on mobility enhancement [10, 11].

b. Steam injection: This method involves the introduction of steam to heat the oil, significantly improving flow characteristics. While it is widely adopted owing to its effectiveness, it is also characterized by high energy demands and substantial operational costs [12–16].

c. In situ combustion: This technique entails igniting the oil within the reservoir, generating heat and gases that facilitate the fluidization of the crude oil. Although it possesses a high recovery potential, its application is often complicated due to the challenges associated with controlling the combustion process [17–21].

Among the aforementioned methods, Cyclic Steam Stimulation (CSS), Steam-Assisted Gravity Drainage (SAGD), and in situ combustion are the most prevalent thermal technologies aimed at enhancing oil mobility to improve overall recovery from geological formations, as well as to ease transportation.

However, it is crucial to acknowledge that the effectiveness of a particular technology may not be universally applicable across all reservoir types. For example, a reservoir characterized by a thickness of 20 feet, a depth of 10,000 feet, and a permeability of 100 millidarcy (mD) may be more suited to in situ combustion methods. Conversely, reservoirs exhibiting higher permeability levels may benefit more from steam injection or hot water flooding. In scenarios involving low permeability (< 100 mD), such methodologies often fail to adequately penetrate the rock formation, leading to uneven heat distribution and suboptimal oil flow.

Consequently, reservoir characterization emerges as a vital component in the selection of appropriate thermal technologies for EOR applications. Additionally, operational considerations and energy requirements should be factored into decision-making processes, given their direct implications on safety and cost.

Another noteworthy thermal method for achieving permanent viscosity reduction is aquathermolysis [11, 22]. This technology employs catalysts to dissociate water vapor molecules, necessitating careful catalyst selection to facilitate the catalytic reaction effectively. Positive outcomes have been reported, including an increase in the saturate and aromatic fractions by up to 35% on average, alongside a viscosity reduction of 58%. Nevertheless, the high energy demand associated with this process—typically requiring temperatures of at least 350°C—poses a significant challenge [23]. Additionally, the practical implementation of aquathermolysis in reservoirs is complicated by the difficulties in achieving the necessary temperature for catalyst activation.

4. Use of light fractions for EOR

Heavy and extra-heavy oils are characterized by their elevated content of asphaltenes and other high molecular weight components. Asphaltenes, in particular, are quasi-insoluble in liquid aliphatic hydrocarbons such as pentane, hexane, and heptane, necessitating the use of aromatic solvents to prevent asphaltenes precipitation within both the wellbore and the transportation pipelines [24]. This precipitation can lead to increased pressure drops along the pipeline, thereby augmenting the energy demands and costs associated with higher power pumping. Furthermore, asphaltene deposition can result in clogging within the well, consequently diminishing flow rates [25].

To address these challenges, mixtures of light and aromatic fractions are commonly injected to enhance oil mobility by reducing viscosity; however, such mixtures may constitute up to 30% by weight, significantly elevating operational costs [26]. These diluents, which are less viscous than heavy crudes, effectively lower the viscosity of the resultant heavy oil mixture [27]. Commonly used diluents comprise light products generated during natural gas production, naphtha, kerosene, and lighter crude oils. In certain scenarios, organic solvents possessing polar entities—such as alcohol, methyl tert-butyl ether, and tert-amyl methyl ether—have also been utilized [28]. Mixtures of hydrocarbons with these organic solvents have proven effective in minimizing the viscosity of heavy crude oil [27].

Light natural gas condensate (C5 or higher) serves as a low-density and less viscous hydrocarbon mixture used to dilute heavy crude oil; however, challenges such as asphaltenes precipitation, segregation, and aggregation lead to instability during both extraction and transportation processes [29]. The incorporation of light hydrocarbons—like kerosene—has also been shown to enhance the recovery and transportation of heavy crude oil. Naphtha, a well-known petroleum fraction, is frequently used as a diluent due to its high API gravity and compatibility with asphaltenes. Thus, blending naphtha with organic solvents could reduce the necessary quantity of diluents to achieve the viscosity required for heavy oil-to-pipeline transportation specifications. It was observed that the relative viscosity of heavy oil diluted with naphtha-organic solvent mixtures was significantly reduced, which can be attributed to the increased polarity or hydrogen bonding capabilities of the solvents, alongside the polar solvent's efficacy in interacting with the asphaltene components in heavy crude oil [27]. Accordingly, utilizing a higher polarity solvent results in a more pronounced reduction in the viscosity of the diluted heavy crude oil.

5. Additives for EOR

During the primary recovery phase, approximately 5–20% of the oil in place is typically recovered. This percentage can be increased to around 40% through the secondary recovery phase, indicating that a significant quantity of oil, roughly 60–70%, remains trapped underground [30, 31]. The extent of recoverable oil largely depends on whether it is light or heavy oil. Among the thermal and non-thermal methods utilized in Enhanced Oil Recovery (EOR), also known as the tertiary recovery phase, chemical or additive flooding is employed.

Additive flooding has two primary objectives: (1) to increase the capillary number (N_C) and (2) to decrease the mobility ratio (M) [32]. The capillary number N_C serves as a useful correlation that compares the ratio of viscous forces to surface tension forces, as expressed in Eq. (1). This correlation is commonly used to evaluate the relative impact of fluid motion against the forces resulting from interfacial tension when a liquid flows across a second fluid layer [33]. Achieving a reduction in interfacial tension to zero results in an infinite capillary number, indicating a displacement efficiency of 100% from the remaining oil.

$$N_C = \frac{\eta \upsilon}{\gamma} = \frac{viscous\ forces}{inter\ facialtension} \tag{1}$$

Where:

η : Viscosity of the displacing fluid.

υ : Velocity.

γ : Interfacial tension.

On the other hand, the water-oil mobility ratio, M, should be less than or equal to unity for a good recovery oil [31]. A favorable displacement, where oil moves faster than water, is given by $M < 1$. If $M > 1$, an unfavorable displacement, where water moves faster than oil, is observed (**Figure 3**) [34]. When $M = 1$, water and oil move at equal speed. This ratio is defined by the mobilities of both phases as defined in Eq. (2).

$$M = \frac{M_w}{M_o} = \frac{mobility\,of\ the\,displacing\ phase\ (water)}{mobility\,of\ the\,displaced\ phase\ (oil)} \tag{2}$$

There are four main additive flooding processes for enhancing oil recovery defined by the type of the injected chemical compound. These are surfactant flooding, polymer flooding, alkaline flooding, and alkali-surfactant-polymer flooding. All of them look to reduce the interfacial tension between the injected fluid and the reservoir oil to improve the oil recuperation.

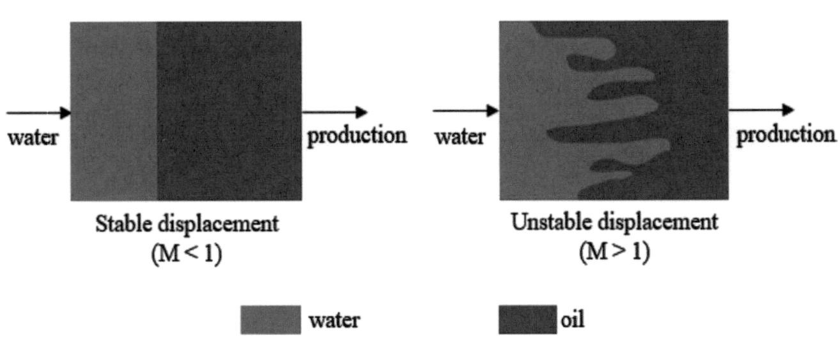

Figure 3.
Difference between water-oil displacement at different values of mobility ratio.

5.1 Surfactant flooding

In carbonate oil reservoirs, a prominent secondary recovery method involves the injection of seawater or brine solutions. This method induces wettability alterations that facilitate the displacement of oil from the rock matrix, thereby enhancing the oil recovery factor from 20 to 50% in optimal scenarios. The success of this method is dependent on the reservoir characteristics such as viscosity, wettability, and permeability [35]. Nevertheless, the efficacy of oil displacement through brine injection is constrained by two significant factors. First, the inherent heterogeneities within the reservoir may lead to water flow in highly permeable zones between the injection point and production points, causing water losses. Second, the high surface tension at the oil-water interface can trap oil within the rock's pore spaces, impeding effective displacement [36].

To further enhance the oil recovery factor, surfactants can be employed in surfactant flooding. Surfactant flooding entails the injection of brine with dissolved surfactants, which act primarily by reducing the interfacial tension (IFT) at the oil-brine interface. This reduction facilitates the mobilization of trapped oil toward the production wells (**Figure 4**) [37]. The decrease of IFT consequently increases the capillary number, thereby promoting oil recovery. Furthermore, surfactants exert additional effects, such as wettability alterations and emulsification, both of which contribute to the effective sweeping of oil. However, it is important to note that wettability can be double-edged; while they may assist in recovery, the adsorption of surfactants onto the rock surface can diminish their concentration and effectiveness [38].

The implementation of surfactant flooding has been shown to yield increments in oil recovery ranging from 16.5 to 23% [38, 39]. Surfactants may be classified as anionic, cationic, non-ionic, or zwitterionic (amphoteric). Nonetheless, the limitations of this EOR method are the expensive cost despite the typical concentration being between 0.1 and 0.5 wt% (<500 ppm) and the potential toxicity of the surfactants.

5.2 Polymer flooding

This method involves the injection of water mixed with polymers to reduce the oil-water mobility ratio and minimize water permeability. The addition of polymers to the water increases the viscosity of the injected fluid, enhancing the stable mobility of oil within the reservoir [40].

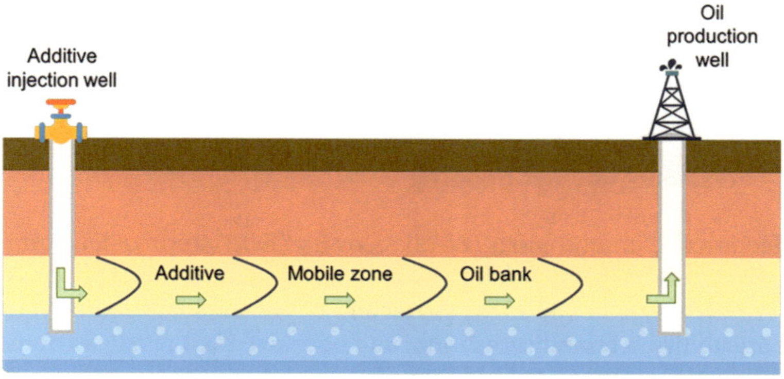

Figure 4.
Diagram showing the EOR with surfactant flooding.

There are two categories of polymers used for injection: synthetic and biopolymers. The most commonly utilized polymers in enhanced oil recovery (EOR) include polyacrylamide (PAM), partially hydrolyzed polyacrylamide (HPAM), polyacrylates, xanthan gum (a polysaccharide), and cellulose derivatives [37, 41]. Typical concentrations of these polymers range from 0.08% to 0.40% (800–4000 ppm), with reported optimal oil recovery results from polymer flooding between 15% and 25% [42–45].

One challenge associated with polymer flooding is polymer degradation. Shear forces are present from the moment of injection until the oil is displaced, during which the polymers are subjected to these forces, leading to degradation. Generally, a water-polymer fluid can lose up to 50% of its viscosity due to shear forces. The effectiveness of polymer flooding for oil recovery is also impacted by salinity—specifically the presence of divalent cations such as Ca^{2+} and Mg^{2+}—which can cause precipitation, as well as high temperatures that may lead to thermal degradation and adsorption on rock surfaces. In the case of biopolymers, degradation can also result from microbial activity. Additionally, the high costs associated with maintenance, elevated concentrations of polymers, and ensuring specific molecular weights to avoid clogging rock pores can pose significant challenges, potentially leading to injectivity issues or increased inaccessible pore volume [39, 46].

5.3 Alkaline flooding

Alkaline flooding involves the injection of an alkaline solution to generate surfactant molecules through the reaction of hydroxyl groups in the solution with acid groups in crude oil via a neutralization reaction. This process can lead to a reduction in interfacial tension, facilitate spontaneous emulsification, or induce wettability alterations in the rock [34]. When surfactants decrease the interfacial tension between the brine and the crude oil phase, microscopic displacement efficiency improves, subsequently enhancing oil recovery [47].

Common chemicals utilized in alkaline flooding include sodium hydroxide, ammonium hydroxide, ammonia, sodium orthosilicate, sodium metasilicate, and sodium carbonate. Typically, these agents are employed at concentrations ranging from 0.2 to 5.0%, making alkaline flooding one of the most cost-effective enhanced oil recovery methods [48].

However, alkaline flooding is not without its drawbacks. Reactions with rock can diminish alkalinity due to the presence of clay. The hydrogen ions on the clay surface react with the hydroxyl ions from the alkaline solution, resulting in a lower pH that reduces effectiveness [49]. Similarly, the salinity of the *in situ* water and the precipitation formed by divalent cations can adversely impact the efficacy of alkaline flooding [50, 51].

To address these challenges, combinations of different additives are used with the goal of maximizing benefits while minimizing drawbacks, thereby increasing the overall effectiveness of oil recovery, as will be discussed shortly.

5.4 Alkali-surfactant-polymer flooding

The selection of a unique enhanced oil recovery (EOR) approach is complicated by the inherent variability of the oil in place—encompassing factors such as viscosity, pH, and permeability—as well as the characteristics of the surrounding rock, which include permeability, wettability, salinity, and pore size. As a result, various alternatives have emerged, as illustrated in **Figure 5**.

These alternatives utilize combinations of the previously mentioned EOR methods, such as alkali, surfactant, and polymer flooding. The classification of these combinations

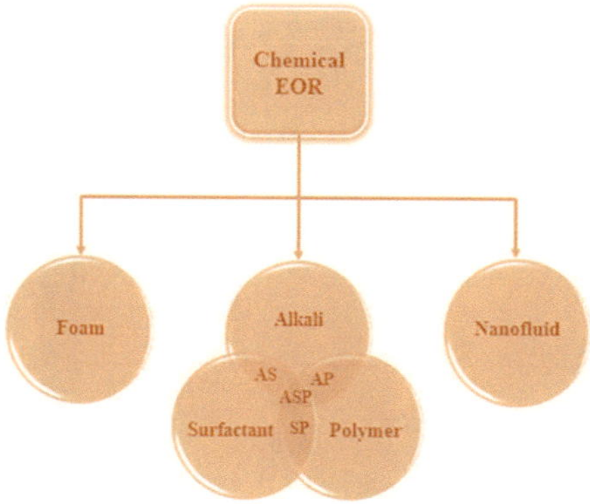

Figure 5.
Enhanced oil recovery technologies.

is determined by their constituents; for example, alkali-polymer flooding is referred to as AP flooding, while alkali-surfactant and surfactant-polymer flooding are designated as AS and SP flooding, respectively. One notable subtype is alkali-surfactant-polymer flooding (ASP flooding), which involves a ternary mixture of chemical additives.

This method significantly influences EOR through several synergistic mechanisms, including the reduction of the mobility ratio between water and oil phases, the lowering of surface tension at the water-oil interface, alterations in rock wettability (from oil-wet to water-wet), and the emulsification of water and oil [52, 53]. These synergistic effects are crucial and stem from the comprehensive application of various chemical agents, each contributing distinct flooding characteristics.

ASP flooding has gained widespread acceptance globally, with many countries—including China, India, Canada, the USA, and various nations in South America—implementing projects that have led to significant recovery improvements of up to 25% over the past two decades [52, 54]. This enhanced oil recovery method is economically attractive, primarily due to the use of alkali chemicals, which are relatively low-cost. Furthermore, this technology enables a reduction in the concentration of expensive surfactants by a factor of 20–70 [32, 55].

Despite these advantages, ASP flooding encounters challenges, particularly in heterogeneous reservoirs where 50–67% of the oil may remain in place. These challenges arise because ASP flooding is often unsuitable for carbonated reservoirs, and the presence of divalent cations can lead to the precipitation of surfactants and polymers. Additionally, high reservoir temperatures may trigger thermal degradation of the fluid components, alongside concerns related to the handling and toxicity of the materials involved [56, 57].

5.5 Imidazoline-based additives. A case of evaluation at lab and on-field evaluation

The imidazoline and derivatives are conventionally used for pipe protection due to their excellent anticorrosive properties. However, there is the possibility to chemically modify such derivatives and, consequently, impart some characteristics that could improve the heavy oil mobilization through the rock formation and also its transportation by reducing the viscosity. The VRBD additive is a proprietary chemical of CEOS (Villahermosa, Mexico), which is able to effectively reduce the viscosity of heavy and extra-heavy oils. **Figure 6** shows the percentage reduction of three different heavy

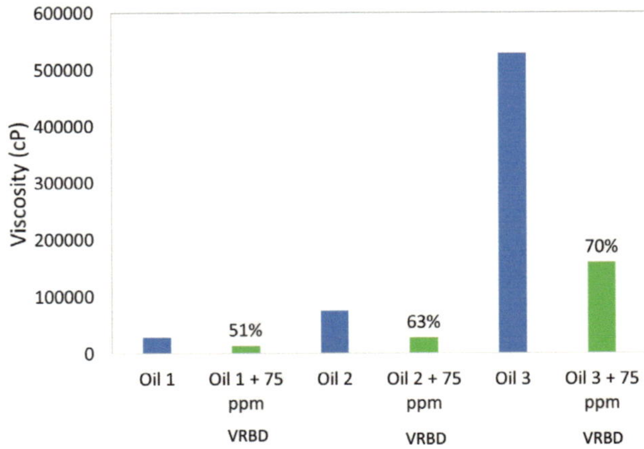

Figure 6.
Viscosity reduction of 3 Mexican heavy and extra-heavy oils using VRBD additive (75 ppm at 25°C).

and extra-heavy oils from a Mexican field, using 75 ppm of VRBD additive. The viscosity was evaluated by Brookfield equipment.

It is clear the great effect of the VRBD additive on the viscosity reduction by using only 75 ppm. Also, the percentage of viscosity reduction depends on the initial oil viscosity being greater, as such, initial viscosity was higher. This could be attributed to a better interaction of the VRBD additive with asphaltene fraction of oils.

In addition, an on-field test was carried out using this additive for 2 months. **Table 1** summarizes the obtained results in terms of production compared with natural decline, the use of mechanical methods, and conventional additives.

The results shown in **Table 2** are meaningful since the VRBD additive not only increases the production but also extends the productive life of the well by 2.3 times compared to the use of conventional additives, which implies an accumulated production 138% higher with VRBD.

These results corroborate that it is possible to use this commercial additive, based on imidazoline, to stimulate wells and increase the oil production by viscosity reduction of oil.

5.6 Ionic liquids

Ionic Liquids (ILs) are typically defined as liquid salts composed of ions with melting points below 100°C. This contrasts with other salts, such as sodium chloride, which has a melting point exceeding 800°C. The first IL, ethyl-ammonium nitrate,

Decline	Productive life (years)	Deferred production (Barrels)	Accumulated production (MMb)
Natural	2.9	—	0.41
Mechanical	1.3	- 229,518	0.18
Conventional additives	2.8	- 23,900	0.39
VRBD additive	6.4	+ 514,918	0.93

Table 1.
Effect of VRBD additive on production and predictive life. Tested on-field (Mexico).

Ionic liquid	CMC / ppm	Brine system	IFT at CMC / mN m-1	Reference
[C12mim][Cl]	2000	—	7.23	[59]
	100	Formation brine	0.81	[59, 60]
	250	Formation brine	0.83	[61]
[C8mim][Cl]	1000	—	16.77	[60]
	500	Formation brine	11.12	[60]
[C12Py][Cl]	500	—	8.72	[60]
	250	Formation brine	7.10	[60]
[C8Py][Cl]	Nd	—	> 21.89	[60]
	750	Formation brine	18.24	[60]
Tetra-alkyl ammonium sulfate	250	20% wt. NaCl and $CaCl_2$	1.65	[62, 63]
Nd: Not detected				

Table 2.
Effect of the different ionic liquids at different CMC on IFT [58].

was reported by Paul Walden in 1994, and since then, interest in ILs has grown significantly. The appeal of ILs lies in their many advantageous properties, including low volatility, non-flammability, a wide electrochemical window, high ionic conductivity, tunable characteristics, excellent solubility, and solvation capabilities, as well as stability in both water and air. These attributes have paved the way for a diverse range of applications for ILs, spanning catalysis, chemistry, analysis, electrochemistry, and their use as additives and solvents (**Figure 7**) [64, 65].

The Ionic liquids (ILs) can be classified in various ways, including categories such as neutral, acidic, basic, functionalized, and supported ionic liquids, as outlined by Verma et al. [66].

In addition, ILs can be categorized based on their chemical entities. When considering cations, ILs may include ammonium, guanidinium, imidazolium,

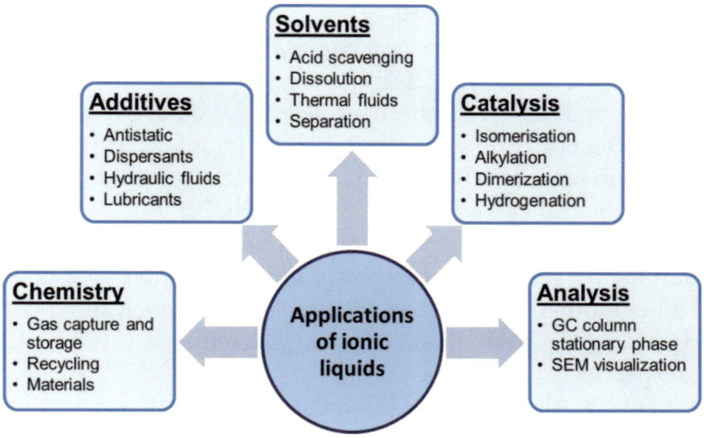

Figure 7.
Some relevant applications of ILs.

morpholinium, phosphonium, piperidinium, pyridinium, pyrrolidinium, and sulfonium. Conversely, classifications based on anions encompass borates, dicyanamide, halides, bis(trifluoromethylsulfonyl)imide, nonaflate, phosphate, sulfate, alkylsulfonate, sulfonate, thiocyanate, and tricyanomethide [67, 68].

Additionally, ILs based on imidazole/imidazolium and pyridiniums have garnered particular attention due to their versatility and anticorrosive properties. For instance, several works have reported the anticorrosive effect of ILs as 1-butyl-3-methyl-1H-benzimidazolium iodide (BMBM), 1-ethyl-3-methylimidazolium dicyanamide (EMID), 1-butyl-3-methylimidazolium chlorides (BMIC), 1-butyl-3-methylimidazolium hexafluorophosphate (BMIHF), and 3-hexadecyl-1-methyl-1H-imidazol-3-ium bromide for the protection of mild steel [66, 69–72].

Furthermore, recent studies have explored the utilization of imidazolium-based ILs in EOR.

The application of these ILs in the petroleum industry has shown promise, particularly in improving the recovery factor. Imidazolium ILs are typically synthesized through the alkylation of imidazole or pyridine with an alkyl halide, resulting in an alkylated imidazolium (quaternary salt). Alternatively, imidazolium can be produced via the heterocyclization of ethylenediamine or diethylenetriamine to yield imidazoline, followed by subsequent quaternization [73–75]. These synthetic pathways allow for the design and optimization of imidazolium/pyridinium-based ILs to enhance their performance for specific applications.

The role of ionic liquids (ILs) in EOR has been documented in various studies. These references highlight that ILs contribute to EOR by reducing the surface tension at the oil/water interface, inhibiting asphaltene precipitation, facilitating demulsification of water/oil emulsions, and altering wettability [76, 77]. **Table 2** summarizes the effects of four different ILs, as compiled by Hanamertani et al. [58]. These include 1-dodecyl-3-methylimidazolium chloride ([C12mim][Cl]), 1-octyl-3-methylimidazolium chloride ([C8mim][Cl]), 1-dodecyl pyridinium chloride ([C12Py][Cl]), and 1-octyl pyridinium chloride ([C8Py][Cl]). The reduction in the IFT between the oil/water phase depends on the alkyl chain and the critical micelle concentration (CMC); as observed, IFT has an increase in CMC of the IL.

This trend is also evident when brine is used during the flooding process, particularly when the alkyl chain of the IL is longer. The longer alkyl chains of ILs facilitate a greater likelihood of intramolecular interactions with oil, indicating that ILs with extended alkyl chains can achieve CMC at lower concentrations. Furthermore, the neutralization of charges from the cations and anions of ILs and brine enhances the contact area between oil and brine [58, 59].

Tafur et al. [78] conducted a comprehensive evaluation of the additional oil recoveries (AOR) effects attributed to three distinct ionic liquids (ILs): 1-dodecyl-3-methylimidazolium bromide ([C12mim][Br]), dodecyl pyridinium chloride ([C12Py][Cl]), and hexadecyl pyridinium chloride ([C16Py][Cl]). Their findings revealed that all three ILs exhibited a synergistic effect when combined with brine (SW0Na), resulting in a reduction of oil/water IFT and an alteration of wettability in chalk cores. This combination increased the capillarity number, significantly enhancing oil recovery by 24.6%, 22.4%, and 11.5% for [C12mim][Br], [C12Py][Cl], and [C16Py][Cl], respectively.

In a similar context, Alarbah et al. [76] documented that the reduction of IFT and modifications in wettability contributed to improved recovery factors when utilizing 1-Benzyl-3-methylimidazolium chloride ([BenzMIM][Cl]), trihexyl tetradecyl phosphonium chloride ([THTDPh][Cl]), and 1-Ethyl-3-methylimidazolium chloride

([EMIM][Cl]), with the latter attaining a remarkable total recovery of 78.36% from the oil in place.

Up to this point, ILs have played a similar role to that of the surfactants, where their main effect is to reduce the IFT, change the wettability, and break oil/water emulsion. However, ILs also demonstrate the properties of interacting with asphaltene fraction. For instance, Liu et al. [79] examined the effect of several ionic liquids, including N-alkylpyridinium chloride ([N-alkyl-py][Cl]; alkyl = ethyl, propyl, or butyl), N-butylisoquinolinium chloride ([C_4isoq][Cl]), and 1-butyl-3-methylimidazolium chloride, hexafluorophosphate, or tetrafluoroborate ([BMIM][Cl/PF_6/BF_4]) in the dissolution of asphaltenes. Their results indicated that, particularly in conjunction with high temperatures, these ILs proved effective in asphaltene dissolution.

This dissolution effect displays a correlation with the length of the alkyl chain and type of anion in such a way that the best performance of ILs was reiterated by [BMIM][Cl] with an asphaltene dissolution of 27.6%. In other words, small anions like $[Cl]^-$ have a better effect on the dissolution due to easy mobility in the medium; consequently, the effectiveness decreases in the order $[Cl]^- > [BF_4] > [PF_6]^-$. Furthermore, the dissolution effectiveness of cations was attributed to their aromatic structures, which allow for significant interaction with the aromatic units of asphaltenes. Conversely, cations with longer alkyl chains were observed to exhibit a diminished effect on asphaltene dissolution. Pereira et al. [80] also documented the efficacy of 1-ethyl-3-methylimidazolium tosylate ([C2mim][OTs]) in enhancing the recovery of aromatic oil, achieving an oil recovery factor of 65.7%. The recovery capabilities of [C2mim][OTs] were postulated to arise from electrostatic interactions and π-π interactions due to the aromatic character of the IL.

On the other hand, Ghanem et al. [81] demonstrated the dispersion of asphaltenes facilitated by ILs. They used imidazolinium-based ILs for the dissolution and observed that ILs interact with the asphaltenes due to the similar properties of ILs with resins. These properties imply interaction such as hydrogen bonding, π-π^* stacking, or Van der Waals forces, which agrees with the results reported by Pereira. Ghanem further noted that an alkyl chain length not exceeding hexadecane contributed to asphaltene dispersion, likely due to its capacity for crystallization and wax-like behavior.

The utilization of ionic liquids (ILs) based on imidazolium with metal-containing anions has garnered significant attention in enhancing the efficiency of enhanced oil recovery (EOR) processes. Nares et al. [82] highlighted the efficacy of ILs synthesized with metal atoms such as iron and molybdenum in treating heavy crude oil. Their findings demonstrated a remarkable reduction in kinematic viscosity from 15,416 to 124 cSt at 15.6°C, alongside a desulfurization rate of 46%, ultimately improving oil quality from 13.5 to 20.2°API. These results can be attributed to the catalytic activity of the ILs, which facilitate the hydrocracking of resins and asphaltenes, thereby liberating heavy fractions from sulfur compounds.

Furthermore, Shaban et al. [83] investigated the specific impact of 1-butyl-3-methylimidazolium chloride ([BMIM][Cl]) and 1-butyl-3-methylimidazolium tetrachloroferrate ([BMIM][$FeCl_4$]) on the recovery of heavy crude oils. Notably, [BMIM][$FeCl_4$] exhibited superior performance concerning viscosity reduction and desulfurization compared to [BMIM][Cl]. The incorporation of Fe in the anion enhanced its capacity to interact with sulfur in the oil, effectively breaking C-S bonds, which resulted in the complexation of sulfur and an improvement in oil flow. This process was found to be optimal at temperatures between 70 and 90°C in the presence of brine, leading to the liberation of hydrogen sulfide. Specifically, the application of [BMIM][$FeCl_4$] resulted in a reduction of sulfur and viscosity by 23.45% and 78.22%, respectively.

In a related study, Fan et al. [84] examined the effects of 1-butyl-3-methylimidazolium tetrachloroaluminate ([BMIM][$AlCl_4$]) on heavy oil recovery. Their experiments indicated that working with [BMIM][$AlCl_4$] at temperatures ranging from 65 to 85°C effectively reduced both viscosity and the content of sulfur and asphaltenes.

The influence of transition metals on ILs employed in EOR can be rationalized by analyzing the various bonding energies present within oil and ILs. In heavy crude oil, notable bonds include C-O, S-H, C-C, and C-S, with bond energies of approximately 360 kJ/mol for C-O and S-H, 248 kJ/mol for C-C, and 272 kJ/mol for C-S. Thus, when ionic liquids interact with heavy oil molecules, the cations predominantly target sulfur atoms due to their relatively weaker bond strength in comparison to carbon atoms. Concurrently, the anionic components of ILs engage with the carbon atoms. These interactions, combined with elevated temperatures, facilitate the cleavage of C-S bonds, enhancing the potential for oil recovery.

In conclusion, the distinct functionalities of ILs, such as interfacial tension reduction, wettability modification, emulsion breaking, sulfur complexation, and asphaltene degradation, significantly contribute to improving the oil recovery factor.

5.7 Importance of rheology

The intricate composition of petroleum, encompassing components such as resins, saturates, aromatics, and asphaltenes, undergoes transformations over time. Consequently, petroleum exhibits viscoelastic behavior, rendering rheology a critical analytical technique for its characterization. Various methodologies have endeavored to correlate specific fraction types with their rheological properties. However, it is essential to recognize that when examining the entire composition, the rheological model may evolve over time [85]. In scenarios where the time frame is relatively brief (spanning months or a few years), rheology assumes considerable importance. The incorporation of additives plays a pivotal role in altering viscosity and, by extension, the rheological characteristics of petroleum products. Notably, the disaggregation or removal of asphaltenes significantly affects the rheology of heavy and extra-heavy oils, with resins also exerting influence due to their interactions with asphaltenes, though asphaltenes are acknowledged to have the most substantial impact [86].

Several rheological parameters are vital for the investigation of heavy oils. Key parameters include the storage modulus (G'), the loss modulus (G"), and their ratio, referred to as the phase angle or damping factor (tan δ). A tan δ value exceeding 1 indicates a predominance of the viscous modulus, while a value below 1 suggests the dominance of the elastic behavior. A tan δ equal to 1 signifies that G" and G' are equivalent, occurring at the pour point or gelation point [87].

It is crucial to emphasize that the viscosity of heavy or extra-heavy oils, which exhibit viscoelastic properties, is contingent upon both temperature and shear rate. Therefore, the formulation of additives for enhanced oil recovery (EOR) must take these dependencies into account. For example, in the context of polymer flooding, the chosen polymer type influences shear thickening and shear thinning characteristics, which directly correlate with polymer concentration, operational temperature, water hardness, and salinity. Consequently, rheological analysis is essential to mitigate pore blockage and polymer losses, ultimately enhancing oil recovery [88–90].

The utilization of surfactants, whether in nanoparticle form or as other chemical additives, often leads to the formation of emulsions that significantly impact the rheological behavior of heavy oil. For example, oil/water (O/W) emulsions at moderate concentrations exhibit shear-thinning properties [91].

Thus, parameters such as temperature, type, and concentration of additives, injection pressure, matrix porosity, initial oil viscosity, and mixing duration must be meticulously considered in the context of EOR. Furthermore, interfacial rheology studies have demonstrated that the viscoelastic properties of O/W emulsions can be adjusted, and the effects on O/W stability at the interfacial scale supersede those at the bulk scale [92]. This insight facilitates a more nuanced understanding of optimal additive selection for the EOR process.

In summary, rheology serves as a sophisticated and powerful analytical tool for elucidating the impacts of additives on enhanced oil recovery.

6. Conclusion

The majority of oil reserves worldwide consist of heavy and extra-heavy oils. However, stimulating wells and transporting oil from these fields presents significant challenges. Enhanced oil recovery techniques are employed to address these issues, with thermal methods currently being widely used due to their effectiveness in reducing heavy oil viscosity and enhancing mobility through rock formations. Nevertheless, the economic and operational aspects of these methods often fail to provide a satisfactory cost/benefit ratio.

The use of chemical additives has created opportunities for more favorable cost/benefit balances, particularly thanks to the lower energy demands associated with chemical injection and the relatively small quantities required. Among these additives, imidazoline-based compounds, typically utilized for their anti-corrosion properties in pipelines, have shown promise. When chemically modified or used as a base for ionic liquids, they exhibit unique characteristics conducive to reducing oil viscosity. This enables improved well stimulation and transportation of heavy and extra-heavy oils.

It is noteworthy that, despite the impressive results achieved with imidazoline-based additives in enhanced oil recovery, the specific interaction mechanisms with rocks and oils remain unclear. Additionally, scaling up the production of these various chemical additives is crucial to meet the demands of the petroleum market.

Furthermore, based on the results obtained with imidazoline-based additives and nanoparticles, the potential for synergy or antagonism between these two could be investigated. There is currently a lack of systematic studies on such mixtures, as well as other potential additives, that might provide multi-effects on viscosity reduction, quality enhancement, and the recovery factor of heavy oils.

Acknowledgements

The authors acknowledge the data provided by CEOS Company (Villahermosa, Mexico) regarding the evaluation of imidazoline-based additives at lab and infield scales. Special acknowledgment and thanks to Mr. Josep Castillo for the discussion of the on-field testing with the VRBD additive.

Author details

Juan Carlos García-Mayorga[1] and Vladimir Escobar-Barrios[2*]

1 CEOS Corporation, Villahermosa, Tabasco, Mexico

2 Instituto Potosino de Investigacion Cientifica y Tecnologica, San Luis Potosi, SLP, Mexico

*Address all correspondence to: vladimir.escobar@ipicyt.edu.mx

References

[1] Wu J, Li S, Li Q, Yan F, Zhou Q, Ma S, et al. Characterization of chemical composition of high viscosity heavy oils: Macroscopic properties, and semi-quantitative analysis of molecular composition using high-resolution mass spectrometry. Petroleum Science. 2024;**21**(59):3612-3620. DOI: 10.1016/j.petsci.2024.02.019

[2] Seidy-Esfahlan M, Tabatabaei-Nezhad S, Khodapanah E. Comprehensive review of enhanced oil recovery strategies for heavy oil and bitumen reservoirs in various countries: Global perspectives, challenges, and solutions. Heliyon. 2024;**10**(18):e37826. DOI: 10.1016/j.heliyon.2024.e37826

[3] Subramanian D, Wu K, Firoozabadi A. Ionic liquids as viscosity modifiers for heavy and extra-heavy crude oils. Fuel. 2015;**143**:519-526. DOI: 10.1016/j.fuel.2014.11.051

[4] Wang C, Gao L, Liu M, Xia S, Han Y. Viscosity reduction mechanism of functionalized silica nanoparticles in heavy oil-water system. Fuel Processing Technology. 2022;**237**:107454. DOI: 10.1016/j.fuproc.2022.107454

[5] Li H, Gao H, Zhao X, Xia Z, Yu B, Sun D. Experimental study on viscosity reduction of heavy oil with water content by synergistic effect of microwave and nano-catalyst. Journal of Petroleum Science and Engineering. 2022;**208**(Part A):109271. DOI: 10.1016/j.petrol.2021.109271

[6] Taborda E, Franco C, Ruiz M, Alvarado V, Cortés F. Experimental and theoretical study of viscosity reduction in heavy crude oils by addition of nanoparticles. Energy & Fuels. 2017;**31**(2):1329-1338. DOI: 10.1021/acs.energyfuels.6b02686

[7] Thangaraj B, Lee Y. Recent progress in catalytic aquathermolysis of heavy oils. Fuel. 2024;**372**:132089. DOI: 10.1016/j.fuel.2024.132089

[8] Subhasis M, Ancheyta J, Marroquín G. Catalytic aquathermolysis used for viscosity reduction of heavy crude oils: A review. Energy & Fuels. 2010;**24**(5):2809-2816. DOI: 10.1021/ef100230k

[9] Duran J, Carbognani L, Scott C, Maini B, Pereira P. In-situ upgrading technology: Nanocatalyst concentration levels effects and hydrocarbons paths in the porous medium. Journal of Petroleum Science and Engineering. 2022;**208**:109594. DOI: 10.1016/j.petrol.2021.109594

[10] Ding Y, Zheng S, Meng X, Yang D. Low salinity hot water injection with addition of nanoparticles for enhancing heavy oil recovery. Journal of Energy Resources Technology. 2019;**141**(7):072904. DOI: 0.1115/1.4042238

[11] Askarova A, Turakhanov A, Markovic S, Popov E, Maksakov K, Usachev G, et al. Thermal enhanced oil recovery in deep heavy oil carbonates: Experimental and numerical study on a hot water injection performance. Journal of Petroleum Science and Engineering. 2020;**194**:107456. DOI: 10.1016/j.petrol.2020.107456

[12] Sun F, Yao Y, Chen M, Li X, Zhao L, Meng Y, et al. Performance analysis of superheated steam injection for heavy oil recovery and modeling of wellbore heat efficiency. Energy. 2017;**125**:795-804. DOI: 10.1016/j.energy.2017.02.114

[13] Xianghong W, Anzhu X, Hailiang F. An integrated evaluation on factors affecting the performance of superheated steam huff and puff in heavy oil reservoirs. Petroleum Exploration and Development. 2010;**37**(5):608-613. DOI: 10.1016/S1876-3804(10)60059-6

[14] Fan J, Yang J, Fan X, Wu L. Experimental study on the mechanism of enhanced oil recovery by non-condensable gas-assisted steam flooding process in extra-heavy oil reservoir energy sources, part a recovery. Utilization, and Environmental Effects. 2021;**43**(4):444-460. DOI: 10.1080/15567036.2019.1624894

[15] Pu W, Gao H, Li Y, Luo Q. Performance and mechanisms of enhanced oil recovery via CO_2 assisted steam flooding technique in high heterogeneity heavy oil reservoir: PVT and 3D experimental studies. Petroleum Science and Technology. 2020;**38**(15):823-835. DOI: 10.1080/10916466.2020.1789658

[16] Gu H, Cheng L, Huang S, Li B, Shen F, Fang W, et al. Steam injection for heavy oil recovery: Modeling of wellbore heat efficiency and analysis of steam injection performance. Energy Conversion and Management. 2015;**97**:166-177. DOI: 10.1016/j.enconman.2015.03.057

[17] Ahmadi M, Masoumi M, Askarinezhad R. Evolving connectionist model to monitor the efficiency of an in situ combustion process: Application to heavy oil recovery. Energy Technology. 2014;**2**(9-10):811-818. DOI: 10.1002/ente.201402043

[18] Pei S, Cui G, Zhang L, Zhang P, Huang L, Ren S. Performance and important engineering aspects of air injection assisted in situ upgrading process for heavy oil recovery. Journal of Petroleum Science and Engineering. 2021;**202**:108554. DOI: 10.1016/j.petrol.2021.108554

[19] Shokrlu Y, Maham Y, Tan X, Babadagli T, Gray M. Enhancement of the efficiency of in situ combustion technique for heavy-oil recovery by application of nickel ions. Fuel. 2013;**105**:397-407. DOI: 10.1016/j.fuel.2012.07.018

[20] Moore R, Laureshen C, Belgrave J, Ursenbach M, Metha S, et al. In situ combustion in Canadian heavy oil reservoirs. Fuel. 1995;**74**(8):1169-1175. DOI: 10.1016/0016-2361(95)00063-B

[21] Zheng W, Tan X, Wang T, Bai Y. Experimental and simulation study of the in situ combustion process in offshore heavy oil reservoirs. Petroleum Science and Technology. 2020;**38**:983-991. DOI: 10.1080/10916466.2020.1808673

[22] Tamanyan B. Aquathermolysis of crude oils and natural bitumen: Chemistry, catalysts and prospects for industrial implementation. Russian Chemical Reviews. 2015;**84**:1145-1175. DOI: 10.1070/RCR4500

[23] Yatimi Y, Mendil J, Marafi M, Alalou H, Al-Dahhan M. Advancement in heavy oil upgrading and sustainable exploration emerging technologies. Arabian Journal of Chemistry. 2024;**17**:105610. DOI: 10.1016/j.arabjc.2024.105610

[24] Hart A. A review of technologies for transporting heavy crude oil and bitumen via pipelines. Journal of Petroleum Exploration and Production Technology. 2014;**4**:327-336. DOI: 10.1007/s13202-013-0086-6

[25] Martinez-Palou R, Mosqueira M, Zapata-Rendón B, Mar-Juárez E,

Bernal-Huicochea C, Clavel-López J, et al. Transportation of heavy and extra-heavy crude oil by pipeline: A review. Journal of Petroleum Science and Engineering. 2011;**75**:274-282. DOI: 10.1016/j.petrol.2010.11.020

[26] Saniere A, Henaut I, Argillier J. Pipeline transportation of heavy oils, a strategic, economic and technological challenge. Oil & Gas Science and Technology—Revue d'IFP Energies Nouvelles. 2004;**59**(5):455-466. DOI: 10.2516/ogst:2004031

[27] Gateau P, Henaut I, Barre L, Argillier J. Heavy oil dilution. Oil & Gas Science and Technology—Revue d'IFP Energies Nouvelles. 2004;**59**(5):503-509. DOI: 10.2516/ogst:2004035

[28] Anhorn JL, Badakhshan A. MTBE: A carrier for heavy oil transportation and viscosity mixing rule applicability. Journal of Canadian Petroleum Technology. 1994;**33**(4):17-21. DOI: 10.2118/94-04-02

[29] Shigemoto N, Al-Maamari R, Jibril B, Hirayama A. A study of the effect of gas condensate on the viscosity and storage stability on Omani heavy crude oil. Energy & Fuels. 2006;**20**(6):2504-2508. DOI: 10.1021/ef060074h

[30] Ragab A, Mansour A. Enhanced oil recovery: Chemical flooding. In: Essa K, Di Risio M, Celli D, Pasquali D, editors. Geophysics and Ocean Waves Studies. London, UK: IntechOpen; 2021. DOI: 10.5772/intechopen.90335

[31] Machale J, Majumder S, Ghosh P, Sen T. Role of chemical additives and their rheological properties in enhanced oil recovery. Reviews in Chemical Engineering. 2020;**36**(7):789-830. DOI: 10.1515/revce-2018-0033

[32] Ahmed T, Meehan D. Introduction to enhanced oil recovery. In: Ahmed T, Meehan D, editors. Advanced Reservoir Management and Engineering. MA, USA: Gulf Professional Publishing; 2012. pp. 541-585. DOI: 10.1016/b978-0-12-385548-0.00006-3

[33] Rapp B. Fluids. In: Microfluidics: Modelling, Mechanics and Mathematics. MA, USA: Elsevier; 2017. pp. 243-263. DOI: 10.1016/B978-1-4557-3141-1.50009-5

[34] Speight J. Exploration and General Methods for Oil Recovery. MA, USA: Gulf Professional Publishing; 2009. DOI: 10.1016/B978-1-933762-25-8.50010-9

[35] Khan M, Piñerez I, Aldeen S, Puntervold T, Strand S. Polysulphate: A new enhanced oil recovery additive to maximize the oil recovery from carbonate reservoirs at high temperature. SPE Reservoir Evaluation and Engineering. 2023;**26**(03):873-887. DOI: 10.2118/211443-PA

[36] Massarweh O, Abushaikha A. The use of surfactants in enhanced oil recovery: A review of recent advances. Energy Reports. 2020;**6**:3150-3178. DOI: 10.1016/j.egyr.2020.11.009

[37] Pothula G, Vij R, Bera A. An overview of chemical enhanced oil recovery and its status in India. Petroleum Science. 2023;**20**(4):2305-2323. DOI: 10.1016/j.petsci.2023.01.001

[38] Austad T, Milter J. Surfactant Flooding in Enhanced Oil Recovery, in Surfactants. Cambridge, GB: Cambridge University Press; 2000. pp. 203-250. DOI: 10.1017/cbo9780511524844.007

[39] Olabode O, Dike H, Olaniyan D, Oni B, Faleye M. Experimental investigation of the effect of surfactant–polymer flooding on enhanced oil recovery for medium crude oil. Polymers.

2024;**16**(12):1674. DOI: 10.3390/polym16121674

[40] Karimov D, Toktarbay Z. Enhanced oil recovery: Techniques. Strategies, and Advances, ES Materials & Manufacturing. 2024;**23**:1005. DOI: 10.30919/esmm1005

[41] Firozjaii A, Saghafi H. Review on chemical enhanced oil recovery using polymer flooding: Fundamentals, experimental and numerical simulation. Petroleum. 2020;**6**(2):115-122. DOI: 10.1016/j.petlm.2019.09.003.: 10.1016/j.petlm.2019.09.003

[42] Malozyomov B, Martyushev N, Kukartsev V, Tynchenko V, Bukhtoyarov V, Wu X, et al. Overview of methods for enhanced oil recovery from conventional and unconventional reservoirs. Energies. 2023;**16**:4907. DOI: 10.3390/en16134907

[43] Gbadamosi A, Junin R, Manan M, Agi A, Yusuff A. An overview of chemical enhanced oil recovery: Recent advances and prospects. International Nano Letters. 2019;**9**:171-202. DOI: 10.1007/s40089-019-0272-8

[44] Druetta P, Raffa P, Picchioni F. Chemical enhanced oil recovery and the role of chemical product design. Applied Energy. 2019;**252**:113480. DOI: 10.1016/j.apenergy.2019.113480

[45] Cheraghian G, Hendraningrat L. A review on applications of nanotechnology in the enhanced oil recovery part A: Effects of nanoparticles on interfacial tension. International Nano Letters. 2016;**6**(2):129-138. DOI: 10.1007/s40089-015-0173-4

[46] Gbadamosi A, Patil S, Shazad M, Adewunmi A, Yussuf A, Agi A, et al. Application of polymers for chemical enhanced oil recovery: A review. Polymers. 2022;**14**(7):1433. DOI: 10.3390/polym14071433

[47] Terry R. Enhanced oil recovery. In: Meyers RA, editor. Encyclopedia of Physical Science and Technology. 3rd ed. MA, USA: Academic Press; 2003. pp. 503-518. DOI: 10.1016/B0-12-227410-5/00868-1

[48] Lyons W. Enhanced oil recovery methods. In: Lyons WC, editor. Working Guide to Reservoir Engineering. MA, USA: Elsevier; 2010. pp. 279-310. DOI: 10.1016/B978-1-85617-824-2.00005-8

[49] Rahimi A, Jami M, Divandari H, Safari M. Alkaline flooding. In: Hemmati-sarapardeh A, series editor. Chemical Methods. Enhanced Oil Recovery Series Elsevier; 2022. pp. 141-220. DOI: 10.1016/B978-0-12-821931-7.00010-9

[50] Sheng J. Alkaline Flooding. In: Sheng JJ, editor. Enhanced Oil Recovery Field Case Studies. MA, USA: Gulf Professional Publishing; 2013. pp. 143-167. DOI: 10.1016/B978-0-12-386545-8.00006-3

[51] Mahdavi E, Zebarjad F. Screening criteria of enhanced oil recovery methods. In: Bahadori A, editor. Fundamentals of Enhanced Oil and Gas Recovery from Conventional and Unconventional Reservoirs. MA, USA: Gulf Professional Publishing; 2018. pp. 41-59. DOI: 10.1016/B978-0-12-813027-8.00002-3

[52] Yang H, Li J, Jiang H, Hu J, Zeng J, Nie C. Alkaline-surfactant-polymer flooding: Where is the enhanced oil exactly? In: SPE EOR Conference at Oil and Gas West Asia, Muscat, Oman. SPE, TX, USA; 2018. DOI: 10.2118/190340-MS2018

[53] Wang X, Wang F, Taleb M, Wen Z, Chen X. A review of the seepage

mechanisms of heavy oil emulsions during chemical flooding. Energies. 2022;**15**:8397. DOI: 10.3390/en15228397

[54] Olajire A. Review of ASP EOR (alkaline surfactant polymer enhanced oil recovery) technology in the petroleum industry: Prospects and challenges. Energy. 2014;**77**:963-982. DOI: 10.1016/j.energy.2014.09.005

[55] Dinesh N, Pavan P, Sivasankar P. Decarbonization of alkali-surfactant-polymer flooding enhanced oil recovery process with hydrogen under uncertainty. Energy and Fuels. 2024;**38**(12):11093-11106. DOI: 10.1021/acs.energyfuels.4c01872

[56] Han D, Yang C, Zhang Z, Lou Z, Chang Y. Recent development of enhanced oil recovery in China. Journal of Petroleum Science and Engineering. 1999;**22**(1-3):181-188. DOI: 10.1016/S0920-4105(98)00067-9

[57] Lei Z, Chen B, Koo Y, MacFarlane D. Introduction: Ionic liquids. Chemical Reviews. 2017;**117**(10):6633-6635. DOI: 10.1021/acs.chemrev.7b00246

[58] Hanamertani A, Pilus R, Irawan S. A Review on the Application of Ionic Liquids for Enhanced Oil Recovery. In: Awang M, Nagash B, Md Akhir N, Lubis L, Rafek A, editors. ICIPEG 2016 (Proceedings of the International Conference on Integrated Petroleum Engineering and Geosciences). Singapore: Springer; 2017. DOI: 10.1007/978-981-10-3650-7

[59] Hezave A, Dorostkar S, Ayatollahi S, Nabipour M, Hemmateenejad B. Investigating the effect of ionic liquid (1-dodecyl-3-methylimidazolium chloride ([C12mim] [Cl])) on the water/oil interfacial tension as a novel surfactant. Colloids and Surfaces, A: Physicochemical and Engineering Aspects. 2013;**421**:63-71. DOI: 10.1016/j.colsurfa.2012.12.008

[60] Hezave A, Dorostkar S, Ayatollahi S, Nabipour M, Hemmateenejad B. Effect of different families (imidazolium and pyridinium) of ionic liquids-based surfactants on interfacial tension of water/crude oil system. Fluid Phase Equilibria. 2013;**360**:139-145. DOI: 10.1016/j.fluid.2013.09.025

[61] Hezave A, Dorostkar S, Ayatollahi S, Nabipour M, Hemmateenejad B. Dynamic interfacial tension behavior between heavy crude oil and ionic liquid solution (1-dodecyl-3-methylimidazolium chloride ([C12mim] [Cl] + distilled or saline water/heavy crude oil)) as a new surfactant. Journal of Molecular Liquids. 2013;**187**:83-89. DOI: 10.1016/j.molliq.2013.05.007

[62] Bin Dahbag M, Hossain M, AlQuraishi A. Efficiency of ionic liquids as an enhanced oil recovery chemical: Simulation approach. Energy and Fuels. 2016;**30**(11):9260-9265. DOI: 10.1021/acs.energyfuels.6b01712

[63] Dahbag M, AlQuraishi A, Benzagouta M. Efficiency of ionic liquids for chemical enhanced oil recovery. Journal of Petroleum Exploration and Production Technologies. 2015;**5**(4):353-361. DOI: 10.1007/s13202-014-0147-5

[64] Shamshina J, Zavgorodnya O, Rogers R. Ionic Liquids. In: Davies P, Manisha K, editors. Encyclopedia of Analytical Science. Chemistry Molecular Sciences and Engineering. 3rd ed. MA, USA: Elsevier; 2019. pp. 218-225. DOI: 10.1016/B978-0-12-409547-2.13931-9

[65] Greer A, Jacquemin J, Hardacre C. Industrial applications of ionic liquids. Molecules. 2020;**25**:5207. DOI: 10.3390/molecules25215207

[66] Verma C, Eno E, Mumtaz A. Ionic liquids as green corrosion inhibitors for industrial metals and alloys. In: Green Chemistry. London, UK: InTech; 2018. DOI: 10.5772/intechopen.70421

[67] Suresh SJ. Recent advances in ionic liquids: Green unconventional solvents of this century: Part I. Green Chemistry Letters and Reviews. 2011;**4**(4):289-310. DOI: 10.1080/17518253.2011.572294

[68] Hajipour A, Rafiee F. Recent progress in ionic liquids and their applications in organic synthesis. Organic Preparations and Procedures International. 2015;**47**(4):249-308. DOI: 10.1080/00304948.2015.1052317

[69] Subasree N, Selvi J. Imidazolium based ionic liquid derivatives; synthesis and evaluation of inhibitory effect on mild steel corrosion in hydrochloric acid solution. Heliyon. 2020;**6**(2):e03498. DOI: 10.1016/j.heliyon.2020.e03498

[70] Zunita M, Kevin Y. Ionic liquids as corrosion inhibitor: From research and development to commercialization. Results in Engineering. 2022;**15**:100562. DOI: 10.1016/j.rineng.2022.100562

[71] Kaya S, Erkan S, Simşek S, Kumar A. Ionic liquids as corrosion inhibitors. In: Hussain C, Verma C, editors. Sustainable Corrosion Inhibitors II: Synthesis, Design, and Practical Applications, ACS Symposium Series. Vol. 1404. Washington, DC, USA: American Chemical Society; 2021. pp. 103-119. DOI: 10.1021/bk-2021-1404.ch004

[72] Gadioli A, de Souza L, Pereira E, Monteiro S, Azevedo A. Imidazolium-based ionic liquids as corrosion inhibitors for stainless steel in different corrosive media: An overview. Journal of Materials Research and Technology. 2024;**29**:803-823. DOI: 10.1016/j.jmrt.2024.01.066

[73] Bajpai D, Tyagi V. Fatty imidazolines: Chemistry, synthesis, properties and their industrial applications. Journal of Oleo Science. 2006;**2006**(55):319-329. DOI: 10.5650/JOS.55.319

[74] Tyagi R, Tyagi V, Pandey S. Imidazoline and its derivatives. An overview. Journal of Oleo Science. 2007;**56**(5):211-222. DOI: 10.5650/jos.56.211

[75] Shamsa A, Barmatov E, Hughes T, Hua Y, Neville A, Barker R. Hydrolysis of imidazoline based corrosion inhibitor and effects on inhibition performance of X65 steel in CO_2 saturated brine. Journal of Petroleum Science and Engineering. 2022;**208**:109235. DOI: 10.1016/j.petrol.2021.109235

[76] Alarbah A, Shirif M, Shirif E. Investigation of different ionic liquids in improving oil recovery factor. Advanced Chemical Engineering Science. 2019;**9**(1):87-98. DOI: 10.4236/aces.2019.91007

[77] Bera A. Applications of ionic liquids as green solvents in enhanced oil recovery. In: Inamuddin AT, editor. Green Sustainable Process for Chemical and Environmental Engineering and Science. MA, USA: Elsevier; 2023. pp. 125-144. DOI: 10.1016/B978-0-323-95156-2.00016-7

[78] Tafur N, Mamonov A, Islam Khan M, Soto A, Puntervold T, Strand S. Evaluation of surface-active ionic liquids in smart water for enhanced oil recovery in carbonate rocks. Energy and Fuels. 2023;**37**(16):11730-11742. DOI: 10.1021/acs.energyfuels.3c01488

[79] Liu Y, Hu Y, Wang H, Xu C, Ji D, Sun Y, et al. Ionic liquids: Novel solvents for petroleum asphaltenes. Chinese Journal of Chemical Engineering. 2005;**13**:546-567

[80] Pereira J, Costa R, Foios N, Coutinho J. Ionic liquid enhanced oil recovery in sand-pack columns. Fuel. 2014;**134**:196-200. DOI: 10.1016/j.fuel.2014.05.055

[81] Ghanem A, Alharthy R, Desouky S, El-Nagar R. Synthesis and characterization of imidazolium-based ionic liquids and evaluating their performance as asphaltene dispersants. Materials. 2022;**15**(4):1600. DOI: 10.3390/ma15041600

[82] Nares H, Schacht-Hernández P, Ramírez-Garnica M, Cabrera-Reyes M. Upgrading heavy and Extraheavy crude oil with ionic liquid. In: International Oil Conference and Exhibition. Veracruz, Mexico. TX, USA: SPE; 2007. DOI: 10.2118/108676-MS

[83] Shaban S, Dessouky S, Badawi A, El Sabagh A, Zahran A, Mousa M. Upgrading and viscosity reduction of heavy oil by catalytic ionic liquid. Energy and Fuels. 2014;**28**(10):6545-6553. DOI: 10.1021/ef500993d

[84] Fan Z, Wang T, He Y. Upgrading and viscosity reducing of heavy oils by [BMIM][$AlCl_4$] ionic liquid. Journal of Fuel Chemistry and Technology. 2009;**37**(6):690-693. DOI: 10.1016/s1872-5813(10)60015-1

[85] Strelets L, Ilyin S. Effect of enhanced oil recovery on the composition and rheological properties of heavy crude oil. Journal of Petroleum Science and Engineering. 2021;**203**:108641. DOI: 10.1016/j.petrol.2021.108641

[86] Murkazov R, Sabanenkov S, Syunyaev Z. Influence of petroleum resins on colloidal stability of asphaltene-containing disperse systems. Chemistry and Technology of Fuels and Oils. 1980;**16**(10):674-677. DOI: 10.1007/BF00726261

[87] Ilyin S, Strelets L. Basic fundamentals of petroleum rheology and their application for the investigation of crude oils of different natures. Energy & Fuels. 2018;**32**:268-278. DOI: 10.1021/acs.energyfuels.7b03058

[88] Hashmet R, Demiral B, Ali M, Kashif M. Rheology of enhanced oil recovery polymers-a review. National Postgraduate Conference. 2011:1-5. DOI: 10.1109/NatPC.2011.6136467

[89] Wei B. Flow characteristics of three enhanced oil recovery polymers in porous media. Journal of Applied Polymer Science. 2015;**132**:41598. DOI: 10.1002/app.41598

[90] Uwaezuoke N. Perspective chapter: Rheological considerations for drilling and enhanced oil recovery fluids. London, UK: IntechOpen; 2023. DOI: 10.5772/intechopen.106897

[91] Arinina M, Zuev K, Kulichikhin V, Malkin A. Effect of composition and interfacial tension on the rheology and morphology of heavy oil-in-water emulsions. ACS Omega. 2020;**5**:16460-16469. DOI: 10.1021/acsomega.0c00769

[92] Kamkar M, Bazazi P, Kannan A, Suja V, Hejazi H, Fuller G, et al. Polymeric-nanofluid stabilized emulsions: Interfacial versus bulk rheology. Journal of Colloid and Interface Science. 2020;**576**:252-263. DOI: 10.1016/j.jcis.2020.04.105

Chapter 8

Hybrid Polymer Injection: A Modular, Factory-Optimised Platform for Burst Polymer Injection, Water-Alternating Polymer and Polymer On-Polymer Off

Juan Ernesto Juri

Abstract

Current studies neglect hybrid approaches that exploit hysteresis phenomena to maximize efficiency and optimize resources. Hysteresis is a fundamental phenomenon arising from irreversible processes within natural systems. This irreversibility reflects nature's inherent tendency toward dissipation, where returning to a prior state requires a different energetic pathway, leading to distinct forward and backward trajectories. We pioneered the concept of hybrid polymer injection in Manantiales Behr, leading to multiple successful pilots and expansions conducted in the Grimbeek field between 2021 and 2023. Throughout these pilots, more than 54,800 tonnes of polymer were injected, demonstrating a clear relationship between injection cycles and consistent, repeatable and scalable production gains, while supporting an efficient operating model. Central to our hybrid polymer injection approach is the Burst and Water-Alternating-Polymer (Burst-WAP) and On-Off strategy, effectively balancing immediate recovery impact, long-term efficiency, cost control and low risk. This technique strategically employs initial high concentration continuous polymer injection (Burst phase) and the alternating cycles of water injection and polymer injection. In zones developed with polymer in secondary mode, we deliberately pause injection to slow the displacement front and widen the advancing oil bank, which maximizes microscopic displacement efficiency. The burst phase creates a robust polymer bank, improving mobility control and forming a high pressure, stable front that promotes efficient three-dimensional polymer crossflow. This reduces premature water breakthrough and polymer dilution along the established high-water saturation paths, leading to immediate and significant improvements in oil rate. The subsequent water-alternating-polymer (WAP) phase sustains the polymer's effectiveness by intermittently alternating polymer injection with water or periods of no injection. This approach activates hysteresis effects in fluid displacement, where the cyclic variation in injection conditions delays the polymer front's arrival at production wells, enhancing cost efficiency. Standardizing the hybrid polymer injection process using modular plants is because this modular scale enables high polymer throughput and concentration, resulting in substantial incremental oil response, making each standalone unit economically attractive on its own at low risk. Modular plants (or the polymer PAD) further enhanced this capability, offering a standardized and flexible platform for implementing and adjusting injection parameters. This flexibility allows us to scale operations effectively, rapid payback and sustained profitability while managing operational costs and polymer consumption. By framing polymer injection as burst-and-water-alternate-polymer strategy combined with modular, short-cycle hybrid scheme, which turned polymer flooding into a repeatable, scalable, and factory-efficient process. By avoiding prolonged polymer injection periods and combining hybrid injection physics with modular-plant logistics, this approach cuts OPEX, delivers early oil, and stabilizes water cut. Even in a fundamentally slow process like polymer injection. This not only enables conceptualization for optimization across diverse reservoir settings but also allows the approach to economically compete with factory-style unconventional development model. Rystad indicates that exploration is declining worldwide and offshore costs are rising; as a result, proven oil reserves will struggle to meet even a standard demand scenario. Hybrid polymer injection can rapidly add proven oil reserves.

Keywords: polymer flooding, alternate, polymer injection, EOR, water-alternate-polymer, WAP, modular polymer injection, on-off polymer injection, surfactant injection

1. Introduction

In our experience, improving efficiency requires abandoning or significantly altering the mechanisms that led to inefficiencies. For instance, if displacing highly viscous oil with water proves inefficient, transitioning to polymer injection must involve polymers whose properties are distinctly different from water; otherwise, one risks replicating or only mildly improving upon the original inefficiencies [1–3].

Historically, polymer floods have often used low polymer concentrations, primarily because of underestimation in numerical simulations. This has resulted in a delayed oil response, marginal economics, and high OPEX, with only modest incremental oil gains. From our perspective, reservoir simulations, although valuable, often fall short of accurately capturing the real behavior of macromolecular polymer solutions. Polymers nominally described as 18 million Daltons represent a molecular-weight distribution that, under the concentrations and viscosities applied in these field implementations, are prone to collective behavior and interactions with porous media that are difficult to capture in reservoir simulation. Additionally, when water displaces more viscous polymer solutions, fingering occurs—another phenomenon challenging to simulate because inherent scale dependencies. Polymer with viscosity in the same order as oil viscosity flows in all directions, while water displacing high-viscosity oil flows mainly in a horizontal direction or the shortest path between injection and production.

Reservoir simulations often assume high polymer residual resistance factors, underestimating required polymer concentrations and causing ineffective, wasted polymer injections.

Polymers uniquely display collective behaviors that are absent in other fluids of comparable viscosity, like glycerin. Polymer solutions, composed of long-chain molecules, have a local mechanism of chain entanglements and interactions with water molecules, creating a network-like structure. Acting on this non-Newtonian behavior is essential for implementing effective hybrid polymer injection strategies [4–8].

Moreover, the hybrid polymer injection strategy explicitly exploits hysteresis effects by alternating polymer concentration and injection rates, optimizing reservoir response through cyclic variations. These alternating cycles create conditions that maximize both microscopic and macroscopic sweep efficiencies, leveraging the natural hysteresis behavior of reservoirs [9–12].

Although finite difference simulations, finite element methods, and analytical models provide valuable insights, ultimate oil recovery results from complex pore-scale displacement processes. While 3D modeling and reservoir simulation remain critical for identifying macroscale injector-producer connectivity, well placements, and production strategies, they do not fully capture the microscale flow processes that govern ultimate recovery efficiency [13, 14].

In our view, new projects can significantly enhance their investment returns by thoroughly understanding target polymer viscosities and concentrations, clarifying multiphase flow implications, and avoiding the temptation to reduce polymer concentration solely to lower costs. At Manantiales Behr, a hybrid, innovative, and globally pioneering strategy has been implemented. It combines high-concentration polymer phases (Burst) with water-polymer alternation (WAP) and intermittent On–Off injection schemes. Establishing optimal slug cycle is crucial for project success [15, 16].

1.1 Organization of the chapter

This chapter is organized into four sections: Section 1 introduces new concepts for polymer flooding design and explains why past projects typically used low polymer concentrations. Sections 2.1 and 2.2 describe the pore network model and the simulation parameters. Section 2.3 addresses polymer injection concentration from molecular concepts and reservoir heterogeneity, and highlights the percolative behavior under different polymer viscosity conditions and injection sequences. Section 2.4 covers laboratory experiments evaluating different sequences of water and polymer injection. Section 2.5 briefly discusses modular deployment, which is crucial for implementing the proposed hybrid polymer flooding strategy at the field scale. Section 3 presents simulation results at the pore network and field-equivalent scales. Detailed descriptions of injection sequences, slug volumes, and zones are omitted here since they are published elsewhere. Section 3.4 provides a dimensionless scaling analysis of reservoir analog zones, validating pore-scale simulations and supporting the effectiveness of the hybrid injection strategies. Section 4 summarizes the significance of hybrid polymer injection with a high concentration initial phase as a promising future method for polymer flooding.

2. Materials and methods

2.1 Pore network model

Incremental oil production in a production well is the result of an aggregated displacement chain, moving from pore to pore and originating at the injector. We conceptualize porous media as a network of pore bodies connected by pore throats. This network is represented as a system of nodes (pore bodies) and bonds (pore throats) with appropriate characteristics to enable flow calculations. The porous medium is modeled using a three-dimensional lattice cubic network of interconnected capillary elements, with the number of nodes in each direction as NxNyNz. Pores and throats are located at the sites (nodes) and bonds (edges) of the lattice. The rules that define if a pore is occupied by water, oil, or polymer are capillary [17, 18] and viscous forces [19].

We use the discrete diffusion-limited aggregation models [19] to simulate the rules that govern the moves of the different fluids inside the pore network model. This method resolves displacement at the pore scale, ensuring a detailed understanding of he movement of all fluids in the network.

We assume all pores and throats have a uniform size of 10 microns, based on common measurements from microtomography and thin sections in the target zone, where absolute permeability exceeds 8000 md [20–23]. We exclude the gas phase, a valid assumption for typical polymer injection cases, and apply periodic boundary conditions perpendicular to the flow direction.

The simulations described later produce realistic and structurally heterogeneous displacement fronts. The algorithm allows fronts to evolve *via* self-organization, capturing how fluid properties differ ahead and behind the front and how the front may break down. We ran thousands of replicate simulations to confirm the repeatability of observed patterns, each lasting until reaching the required injected pore volume. Comparisons between theoretical results, laboratory experiments, and field pilot data continuously enhance our understanding and optimization of polymer flooding.

We performed simulations on progressively larger networks to confirm pattern coherence. At a constant pressure drop between the inlet and outlet, viscosity ratios determine fluid occupancy. The displaced-to-displacing viscosity ratio controls the number of pores contributing to displacement at the inlet. An essential displacement mechanism observed is oil trapping, which occurs as advancing water surrounds isolated oil regions. These simulations primarily illustrate general displacement phenomena and demonstrate the impact of polymer and hybrid injection strategies.

2.2 Simulation approach and parameters

We simulated pore-scale displacement under various scenarios by changing polymer viscosity, injection sequences, and mobility conditions to study fluid distribution and recovery behavior. Each case included thousands of replicates to ensure consistent pattern development.

In all cases, we set oil viscosity to 100 cp and water to 1 cp, and adjusted polymer viscosity based on the scenario. The shape of the advancing water front strongly depended on the mobility ratio and injection strategy. A number of images will be displayed in the coming sections to illustrate the resulting displacement patterns.

We projected the simulated flow on a 2D 100 × 100 pore lattice, with flow from left to right to represent injection from the left boundary. Periodic boundary conditions in the y and z directions modeled an infinite porous medium in those planes. Each simulation began with 100% oil saturation in uniformly sized pores. We tested several field-relevant injection strategies. In one case, polymer followed water (tertiary injection). In another, we applied a Water-Alternating-Polymer (WAP) strategy after reaching the oil production peak during continuous polymer injection. In a modified WAP, polymer runs intermittently (On-Off), leaving out the water period so no injection takes place during the off cycles.

2.3 Polymer specifications

In this study, we use the polymer with the properties described in **Table 1**. The formation water salinity is 11,000–15,000 ppm total dissolved solids (TDS), and the formation permeability is greater than 8000 mD [23].

Polymer type	Partially hydrolyzed polyacrylamide (HPAM)
Molecular weight (million Daltons)	18–30
Concentration (ppm)	2500–5500
Viscosity (cp) (60°C) and 0.5 1/s	60–180 cP
Reservoir temperature (°C)	60
Hydrolysis degree (%)	18–32
Water salinity	11,000–15,000
Absolute liquid permeability	8000–28,000 md
Pore size (microns)	1–10 μm

Table 1.
Polymer, water, and rock properties.

2.4 Argument of burst polymer concentration: High-concentration polymer injection

Selecting the appropriate polymer concentration is crucial for effective polymer flooding, as it directly impacts flow control and pressure management. At higher polymer concentrations, the primary effect is the activation of high-pressure viscous crossflow fronts, previously detailed in Refs. [15, 16, 24], key factor to generate early, increase in oil production. A secondary effect involves local molecular entanglement, particularly significant for high-molecular-weight HPAM polymers, such as those averaging around 18 million Daltons.

2.4.1 Molecular behavior justification

Using a polymer with an average molecular weight near 10 million Daltons provides an optimal balance: sufficient polymer molecules per pore and adequate molecular size to effectively mobilize oil, enhancing sweep efficiency without significantly reducing injectivity.

The actual hydrodynamic radius (R_h) of an 18-million-Dalton polymer in solution depends on factors including molecular structure, hydration degree, and solvent characteristics. Typically, this radius is estimated using empirical power-law models for dilute polymer solutions [4].

$$R_h = k \cdot M_w^{\,v}$$

where:

M_w is the average molecular weight in Daltons (18×10^6).

k and ν are empirical constants that depend on the polymer type and solvent. For polyacrylamide in water:

$$k \approx 0.027\, nm / \text{Dalton}$$

$\nu \approx 0.5$ (for polymers in good solvents).

The estimated hydrodynamic radius is 0.1145 μm, and the diameter in solution is 0.229 μm. This range will vary depending on the water salinity, in these applications, the water injection salinity is below 15,000 ppm. High-permeability reservoirs with viscous oil in the range of 100–1000 cp have pore sizes around 1–10 micros. In this range of sizes, we could expect molecular interaction with the pore space that will affect the sweep.

The polymer molecule interaction inside the pore space can induce polymer entanglement and collective size in solution to exploit the possibility that locally, polymers increase viscosity by entanglement when long polymer chains physically interlock (note that it is not gel cross-linking) with each other, creating a local network that resists flow.

The number of polymer molecules entangled depends on the overlap polymer concentration (C^*), which marks the transition from dilute to semi-dilute regimes, where chains begin to interact strongly.

Using the overlap concentration formula [5]:

$$C^* \approx M_w / \left(NA\, Rg^3 \right)$$

where:

- C^* is the overlap concentration.
- Mw is the polymer's molecular weight (18 million Daltons = 18×10^6).
- NA is Avogadro's number (6.022×10^{23}).
- Rg is the radius of gyration, which we estimate from the relationship:

$$Rg \approx 0.66 \cdot R_h$$

Since $R_h \approx 114.55$ nm, then:

$$Rg \approx 0.66 \times 114.55\, nm = 75.6\, nm$$

The concentration where polymer molecules begin to entangle is ~0.069 g/cm^3, which is equivalent to 69,000 ppm (or 69,000 mg/L) of polymer concentration in solution. This means that entanglement starts occurring when polymer concentration reaches approximately 69,000 ppm.

Once polymer concentration exceeds C^*, molecules start forming an entangled network. The number of molecules in an entangled structure can be estimated using [6]:

Number of molecules per entanglement = Entanglement Factor/C^* [6–8].

The entanglement factor is typically between 2 and 5 in semi-dilute solutions.

We assume 3 (a common empirical value for polymer physics).

Since C^* is in g/cm^3, we take the inverse to estimate how many molecules fit into a given volume.

Number of molecules per entanglement

$$= 3 \times (1 / 0.069)$$

$$= 3 \times 14.47$$

$$= 43.37 \approx 43\, \text{molecules}$$

Most polymer flooding operations in the field occur below the entanglement threshold. Consequently, the observed increase in viscosity typically results from polymer coil expansion rather than true molecular entanglement. However, localized entanglement can still occur due to the irregular flow paths within porous media and the converging and diverging flow conduits in the pore space.

2.4.2 Localized concentration variations

Polymer molecules can accumulate on pore walls or other surfaces, creating microenvironments where the effective local polymer concentration temporarily rises above 5000 ppm. While high shear often reduces average molecular weight, transient shear zones can also concentrate macromolecules in particular regions. Larger polymer molecules may get preferentially retained, locally elevating their concentration in low-flow "stagnant" pockets.

2.4.3 Semi-dilute regime and coil overlap

Even below C^*, polymer coils can overlap enough to increase viscosity *via* intermittent entanglements—particularly in heterogeneous reservoirs. At high concentration, for example, 5000 ppm, coils still interact to a degree, forming transitory entangled clusters that strengthen viscosity without establishing a fully entangled "mesh." As the flood advances, mixing with *in situ* fluids may weaken these overlaps, but local conditions can still favor short-term entanglements.

2.4.4 Mechanical and hydrodynamic factors

High shear zones (e.g., pore throats, flow convergences) can stretch chains, making them more likely to intersect and bind. The presence of converging and diverging local flow-path constrictions may cause polymer concentration to spike because of the reduced fluid volume, enabling entanglement events even at lower bulk ppm levels. Beyond the simple bulk calculations provided above, the field conditions rarely match ideal laboratory setups; thus, even at around 5000 ppm, they can generate localized entanglement that boosts viscosity in key regions. Where partial entanglements arise, the polymer flood may maintain better mobility control than expected from bulk concentration estimates. And favoring this condition only is achieved with high-concentration floods.

2.4.5 High-concentration polymer for activating crossflow in fluvial systems: Geological justification

Another geological aspect to consider in polymer flooding is its correlation with the use of high polymer concentrations. By linking polymer viscosity and reservoir architecture, this section underscores the necessity of choosing a sufficiently high polymer concentration to activate crossflow across the distinct flow units and in all 3D directions found in fluvial settings. The following discussion expands on these findings, exploring the interplay of vertical layering, fluid mobility, and polymer-induced flow diversion, and how fine-tuning each factor can lead to a more successful polymer strategy.

Particularly in fluvial reservoirs, higher polymer concentrations play a crucial role beyond traditional mobility control, directly affecting crossflow between reservoir layers of varying permeabilities and porosities in Refs. [2, 3, 16, 24–26]. In fluvial reservoirs, vertical and lateral heterogeneities often result in uneven displacement and reduced sweep efficiency when conventional waterflooding is applied. One way to overcome these challenges is by injecting higher-viscosity polymer and high concentration to ensure enought margin above adsorption level. The increased viscosity and subsequent mobility control can drive more effective crossflow between reservoir layers and within the internal bed structure exhibiting variations in texture and permeability and porosity, thereby improving overall oil recovery.

Figure 1 illustrates how vertical heterogeneity (represented by 2, 3, and 4 subdivisions) correlates with the polymer-to-oil viscosity ratio to affect the recovery factor. As polymer viscosity rises, recovery increases—albeit with diminishing returns beyond a ratio of approximately 2 to 3. Reservoirs modeled with fewer subdivisions exhibit higher apparent recovery, partly because vertical permeability contrasts are less rigorously captured. More detailed subdivision (3 or 4 divisions) better reflects the complex layering of fluvial deposits, giving a more realistic picture of how high-concentration polymer floods redistribute flow and enhance oil displacement.

The number of divisions (2, 3, or 4) represents the level of detail in capturing vertical interfaces of reservoir quality. More divisions (4 vs. 2) provide a better description of changes in petrophysical properties, allowing for: (i) more accurate crossflow between layers and (ii) better representation of vertical permeability contrasts. Thus, as the number of divisions increases, the model better represents the connectivity between sand bodies, leading to improved polymer sweep efficiency

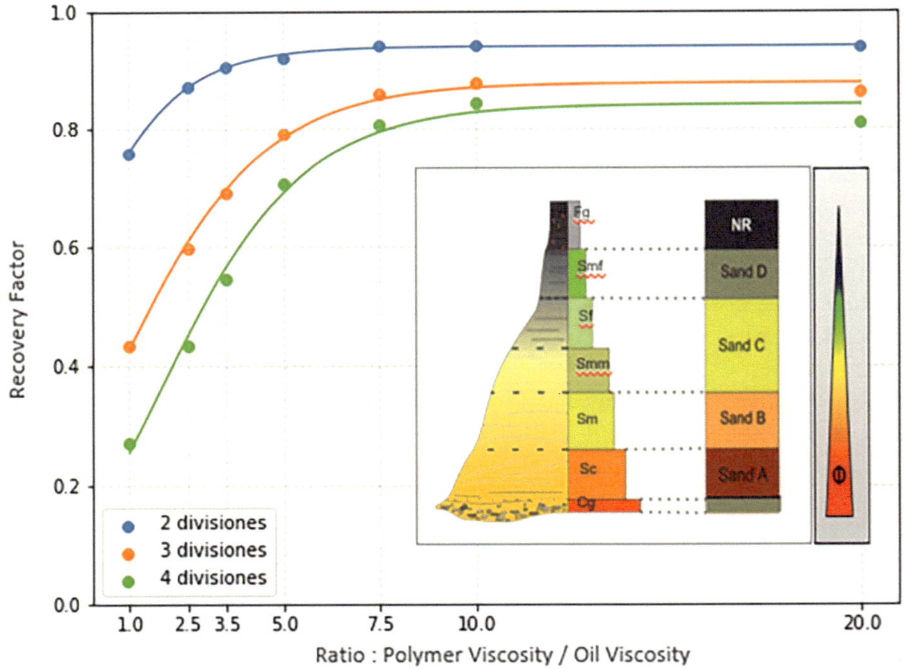

Figure 1.
Impact of vertical heterogeneity and polymer viscosity on oil recovery factor. The plot displays the recovery factor as a function of the polymer-to-oil viscosity ratio under different levels of vertical reservoir subdivision: 2, 3, and 4 divisions, representing increasing heterogeneity. As polymer viscosity increases, recovery improves across all cases, but with diminishing returns beyond a ratio of ~5. Reservoirs with fewer subdivisions (blue line, 2 divisions) exhibit higher recovery because of more uniform vertical sweep. Increased heterogeneity (3 or 4 divisions) reduces polymer efficiency because of flow diversion into high-permeability layers. The inset shows the stratigraphic architecture of the reservoir used in the model, with colored facies and porosity distribution.

and a higher recovery factor. With more divisions, the model better captures flow redistribution, ensuring polymer moves into all sand layers rather than preferentially flowing through high-permeability zones. In simpler (2-division) models, the polymer sweep appears more efficient, but this might be an overestimation because the coarse representation since some low-permeability layers might not be fully captured. More divisions enable better crossflow representation, enhancing polymer efficiency. Higher polymer viscosity increases recovery, but the effect is more accurately captured when the reservoir is better divided into its actual fine-scale flow units.

2.5 Experimental setup

In this section, we present the design and outcomes of four coreflood experiments conducted on a segmented core sample (**Figure 2**) [23, 27, 28]. Each experiment follows a distinct sequence of pore volumes and injection velocities, enabling us to evaluate secondary polymer injection against tertiary polymer injection. Through these experiments, we aim to establish whether injecting polymer earlier (secondary injection) yields more efficient displacement and higher recovery factors (RF) compared to late-stage (tertiary) polymer injection. This setup allows for comparing the impact of initial waterflooding under different pore volumes injected and under different velocities on polymer flood performance.

Figure 3 shows the recovery factor (RF) for each composite core, along with the pore volumes injected (PV) in **Figure 4** and flow velocity in **Figure 5**.

Composite Core A:

Waterflood (WF) recovery: 0.583.

Polymer flood (PF) recovery: 0.116.

Total recovery: ~0.70.

Figure 2.
Photograph (left) and schematic diagram (right) of a segmented core sample used for flow experiments. The core, measuring 22.54 cm in length, is divided into four equal sections (Section 1 to Section 4), each equipped with a pressure tap to monitor differential pressure (ΔP_1 to ΔP_4) during fluid injection. This setup allows for detailed analysis of pressure distribution and flow behavior along the core, supporting the evaluation of heterogeneity and permeability variations within the reservoir rock [23, 27, 28].

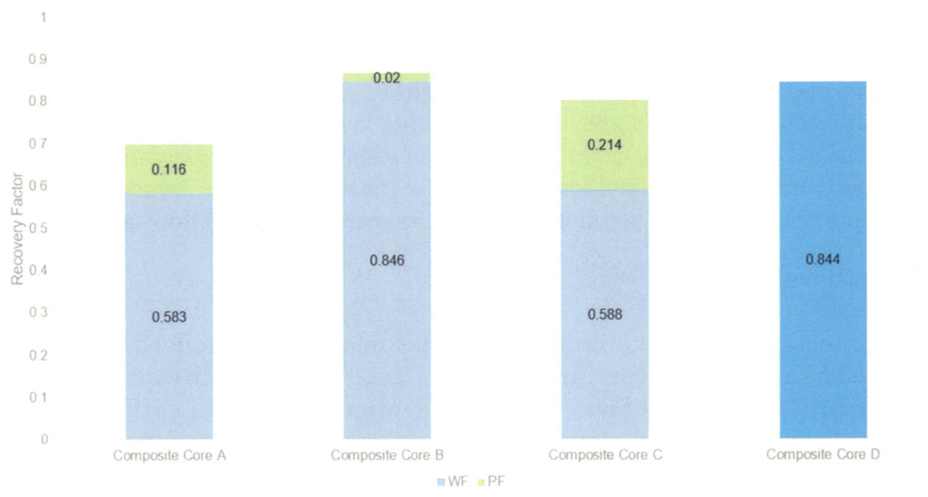

Figure 3.
Oil recovery factors achieved in coreflood experiments by the injection phase. This figure shows the contribution of waterflood (WF) and polymer flood (PF) to the total oil recovery factor for each composite core. Cores A, B, and C underwent water injection followed by polymer injection, while Core D received polymer injection from the start. Core B achieved the highest total recovery (0.87), mainly because of waterflooding (0.846), with minimal polymer contribution (0.02). Core C exhibited the highest incremental recovery from polymer (0.214). Notably, Core D achieved a 0.844 recovery factor using polymer alone, demonstrating the efficiency of early polymer application [23, 27, 28].

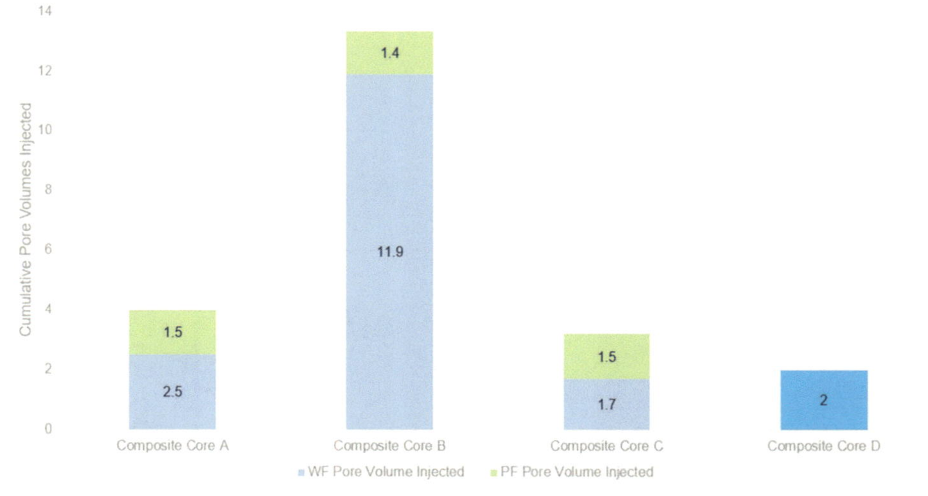

Figure 4.
Cumulative pore volumes injected during coreflood experiments. The figure presents the total volumes of waterflood (WF) and polymer flood (PF) injected, expressed in pore volumes (PV), for each composite core. Cores A, B, and C underwent sequential WF followed by PF, whereas Core D received only polymer injection. Core B had the highest cumulative injection (13.3 PV), primarily because of an extensive waterflood phase (11.9 PV). In contrast, Core D achieved injection with only 2 PV of polymer, highlighting a more efficient approach. PF injection volumes were kept relatively consistent across cores (~1.4–2.0 PV) [23, 27, 28].

This core had a high initial waterflood velocity (10.6 ft./day) and moderate polymer flood velocity (2 ft./day). The polymer flooding contributed significantly to additional recovery.

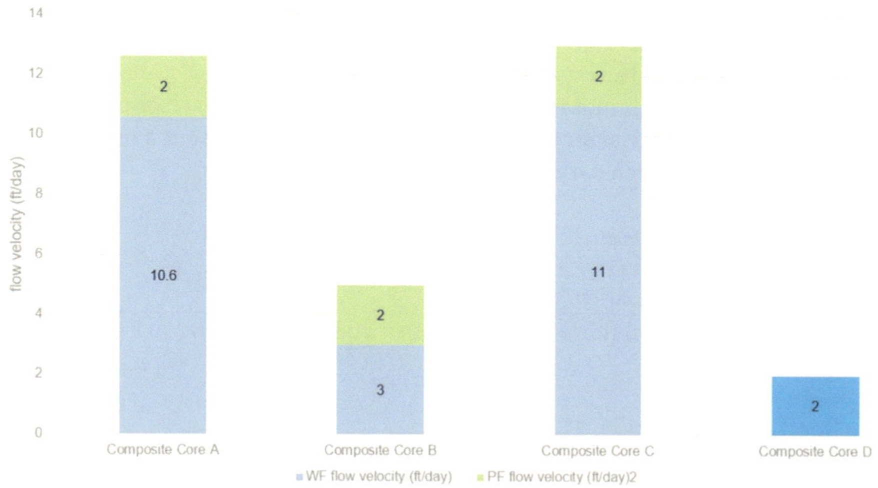

Figure 5.
Flow velocities applied in coreflood experiments. Each composite core was subjected to an initial waterflood (WF) phase followed by polymer flood (PF), except for Composite Core D, which received only polymer injection from the start. The figure shows that Cores A and C were injected at higher WF velocities (10.6 and 11 ft./day, respectively), while Core B experienced a lower WF velocity (3 ft./day). Polymer injection was performed at a consistent flow velocity of 2 ft./day across all tests [23, 27, 28].

Composite Core B:

Waterflood (WF) recovery: 0.846.

Polymer flood (PF) recovery: 0.02.

Total recovery: ~0.87.

Despite the lower waterflood velocity (3 ft./day), this core achieved the highest waterflood recovery, leaving little mobile remaining oil for polymer flood to contribute.

Composite Core C:

Waterflood (WF) recovery: 0.588.

Polymer flood (PF) recovery: 0.214.

Total recovery: ~0.80.

With another high waterflood velocity (11 ft./day) and polymer flood at 2 ft./day, this case shows a noticiable polymer response, indicating the polymer effectively mobilized the remaining oil.

Composite Core D:

Polymer flood (PF) recovery: 0.844.

No waterflood phase.

This core, which only experienced polymer flooding from the start, achieved a recovery virtually identical to Core B (0.844 vs. 0.846), suggesting that polymer flooding alone can reach the same recovery as a laboratory-only 11.9 PV waterflood, but with six times lower, field-feasible injection volume. Waterflood effectiveness varies: Core B's high WF recovery (0.846) suggests that lower velocity waterflooding (3 ft./day) was effective in achieving a high oil recovery without much need for polymer but required six times more water—an injection volume that would take geological timescales at field scale and is therefore unfeasible.

Polymer effectiveness depends on residual remaining oil saturation: Core C had a strong polymer response (0.214 additional RF), likely because of higher remaining oil saturation after WF.

Direct polymer injection (Core D) performed similar to a highly effective 11.9PV WF although not feasible at field scale (Core B), reinforcing the potential of polymer flooding as an initial recovery strategy.

Figure 4 illustrates the cumulative pore volumes injected for each composite core, connected to the previous recovery data (**Figure 3**). Here, we present a breakdown of the injection strategy for each core:

Composite Core A:

Waterflood (WF) injection: 2.5 PV.

Polymer flood (PF) injection: 1.5 PV.

Total injected volume: 4.0 PV.

Moderate water injection followed by polymer injection, leading to a 0.70 recovery factor.

Composite Core B:

Waterflood (WF) injection: 11.9 PV.

Polymer flood (PF) injection: 1.4 PV.

Total injected volume: 13.3 PV.

A much higher WF volume was injected compared to the other cores, yielding a 0.87 recovery factor, but with minimal polymer flood response.

Composite Core C:

Waterflood (WF) injection: 1.7 PV.

Polymer flood (PF) injection: 1.5 PV.

Total injected volume: 3.2 PV.

Despite a waterflood of just 1.7 PV (versus 2.5 PV previously), this case doubled recovery with the same polymer PV, resulting in 0.8 RF.

Composite Core D:

Only Polymer flood (PF) injection: 2.0 PV.

No prior water injection.

This direct polymer injection strategy achieved 0.844 RF, demonstrating effectiveness comparable to a high 11.9 PV WF injection (Core B) or WF+PF Core A,C.

Core B had the highest WF volume injected (11.9 PV), which led to very high oil recovery (0.87 RF), but polymer had little impact, meaning most oil was recovered before polymer injection.

Core D (polymer-only) achieved a similar recovery (0.844 RF) with just 2 PV polymer injections, reinforcing the idea that direct polymer injection can be an efficient recovery strategy.

Core C showed a strong polymer response after injecting 30% less WF PV than Core A, highlighting that early transition to polymer flood may be advantageous in some cases.

Figure 5 shows that the four composite cores have different waterflood (WF) and polymer flood (PF) flow velocities. Here is a description of the corefloods:

Composite Core A:

Waterflood (WF) velocity: 10.6 ft./day (shown in blue).

Polymer flood (PF) velocity: 2 ft./day (shown in green).

This core had a high initial waterflood injection rate followed by a lower polymer flood velocity.

Composite Core B:

Waterflood (WF) velocity: 3 ft./day.

Polymer flood (PF) velocity: 2 ft./day.

This core had a lower initial waterflood velocity compared to Core A and received the same polymer flood velocity.

Composite Core C:

Waterflood (WF) velocity: 11 ft./day.

Polymer flood (PF) velocity: 2 ft./day.

Similar to Core A, this core had a high initial waterflood velocity followed by polymer injection at a lower velocity.

Composite Core D:

No initial waterflood.

Polymer flood (PF) velocity: 2 ft./day from the beginning.

This core only experienced polymer flooding without a prior water injection phase. This setup allows for comparing the impact of initial waterflooding on polymer flood performance under different velocities velocities WF or pressure drop.

Tables 2 and **3** summarize the results across the three parameters—flow velocity, recovery factor, and cumulative pore volumes injected.

Core B (3 ft./day WF) had the highest total recovery (0.87 RF), but polymer had little additional impact.

Core C (11 ft./day WF) had the highest polymer flood contribution (0.214 RF), showing polymer can be very effective after a high-throughput WF conducted at 11 ft/d velocity with a substantial pressure drop.

Core D (polymer-only) reached nearly the same recovery (0.844 RF) as Core B, demonstrating that polymer flood alone can match an optimal waterflood + polymer sequence—or even the theoretical recovery from injecting a combined 11.9 + 1.4 PV.

Cores A and C (both with high WF velocity) had significant additional recovery from polymer, unlike Core B.

Core B required the highest total PV injected (13.3 PV) for only a slight RF advantage (~0.87 vs. 0.844 in Core D).

Core D (polymer-only) was the most efficient (0.422 RF/PV), proving that skipping WF can significantly reduce injection volume while achieving comparable recovery.

Core C had the second highest efficiency (0.25 RF/PV), indicating that a balanced WF + PF approach could be effective although it achieved only about half the efficiency of polymer flooding alone.

Core	WF flow velocity (ft/day)	PF flow velocity (ft/day)	Total recovery factor (RF)	WF RF contribution
A	10.6	2.0	0.70	0.583
B	3.0	2.0	0.87	0.846
C	11.0	2.0	0.80	0.588
D	N/A	2.0	0.844	0.000

Table 2.
Flow velocity vs. recovery performance.

Core	WF PV injected	PF PV injected	Total PV injected	Recovery factor per PV injected
A	2.5	1.5	4.0	0.175 RF/PV
B	11.9	1.4	13.3	0.065 RF/PV
C	1.7	1.5	3.2	0.25 RF/PV
D	0.0	2.0	2.0	0.422 RF/PV

Table 3.
Pore volumes injected vs. efficiency.

Core A, while achieving moderate recovery, was less efficient than Core C because of a lower polymer impact.

Waterflood effectiveness depends on initial velocity and pore volume injected. Core B (low WF velocity, high PV injected) outperformed Core A (high WF velocity, moderate PV injected). At the coreflood scale, polymer flooding proves most effective when applied after moderate waterflooding. Core C showed the strongest polymer response (0.214 RF), while Core B achieved only 0.020 RF. Skipping waterflooding and injecting polymer from the start can also be highly efficient—Core D matched or outperformed combined WF + PF strategies in terms of RF per PV injected. Excessive waterflooding, as seen in Core B, leads to diminishing returns. Despite injecting 11.9 PV of water, recovery only slightly exceeded (within experimental error) that of polymer-only injection. Lower total injected volumes consistently yielded better efficiency. Higher pressure drops, as observed in Core A, C and D, resulted in improved recovery; this highlights that elevated pressures are an additional benefit of high polymer concentrations, helping to drive a more encapsulated front, as later sections will show.

2.6 Injection methodology at field scale: Modular injection

Polymer solutions were injected using modular plants with static mixers at injection rates of 75–150 m^3/day, maintaining injection pressures below fracture pressure to prevent channeling. The modular approach enables precise control of polymer concentration and injection schedules, significantly enhancing operational flexibility and agility in implementing hybrid injection schemes [1]. Velocity wins against scale of the facility. We deploy movable units to where it pays first, then redeploy as response matures (lower capital per step, front-loaded cash flow). This is historically different than polymer flooding traditional view. This strategy limits complications of polymer being produced which affects produced-water separation.

Figure 6.
Modular mobile Polymer Injection Unit (PIU) used for Enhanced Oil Recovery (EOR) operations in remote locations. The PIU is composed of stackable, interlinked container modules, enabling rapid deployment, flexible relocation, and minimized site preparation. This setup supports polymer preparation, dosing, and injection without the need for permanent infrastructure [1, 29].

Figure 6 shows a modular mobile Polymer Injection Unit (PIU) installed in a remote oilfield location [1, 29]. The system is constructed using intermodal containers (commonly known as shipping containers), which are adapted and interconnected to function as a cohesive polymer injection facility. This modular design allows for.

Mobility: The entire unit can be disassembled, relocated, and reassembled quickly at a new site.

Compact footprint: Efficient use of space through vertical stacking and lateral modular expansion.

Autonomy: These units are typically equipped with blending tanks, dosing systems, and pumping modules.

Rapid deployment: Ideal for injecting high concentrations in localized injection points for a few pore volumes to accelerate recovery.

3. Results

In the following simulations, black shows the remaining oil that has not been displaced *inside pores*. Blue marks *pores* swept by water before polymer injection. Green highlights *pores* displaced by polymer. Cyan indicates pores where fluid occupancy changed, signaling a transition between oil, water, and polymer.

3.1 Water injection performance

3.1.1 Inefficiencies of water flooding in oil viscosities below 1000 centipoise

The sequence of images illustrates the progression of water injection (1 cP water displacing 100 cP oil) through a 2D pore network (100 × 100 lattice) as a function of increasing pore volumes injected (PV). The flow direction is left to right, and the oil-filled pores are black, while invaded pores with water are blue.

Figure 7.
Early-stage invasion at 0.1 PV shows minimal pore filling, with only a few preferential paths developing.

Figure 8.
By 0.2 PV, water fingering becomes more evident, still largely confined to limited paths.

Figure 9.
At 0.3 PV, the invasion front thickens, but many central areas remain oil-filled due to a poor mobility ratio.

Figure 10.
Water breakthrough becomes imminent at 0.32 PV, with main conductive channels established.

3.1.1.1 Description of the sequence and water injection inefficiencies

Figures 7 and **8** show early injection (0.1–0.2 pore volumes); only a small amount of water has entered the medium from the left. Water begins to invade through a small number of preferential flow paths or fingers, primarily along high-permeability channels. A few thin "fingers" of water start to grow into the oil. The displacement front is highly irregular due to the large viscosity contrast (100 cP oil vs. 1 cP water). The unfavorable mobility ratio ($\mu\_oil/\mu\_water = 100$) makes the displacement unstable, leading to viscous fingering. Water bypasses much of the oil, and large regions remain upswept. Approaching the water breakthrough, more branches form, small pockets of trapped oil appear, and the displacement front (which is not a singular front but rather a zone of changing saturation) advances. Water channels dominate the medium, but significant oil remains unswept. The waterfront begins to link up across larger portions of the medium, moving ever closer to the right edge.

Figures 9 and **10** show the intermediate stages (0.2–0.4 pore volumes); as more water is injected, existing fingers become longer and slightly thicker.

The water fingers extend further to the right, branching out in multiple directions. Because the oil is more viscous, the water tends to bypass it in narrow, branching channels rather than pushing it evenly. This creates a branching pattern of water channels surrounded by oil. This process creates a pattern of interdigitated water channels surrounded by oil. As injection proceeds to approximately 0.32–0.35 pore volumes, the fingers grow further, forming additional branches that spread, interconnect, and sometimes merge. Small pockets of trapped oil appear between the water channels. The fingers then become longer and more numerous, continuously advancing from left to right.

Breakthrough occurs early (around 0.32 PV), meaning water reaches the outlet before much of the oil is displaced. This early breakthrough is a direct symptom of poor mobility control and heterogeneous pore occupation.

Figures 11 and **12** show that, post-breakthrough (above 0.3), additional water injection after breakthrough contributes very little to oil recovery.

Water recirculates through established paths, bypassing oil trapped in less connected regions. It is equivalent to the results in the coreflood for 11 ft./d; increasing flow speed

Figure 11.
At 0.35 PV, water penetrates further, but the displacement is still inefficient in central pore zones.

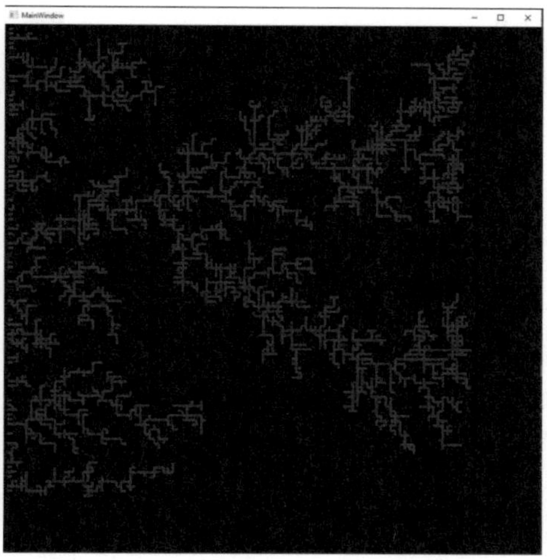

Figure 12.
Injection of 0.4 PV reveals stagnation in additional oil recovery; water flows mainly through the established fingers.

does not improve recovery. The number of fingers and their extension highlight how rapidly drawdown pressure is dampened from the injection side to the production side.

In **Figures 13** and **14**, even at 0.8 PV injected, the sweep efficiency remains low—a large portion of the pore space (central and bottom areas) remains oil-filled (black). The progress of the flow does not lead to meaningful oil recovery as production continues to come from already-invaded paths. Large portions of the oil are swept out, but many isolated oil pockets remain in low-flow zones. The displacement is highly nonuniform, showing how the large viscosity contrast and unstable front lead to incomplete, fingering-dominated displacement.

Overall, this sequence of snapshots of the flow demonstrates how an unfavorable mobility ratio (water much less viscous than oil) promotes a highly unstable front. Instead of sweeping the oil uniformly, water creates narrow, branching channels, leaving behind trapped oil and showing a pronounced viscous fingering pattern. A qualitative analysis suggests that approximately 30–50% of the porous medium remains occupied by oil at later stages because viscous fingering. Nevertheless, these images clearly show that a significant fraction of oil remains trapped in the unswept zones between the water "fingers," highlighting how strong viscous fingering leaves behind a large portion of the original oil in place.

3.1.2 Water and polymer injection in oil viscosities above 1000 centipoise: A heavy oil case

Heavy oils have a viscosity above 1000 cp, normally in the range of 6000–10,000 cp; therefore, in this condition, the fluid flow does not operate like in a conventional reservoir instead, polymers are injected to carve alternative pathways through the oil-filled pores and transmit the pressure drop. Polymer aims to create more fingers; a greater number, rather than a single finger, increase recovery. Therefore, lower viscosity provides better recovery because in diluted solutions of hundreds of ppm, the polymer molecules do not interact; therefore, they are more efficient in creating new viscous pathways. On the contrary, higher viscosity makes the molecules interact, and they tend to generate fewer fingers, but they are never able to make a stable front.

Figure 13.
Injection of 0.6 PV leads to a slight lateral expansion of fingers but with limited improvement in overall sweep.

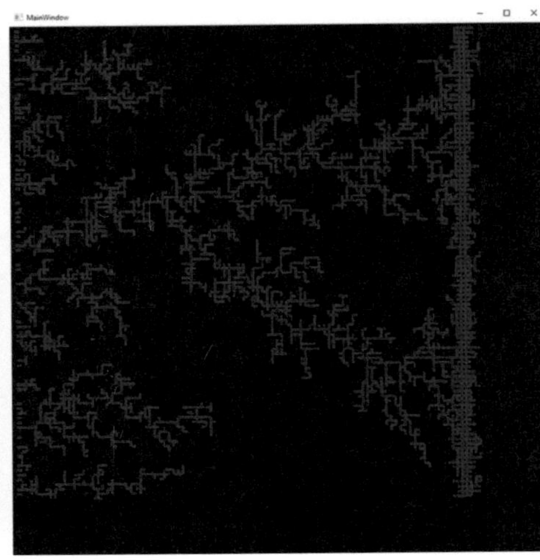

Figure 14.
By 0.8 PV, channeling dominates, and water has bypassed large oil-filled regions, indicating the need for mobility control methods like polymer flooding.

Even in heavy oil environments where matching polymer and oil viscosities is impossible, a small amount of polymer still has a profound impact compared to water alone. Our primary goal in these settings is not to achieve a stable, uniform front—this would be nearly impossible given the oil's extremely high viscosity. Instead, we take advantage of how polymer molecules, even at low concentrations, alter water flow through the pore space. This change in flow behavior creates multiple flow channels—or "fingers"—that sweep more oil than a single dominant water path ever could. Thus, although we cannot realistically match the oil's high viscosity, introducing polymer—even at just a few hundred parts per million—promotes a flow regime that breaks up the otherwise unproductive, single-channel water flow. This strategy leads to greater overall oil recovery despite the significant mismatch in fluid viscosities.

3.2 Polymer injection performance

3.2.1 Burst polymer: High polymer concentration and viscosity

We regard the choice of polymer viscosity, concentration and molecular weight is a critical factor in injection strategy. **Figures 14–17** compare the impact of polymer concentration—all within the Newtonian plateau, taken as reference—after 0.8 pore volumes of water were injected into a 2D, 100 × 100 pore-lattice model. Flow proceeds from left to right, with black indicating oil, blue indicating water, and green indicating polymer. We initiate polymer injection only after water has already broken through, so this process functions as a tertiary recovery method.

We define *burst polymer injection* as any polymer injection that pushes the polymer's viscosity above the resident oil viscosity (here, oil = 100 cP), which in these field applications also implies injecting at high concentration within the water composition. Under these conditions, high-viscosity polymer solutions (within the Newtonian plateau) confines the displacement front within a more limited volume. This restriction concentrates pressure energy near the front itself. Pressure energy dissipates more efficient through pore-level interactions near the confined front, effectively

DOI: http://dx.doi.org/10.5772/intechopen.1011022

mobilizing oil along multiple local displacement chains in the immediate vicinity , this accelerates the production response - a longstanding issue inpolymer flooding that has beenlinked toapparent injectivity losses inreservoir simulation.

3.2.1.1 Analysis of the displacements @polymer viscosity is 0.5 times oil viscosity (polymer viscosity: 50 cP, oil viscosity: 100 cP)

The water flood (blue) has advanced quickly because of high water mobility, creating strong viscous fingering (**Figure 15**). These branched patterns reveal unstable displacement fronts where water bypasses large portions of oil, leaving substantial unswept zones. The polymer flood (polymer displacement into pores shown by green pores) follows existing water pathways but also penetrates new areas, thereby improving sweep. The green zones show more continuous, less branched patterns than the blue regions, indicating enhanced mobility control. By stabilizing the displacement front, the polymer leads to a more uniform oil displacement. The remaining oil (black regions) and large clusters of trapped oil remain in areas not effectively reached by water or polymer. Isolated black patches suggest poor connectivity to the displacement pathways. Compared to water alone, polymer injection demonstrates superior displacement efficiency, mitigating fingering and widening the swept zone. Despite improved displacement from polymer injection, some oil remains stranded because of pore-scale trapping and limited connectivity.

3.2.1.2 Analysis of the displacements @polymer viscosity is 1.5 times oil viscosity (polymer viscosity: 150 cP, oil viscosity: 100 cP)

In **Figure 16**, we observe that the green polymer-displaced regions are spread more uniformly throughout the medium, extending more deeply than in the previous case. Compared to the lower-viscosity polymer scenario (**Figure 15**), the displacement front appears less branched and significantly more stable, demonstrating better mobility control with minimal viscous fingering. We see fewer blue (water) zones, indicating that the polymer has successfully displaced water and remobilized oil that was previously bypassed. In contrast to the earlier example, where large water-filled

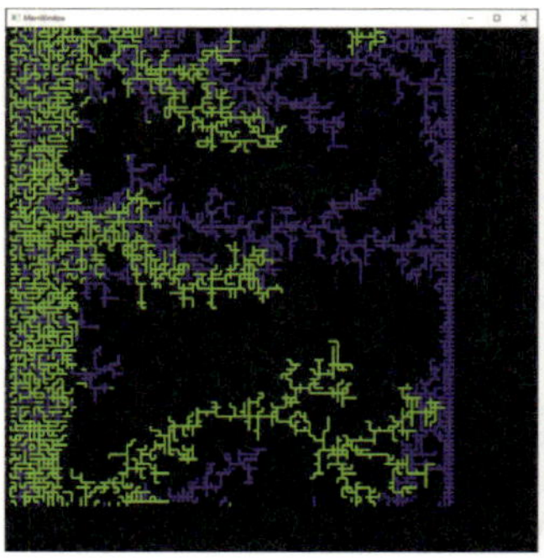

Figure 15.
Polymer injection at 2500 ppm (50 cP viscosity) started after 0.8 PV of water injection. Polymer viscosity remains below oil viscosity, so burst injection is not achieved. Sweep improves modestly but leaves many oil zones untouched.

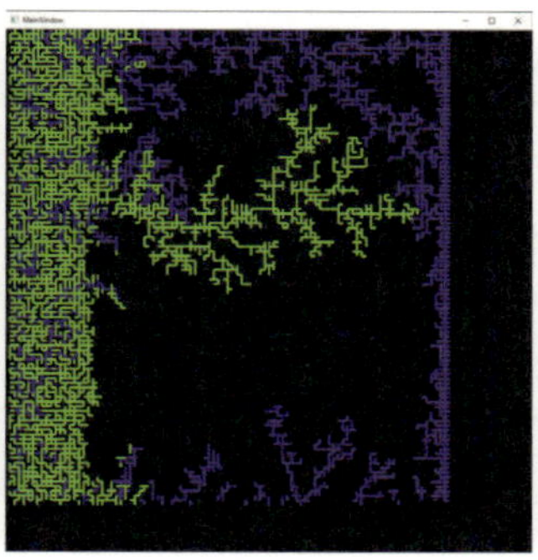

Figure 16.
Polymer injection at 5000 ppm (150 cP), initiated after 0.8 PV of water injection. Polymer viscosity exceeds oil viscosity, achieving burst injection. This results in effective flow redistribution and improved sweep of previously bypassed oil.

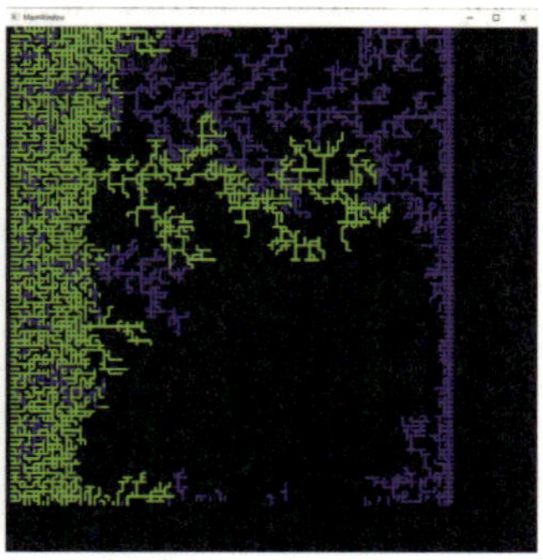

Figure 17.
*Polymer injection at 5500 ppm (200 cP) started after 0.8 PV of water injection. Although polymer viscosity is doubled relative to oil, the incremental sweep gain over the 150 cP case (**Figure 16**) is modest.*

channels remained disconnected and inefficient, this higher-viscosity polymer (150 cP) has "fixed" water fingering to a greater extent. Because the polymer in this case has a viscosity exceeding the oil's (150 vs. 100 cP), it exerts stronger pressure, reducing channeling while improving conformance. Consequently, it accesses more pathways than the lower-viscosity polymer does. The flow now spreads more evenly across the model. Nevertheless, we still observe some pockets of trapped oil, likely because of capillary effects or insufficient pressure to fully mobilize those isolated regions. The implications of using a 150 cP polymer are that it generates a more uniform displacement. The higher-viscosity polymer suppresses viscous fingering and promotes a steadier flood front. This enhances the flow diversion by preventing early polymer breakthrough; the flood can move deeper into unswept areas, thereby improving overall sweep efficiency. By encapsulating a high-pressure front within a confined volume, the polymer induces all-around crossflow, remobilizing trapped water and oil and delivering higher recovery than a lower-viscosity polymer. The increased pressure behind the front continues driving the flood through previously unswept regions, further enhancing displacement leading to early production gains.

3.2.1.3 Analysis of the displacements @polymer viscosity is 2.0 times oil viscosity (polymer viscosity: 200 cP, oil viscosity: 100 cP)

In **Figure 17**, we observe that the polymer flood continues to display strong mobility control, surpassing the performance seen with the 150 cP polymer in **Figure 16**. Despite this improvement, however, some oil remains stranded in disconnected or poorly connected zone. While the overall displacement structure is improved, excessively high polymer viscosities may not entirely solve displacement inefficiencies. We see evidence of potential diminishing returns, indicating that beyond a certain point, increasing polymer viscosity does not guarantee proportionally better oil recovery.

Table 4 compares the impact of polymer viscosity during post-waterflood polymer flooding for three viscosity values—50, 150, and 200 cP—in a porous medium initially saturated with 100 cP oil. The polymer was injected after 0.8 pore volumes of water, and the injection process occurred within the Newtonian plateau for each polymer case. The 50 cP polymer shows a highly branched and unstable displacement front, characteristic of low-viscosity polymer floods. Water bypassing is significant, leading to inefficient oil displacement and low recovery, though the injection pressure remains low. At 150 cP, the polymer enters a more favorable conformance regime with

Feature	50 cP polymer	150 cP polymer	200 cP polymer
Polymer flood efficiency	Highly branched, unstable front	More structured, better conformance	Most uniform, least fingering
Water trapping (Blue regions)	Significant water bypassing	Reduced bypassing, better displacement	Water is further constrained, improving sweep
Oil recovery potential	Lower because of inefficient displacement	Improved, but some areas remain unswept	Best recovery so far, but oil trapping still occurs
Injection pressure needs	Low	Moderate	Moderate to high

Table 4.
Summary of polymer flooding performance at three viscosity levels after 0.8 PV of water injection in a 2D porous medium filled with 100 cP oil. The burst polymer injection condition (polymer viscosity > oil viscosity) at 200 cP results in the most stable front and best sweep efficiency, though at the cost of higher injection pressures.

a more structured front, better sweep, and reduced water bypassing. Oil recovery improves, though some unswept pockets persist. Injection pressure is moderate. This case represents near-balanced viscosity injection, approaching the oil viscosity. At 200 cP, the polymer viscosity exceeds the oil viscosity, entering what is referred to as burst polymer injection. This mode creates a highly uniform front, minimizes fingering, and most effectively constrains water movement, thereby optimizing sweep efficiency. Although oil recovery reaches its peak among the three cases, some oil trapping is still observed because of pore-scale heterogeneities. Injection pressure needs are moderate to high because of the increased viscosity.

3.3 Hybrid polymer injection: Water-Alternating-Polymer and Polymer On-Polymer Off

Projects in China and elsewhere recommend injecting more than one pore volume of polymer to boost recovery factor intended to sustain marginal project economics. Field experience with CO_2 displacement, miscible gas, and steam flooding also show that extended injection beyond one pore volume increases long term economic returns. However, overextended polymer injection periods are one of the pain points of polymer flooding and increase opex continuously. These observations suggest that initial designs were suboptimal. Extended injection leads to prolonged production periods causing high residual polymer concentrations in produced fluids and complicating surface processing. The saturation front does not form a sharp boundary; instead, it results from two overlapping processes: (a) a phase transition that creates a discontinuity in the volume-weighted average saturation, forming a belt-like transition zone, and (b) the rate at which this average saturation evolves.

Alternating polymer injection and water injection at a higher concentration (>5500 ppm) mobilize more oil in poorly swept areas than continuous polymer injection at a lower concentration. Polymer alternation helps because: (i) flow diversion, in which high-concentration polymer temporarily blocks zones, redirecting flow into other regions; and (ii) Pressure pulses caused by alternating polymer slugs create transient pressure gradients that mobilize stagnant oil.

Every time the injectant is modified, the displacement sequence changes, triggering multiple displacement chains that create new flow paths and enhance oil displacement. Factors such as microrugosity, differential adsorption, and dynamic adsorption equilibrium further influence this process. Augmentation of branches retards the velocity profiles significantly, Direct examination of the saturation front shows that even at high flow rates, alternating polymer and water (or cycling polymer on and off) creates tortuous flow paths that reduce water fingering and slow front velocity, while activating flow in the rear and core of the belt-shaped front to better regulate its advancement.

Figures 18 and **19** show the displacement before and after WAP. The displacement sequence is as follows: Waterflooding (blue) of 1 cP water injected into a 100 cP oil-filled medium (black) for 0.4 PV. Because of an unfavorable mobility ratio, water advances in unstable fingers, bypassing significant volumes of oil. Then, polymer flooding (green) of 150 cp polymer (in the Newtonian plateau) displaces the remaining oil. Green zones indicate improved mobility control and deeper oil penetration. Then, Water-Alternating-Polymer (WAP) (cyan) occurs for 0.4 PV, with alternating slugs of polymer and water in equal volumes. Cyan regions represent areas invaded during WAP, where trapped fluids are remobilized through improved sweep and dynamic pressure redistribution.

The pressure contours are more closely spaced in the intermediate saturation zone near the saturation front. This indicates a reduced front mobility because the pressure

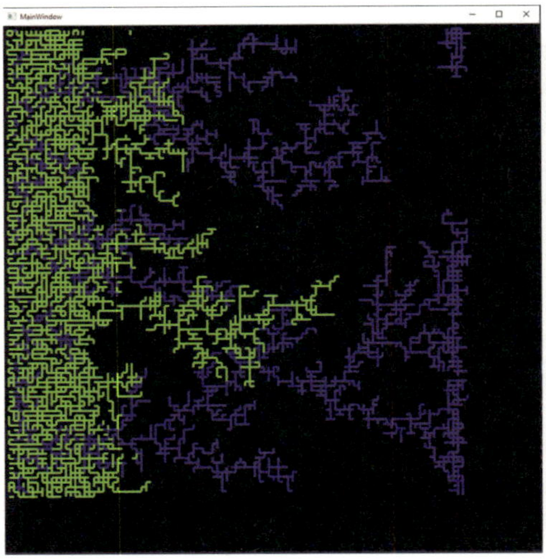

Figure 18.
Final fluid distribution after injecting water (1 cP) for 0.4 pore volumes, followed by continuous polymer injection at 150 cP (Newtonian plateau). This represents the last state before starting WAP.

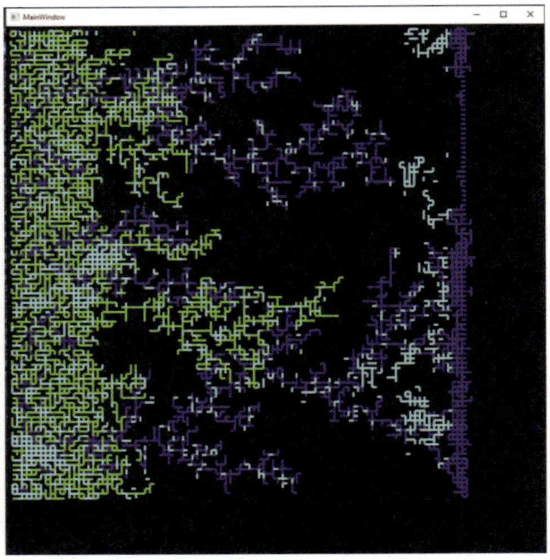

Figure 19.
Fluid distribution after WAP implementation. Cyan pores represent locations where fluid occupancy changed during WAP, highlighting pore-scale redistribution and improved sweep because of dynamic mobility fluctuations. These changes suggest enhanced connectivity and access to previously unswept zones.

gradient is inversely proportional to the total mobility. The fact that there are multiple fingers increases the total mobility (oil mobility plus water mobility), enhancing the oil recovery. The displacement front, taken as a band or belt-like zone, is uniform. Therefore, the Water-Alternating-Polymer reduced the total mobility of the front, inducing a stabilizing effect on the front band or belt-like zone. Similar to this is the Polymer On-Polymer Off [26].

The gradual advance of the belt-shaped front shows a decreasing speed of the saturation front relative to the characteristic of water velocity fingering through the oil. Thus, the polymer front reduced its speed by 2.5-fold because of the Water-Alternating-Polymer.

The Water-Alternating-Polymer revealed a discontinuous mode of front propagation, characterized by having more distributed branches and growth retardation when the branches transitioned from the water-displacing branches. This retardation does not correlate with observations from earlier works on Water-Alternating-Polymer in finite difference simulators. In finite difference simulators, the front brakes rapidly, and the recovery with alternating polymer is equal to the chase water tapering strategy.

Figures 20 and **21** show a snapshot of pore-scale displacement simulation comparing fluid configuration before and after the Water-Alternating-Polymer (WAP) phase for another simulation, but in this case, the WAP takes place at 100 cp polymer. The extent of pore occupancy changes (cyan regions) shows that while in the first image, the polymer front (green) is more continuous and structured, showing a more uniform displacement with minimal readjustments; in the second image, the higher presence of cyan suggests greater interactions between water and polymer, leading to more fluid exchange and redistribution within the pore network. The displacement front has a more compact polymer displacement in the first image, meaning the flow path was more predictable and stable. **Figure 22** suggests a more evolving displacement front, possibly because of a lower-viscosity polymer. In **Figure 22**, the presence of more cyan-colored pores suggests greater local redisplacement,

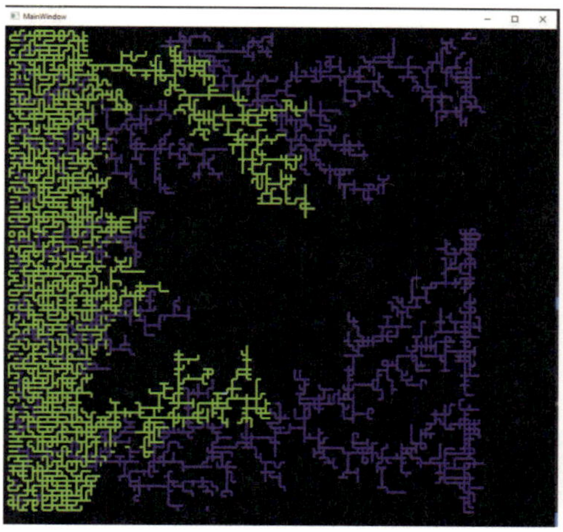

Figure 20.
Before WAP. Comparison of pore-scale fluid occupancy before Water-Alternating-Polymer (WAP) injection. The left panel shows the fluid configuration at the end of polymer flooding (viscosity = 100 cP) following 0.4 pore volumes of initial water injection displacing oil. The right panel represents the pore occupancy after an additional 0.4 pore volumes of WAP injection. Cyan pores indicate locations where fluid occupancy changed during WAP, highlighting enhanced redistribution and dynamic displacement behavior because of alternating water-polymer sequences. The flow direction is from left to right, and the 2D lattice is 100 × 100 pores with periodic boundary conditions in both the y and z directions.

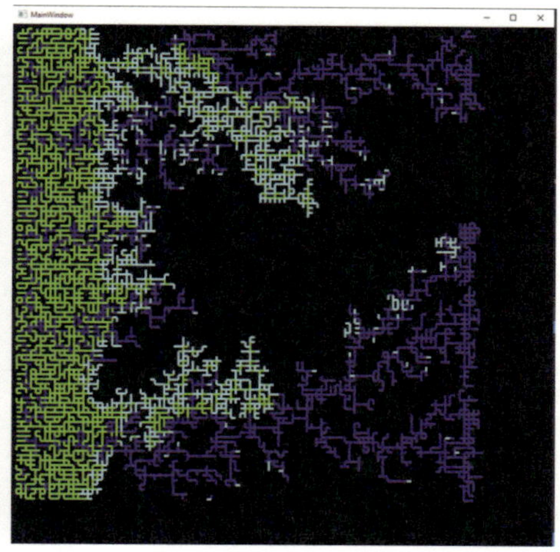

Figure 21.
After WAP. Comparison of pore-scale fluid occupancy after Water-Alternating-Polymer (WAP) injection. The left panel shows the fluid configuration at the end of polymer flooding (viscosity = 100 cP) following 0.4 pore volumes of initial water injection displacing oil. The right panel represents the pore occupancy after an additional 0.4 pore volumes of WAP injection. Cyan pores indicate locations where fluid occupancy changed during WAP, highlighting enhanced redistribution and dynamic displacement behavior because of alternating water-polymer sequences. The flow direction is from left to right, and the 2D lattice is 100 × 100 pores with periodic boundary conditions in both the y and z directions.

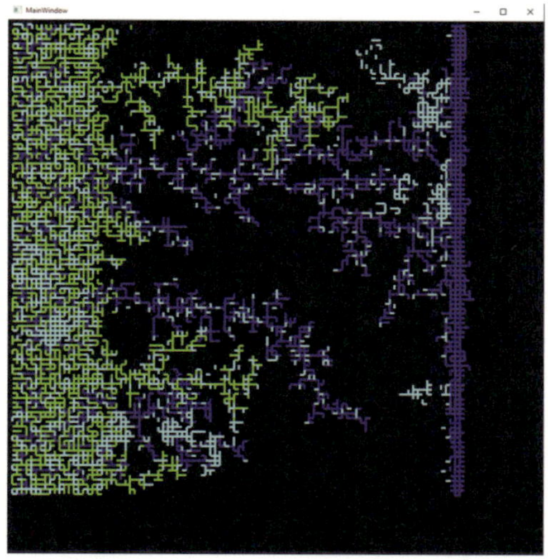

Figure 22.
2D pore network displacement map under sequential water, polymer, and Water-Alternating-Polymer injection (left-to-right flow). The image illustrates fluid movement through a 100 × 100 2D pore lattice under tertiary recovery. Initial water injection (blue) leads to poor sweep because of the mobility contrast with the oil (black). Polymer injection (green) improves displacement. The final Water-Alternating-Polymer (WAP) phase reactivates flow pathways and sweeps bypassed zones. Cyan pores mark regions where occupancy changed during WAP, revealing successful redistribution. Periodic boundary conditions in y and z enhance the representativeness of reservoir-scale flow continuity.

Factor	Before WAP	After WAP
Polymer viscosity	Higher (leading to a stable front)	Lower (allowing more fluid redistribution)
Mobility control	Stronger, with fewer occupancy changes	Weaker, with more dynamic fluid exchanges
Pore-scale instabilities	Minimal, suggesting better conformance	More significant, indicating ongoing redistributions
Sweep efficiency	More structured polymer effectively pushing oil	More readjustment, meaning the polymer is still interacting with water pockets

Table 5.
Comparison of pore-scale behavior before and after Water-Alternating-Polymer (WAP) injection. The table highlights how WAP transitions the system from a stable, structured polymer flood to a more dynamic, redistributive regime, improving sweep in less accessible zones. This reflects the adaptive role of WAP in enhancing conformance and targeting residual heterogeneities.

which could mean a more complex flow path. The extent of pore occupancy changes (cyan regions) in **Figure 22** shows a significant increase in cyan pores, indicating more dynamic fluid changes because of additional remobilization of oil or polymer redistribution. In **Figure 20**, the green polymer-displaced area appears more structured, showing a stable front with minimal fluid exchange. In **Figure 22**, the polymer displacement is more dispersed and interacts more with previously occupied water-filled pores.

Table 5 provides a comparative analysis of pore-scale dynamics before and after Water-Alternating-Polymer (WAP) injection in a polymer flood process. It examines how key parameters such as polymer viscosity, pore-scale instabilities, and sweep efficiency evolve during the transition. Post-WAP, sweep becomes more dynamic and redistributive, where polymer continues interacting with residual trapped water and oil pockets.

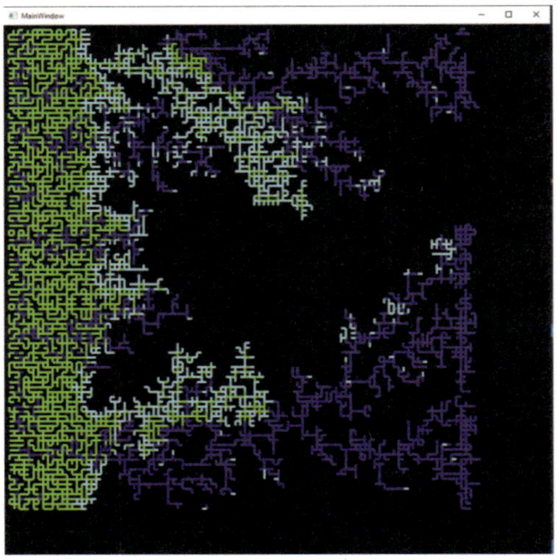

Figure 23.
2D pore network simulation of sequential water, polymer, and Water-Alternating-Polymer injection (flow from left to right). The figure illustrates fluid displacement across a 100 × 100 lattice with periodic boundary conditions. The initial waterflood (blue) shows an uneven sweep because of an unfavorable mobility ratio. Polymer injection (green) enhances front stability. During the WAP stage, cyan pores reveal locations where dynamic occupancy changes occurred, suggesting resaturation and flow reorganization. These redistributions are critical for maximizing sweep efficiency in heterogeneous porous media.

Factor	First image (Figure 22)	Second image (Figure 23)
Polymer viscosity	Likely higher (more stable displacement)	Moderate (more interactions)
Fluid redistribution	Minimal, stable polymer front	Moderate, showing interactions between polymer and water
Water retention (Blue)	Well-separated from polymer	More mixed, suggesting redisplacement
Sweep efficiency	More structured displacement	Fluid redistribution in progress

Table 6.
Comparison of pore-scale displacement behavior between two polymer flooding scenarios. The table highlights differences in viscosity effects, fluid redistribution, and water retention between the two visualizations of polymer injection. The first case shows a more stable and structured front, while the second reveals dynamic interactions and remobilization processes, indicative of lower viscosity and active redistribution in the pore network.

Figures 22 and **23** compare two WAP scenarios: **Figure 22** shows WAP performed after polymer injection at 1.5 times the oil viscosity, while **Figure 23** shows WAP performed after polymer injection at 1.0 times the oil viscosity.

In **Figure 22**, the polymer viscosity is 1.5 times the oil viscosity before Water-Alternating-Polymer and during the polymer cycle in the water-alternating scheme, which supports a more stable and structured displacement front. The polymer front is minimal and appears stable, with clear separation from the previously injected water (blue), resulting in a well-defined displacement pattern and higher sweep efficiency. **Figure 22** shows clearer separation between blue and green zones, meaning that water and polymer did not mix significantly.

In **Figure 23**, polymer viscosity is equal to oil viscosity, leading to more visible interactions between the polymer and water. It shows a more fragmented polymer front, with some green regions interspersed within blue areas, suggesting partial displacement and fluid mixing. It shows a larger amount of blue within polymer-displaced areas, which suggests more interaction between water and polymer and possibly some water retention. It has the least defined separation between water and polymer, indicating a more complex and unstable interaction between the fluids. **Figure 23** shows a more dynamic and complex interaction. Fluid redistribution is underway, with the polymer and residual water becoming increasingly mixed, which implies the remobilization of trapped water. The sweep efficiency is lower than in **Figure 22**, as the displacement front is less organized because of ongoing redistribution.

Table 6 compares two pore-scale displacement images based on key flow behavior factors observed during polymer flooding. It analyzes polymer viscosity, fluid redistribution, water retention, and sweep efficiency between the two scenarios.

3.4 Field-scale hybrid polymer injection: Comparison across different zones

In the previous sections, we examined how Water-Alternating-Polymer (WAP) injection affects fluid flow and oil displacement at the pore scales. We now extend those insights to the field scale, integrating both numerical modeling and real-field performance to assess the broader implications of hybrid strategies in an operational environment. By translating pore-scale phenomena—such as flow diversion, pressure pulsing, and displacement front stabilization—into a full-field context, we can better predict how these mechanisms will influence oil rate response, ultimate recovery, surface facility demands, and overall project economics. The following sections present the key findings and highlight the critical takeaways for reservoir management, including the control knobs to steer injectivity, high pressure oil drive to rapid oil response.

To compare different reservoir zones and injection strategies, we normalized the oil production rate to the maximum rate observed during the first 12 months of injection, aligning with typical economic return periods. We also normalized the x-axis to the estimated ultimate recovery (EUR) for each zone. The y-axis represents the normalized production rate ($QoD = Qo/Qo_{max}$ from continuous polymer injection during the first year), and the x-axis represents cumulative production as a fraction of EUR ($NpD(t) = Np(t)/EUR$). The following sections present data from various zones, each with different pore volumes of water injected prior to polymer injection. They also include cases with varying polymer injection volumes and different combinations of Water-Alternating-Polymer strategies.

3.4.1 Zone with tertiary polymer injection initiating above 0.6 PV water injected, then continuous polymer injection for 0.3 PV, and then Water-Alternating-Polymer for 0.3 PV

Figure 24 shows the production performance of Zone 7, where continuous tertiary polymer injection is initiated after 0.6 PV of water injection. **Figure 24** highlights the main events: Continuous polymer injection, production downtime from operational issues (unrelated to reservoir response), and Water-Alternating-Polymer. The process

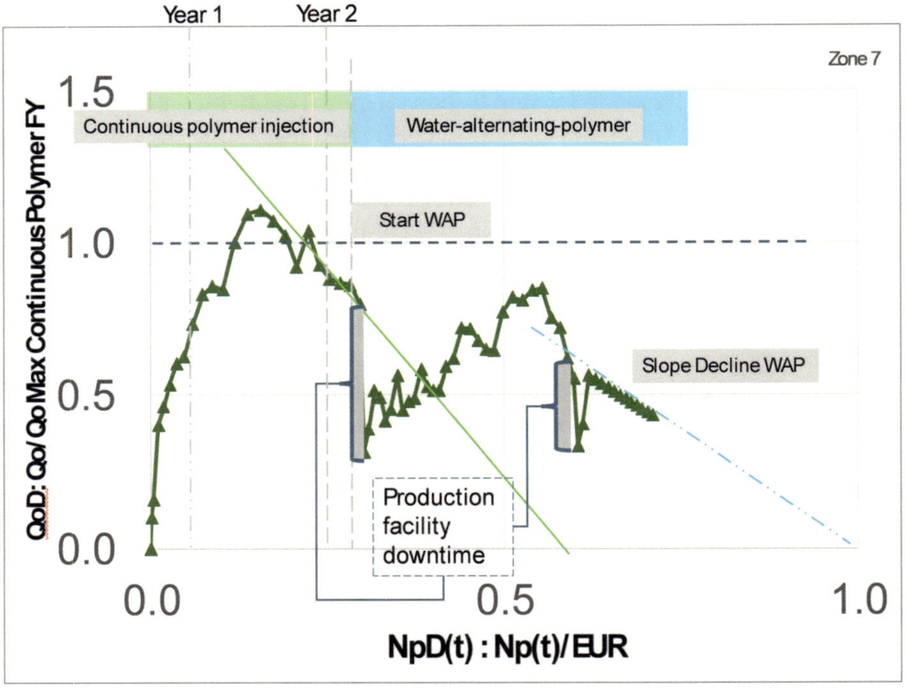

Figure 24.
Production response in Zone 7 during continuous polymer and Water-Alternating-Polymer injection. This plot illustrates the production performance transition from continuous polymer injection to WAP in Zone 7. The continuous polymer phase yields a strong initial response followed by a decline, while WAP implementation leads to a smoother production trend with remobilization peaks and a slower decline rate. Downtime events momentarily disrupt output but are followed by recovery, reflecting system robustness. The figure highlights how WAP can extend recovery and mitigate polymer degradation effects after [30].

follows these sequential steps: Continuous polymer injection for less than 0.3 PV (green region). Water-Alternating-Polymer (WAP) Injection for 0.3 PV (blue-striped region).

Continuous polymer injection phase (green-shaded area): Production increases steadily, reaching a peak slightly above 1.2 relative units. Over time, polymer efficiency declines, leading to a gradual production drop as the displacement process becomes less effective. Two production facility downtimes interrupt this trend, causing sharp, non-reservoir-related drops in production. Transition to WAP Phase (blue-striped area): The "Start WAP" marker indicates the shift from continuous polymer to Water-Alternating-Polymer (WAP) injection. Production stabilizes briefly, with several uplift events linked to the remobilization of oil in previously unswept zones. The "Slope Decline WAP" line shows a gentler decline compared to the continuous polymer phase, indicating improved sweep efficiency and extended recovery performance under the alternating injection strategy.

Table 7 provides a structured comparison of reservoir performance across three distinct operational phases: Continuous polymer injection, transition period, and WAP implementation. During continuous polymer injection, a high production peak is achieved because of effective initial mobility control. However, this is followed by a steep production decline, signaling that polymer efficiency deteriorates over time.

With WAP implementation, production peaks become more stabilized, though lower than the initial peak observed during the continuous polymer phase. The decline trend during this phase is more gradual, suggesting better flow control and improved sweep efficiency. The key feature is that WAP extends the duration of enhanced oil recovery, freeing up polymer resources for redeployment to other injection points and reducing OPEX, while mitigating the sharp declines seen in earlier phases.

Figure 25 presents the production response in Zone 6 during continuous polymer injection followed by a Water-Alternating-Polymer (WAP) strategy.

Early continuous polymer injection:

The green-shaded area indicates continuous polymer injection.

The initial response shows a moderate rise in production with a local maximum, followed by a sharp decline (green trendline).

Operational disruption:

A production facility downtime event (highlighted in the dashed box) creates a sharp drop and temporary halt in production.

Phase	Peak behavior	Decline trend	Key learning
Continuous polymer injection	High peak because of initial mobility control	Steep decline (polymer retention, shear-thinning effects)	Polymer efficiency reduces over time
Transition period	Fluctuating peaks because of injection variability	Short recovery boosts but no sustained gains	Indicates polymer remobilization and reservoir heterogeneity
WAP implementation	Stabilized peaks, but lower than initial peak	Gradual decline, less steep than initial drop	WAP improves oil recovery longevity

Table 7.
Comparative assessment of production dynamics across EOR phases: Continuous polymer, transition, and WAP. The table highlights how reservoir response evolves with different injection strategies. Initial polymer flooding achieves strong performance but suffers from early decline. Transition phases show instability linked to injection dynamics and heterogeneity. WAP stabilizes performance and extends recovery benefits, supporting its implementation in mature fields.

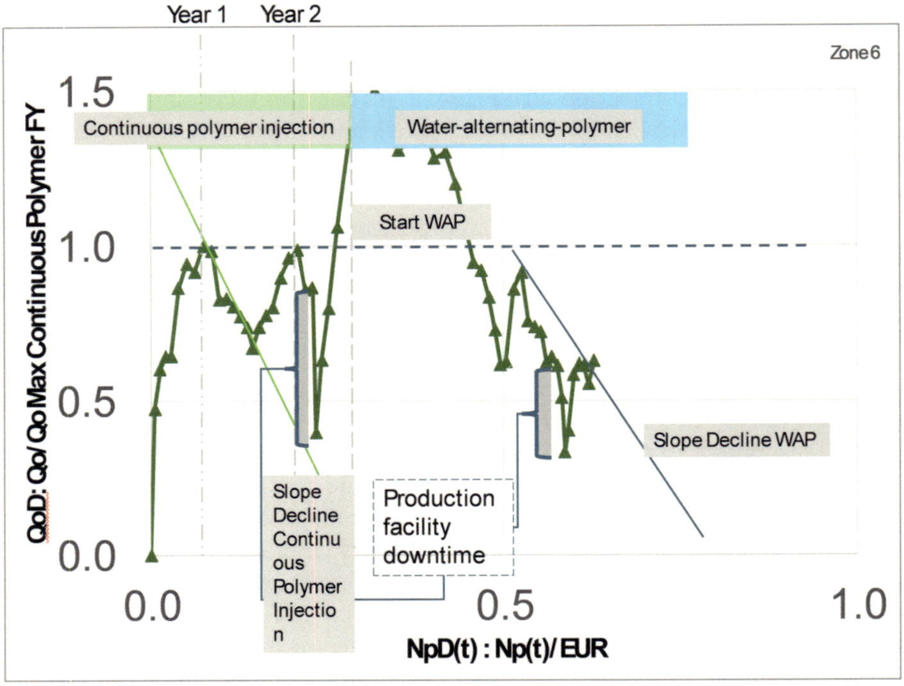

Figure 25.
Production dynamics in Zone 6 during continuous and alternating polymer injection. The plot highlights the performance transition from continuous polymer injection to Water-Alternating-Polymer (WAP) in Zone 6. Following an early production peak and sharp decline because of polymer inefficiencies, WAP implementation temporarily stabilizes production, though overall decline resumes. Facility downtime and alternating injection cycles also contribute to short-term fluctuations, with overall response suggesting moderate WAP improvement over continuous polymer decline after [30].

Post-downtime, production attempts to recover but stabilizes below the initial peak.

Switch to Water-Alternating-Polymer (WAP):

The transition point is indicated by a vertical gray band labeled "Start WAP."

Following the start of WAP, there is a second production peak, slightly higher than the previous.

This is followed by a more controlled decline (blue trendline), representing WAP performance.

Overall decline pattern:

The two key declines—one following continuous polymer, the other post-WAP—illustrate distinct performance regimes.

The WAP decline is less steep, suggesting improved sweep efficiency and polymer effectiveness during this phase.

Figure 25 illustrates the production performance over time for Zone 6, showing the impact of continuous polymer injection followed by a transition to Water-Alternating-Polymer (WAP) injection.

Initial polymer phase: The green-shaded area indicates continuous polymer injection. The production rate initially increases, reaching a peak above 1.4. However, shortly after the peak, a sharp decline in slope (light green line) reflects reduced efficiency. Operational Downtime: Midway through the polymer phase, a production facility downtime occurs, marked in dashed gray. This creates a clear fall in production before a partial recovery resumes.

Transition to WAP: The system transitions into WAP injection, signaled by the dashed blue vertical line and shaded blue/striped bar. A short production increase follows, suggesting pressure improvement and sweep efficiency in previously unswept zones.

Post-WAP behavior: A second decline slope (solid blue line) represents WAP-induced decline. It is less steep than the initial decline, showing moderate performance stabilization, albeit with continued reduction in production.

Figure 25 for Zone 6 follows the same injection scheme as the previous Zone 7 graph: Continuous polymer injection (green region), Water-Alternating-Polymer (WAP) injection (blue-striped region), and it was also affected for production facility downtime effects (blue boxes), which describe the inertia of the process to recover the production as before.

Zone 6 has the same injection sequence and events as Zone 7. However, Zone 6 presents some distinct differences in the production response, which will be discussed below.

A sharp production increase is observed during continuous polymer injection. The peak is higher than the first peak in Zone 7. The peak occurs slightly earlier compared to Zone 7, indicating a faster response to polymer injection. After reaching the first peak, production declines more steeply (green straight line) compared to Zone 7. The faster drop-off indicates that continuous polymer injection did not sustain oil recovery as long as in Zone 7. Then, there are significant production disruptions (blue boxes). The recovery after downtime is stronger than in Zone 7, showing that the reservoir has some resilience to interruptions. It implies that downtime caused temporary setbacks, but polymer-WAP transition helped re-establish production gains. The start of WAP shows another production peak after the decline. However, compared to Zone 7, the WAP transition in Zone 6 has a sharper production drop after the second peak. WAP still provides production uplift, but the decline afterward suggests that WAP in Zone 6 was less effective at stabilizing production compared to Zone 7. A strong production drop follows WAP, which is steeper than in Zone 7, suggesting that the WAP mechanism had a weaker long-term effect in Zone 6 compared to Zone 7.

Table 4 provides a detailed comparison of production dynamics between Zone 6 and Zone 7, highlighting differences in polymer flooding performance and the transition into Water-Alternating-Polymer (WAP) strategies. In contrast, Zone 7 shows a more gradual rise, reflecting a smoother buildup and possibly a more controlled displacement front, which is consistent with a smoother, more gradual decline, suggesting better polymer sweep efficiency in Zone 7.

Zone 6 shows a strong production rebound after downtime, possibly because of pressure buildup and remobilization of fluids. Zone 7 also recovers, but the recovery is moderate and less pronounced, indicating a more stable system less influenced by shutdown periods. In Zone 6, the WAP transition leads to a temporary production uplift, but the effect is short-lived, likely because of residual polymer instability or poor conformance. Zone 7 demonstrates a more stable post-transition, highlighting the importance of timing. After WAP, Zone 7 maintains a slower decline, reinforcing the idea of higher polymer efficiency and delayed water invasion (**Table 8**).

3.4.2 Zone with tertiary polymer injection initiating below 0.5 PV water injected, then continuous polymer injection for less than 0.3 PV, then a water slug for 0.05 PV, and then Water-Alternating-Polymer for 0.3 PV

Figure 26 presents the normalized production performance of Zone 8–9. It follows a slightly different tertiary polymer injection sequence compared to Zones 6 and 7.

Feature	Zone 6 (Figure 25)	Zone 7 (Figure 24)
First production peak	Higher and sharper	More gradual rise
Initial decline after polymer injection	Steeper decline (polymer retention issues)	Smoother, more gradual decline
Effect of facility downtime	Strong recovery after downtime	Moderate recovery, but less pronounced
WAP transition	Provides production uplift, but short-lived	More stable post-transition
Decline after WAP	Rapid decline (water breakthrough)	Slower decline (better polymer efficiency)

Table 8.
Comparison of polymer and WAP injection performance between Zone 6 and Zone 7. Zone 6 shows a sharper peak and stronger but short-lived responses, with production declines driven by polymer retention and water breakthrough. In contrast, Zone 7 achieves more controlled performance, with smoother transitions and better polymer efficiency leading to prolonged oil recovery.

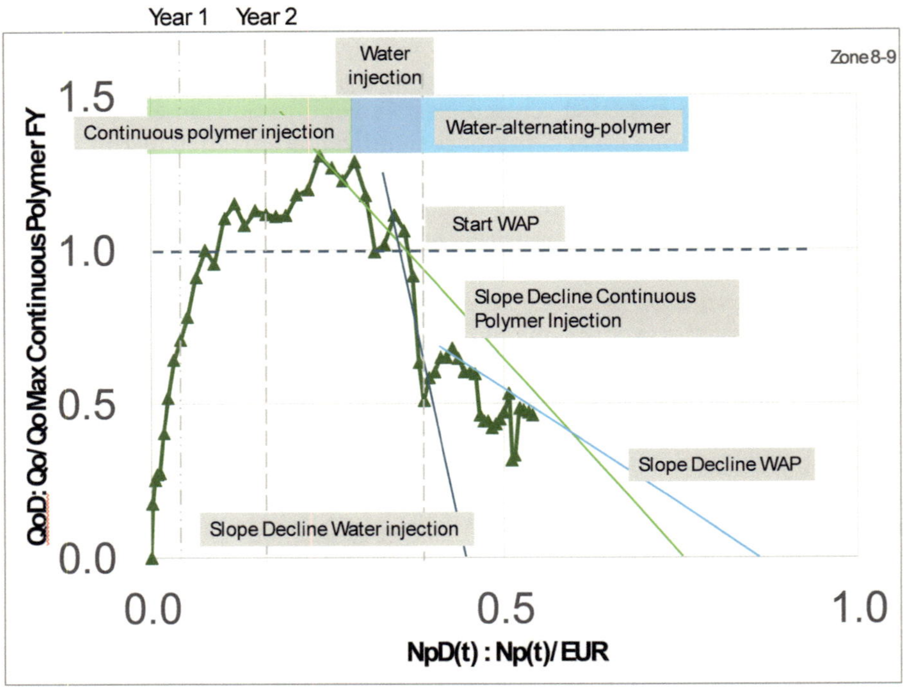

Figure 26.
Normalized oil production response in Zone 8–9 under sequential injection strategies. Following an early peak (~1.2), a water slug disrupts polymer injection and causes a sharp production drop. The onset of Water-Alternating-Polymer (WAP) improves recovery slightly and flattens the decline trend. Each segment's slope highlights the relative performance: steepest underwater, moderate during continuous polymer, and most stable under WAP, emphasizing WAP's role in mobility control and late-time recovery efficiency after [30].

The tertiary polymer injection starts below 0.5 PV of water injection (earlier than in previous zones). Then continuous polymer injection lasts for less than 0.3 PV (similar duration as in previous zones). A short water slug (0.05 PV) is introduced, confirming the polymer front disruption seen in the pore-scale simulations. Then, WAP (blue-striped area) for 0.3 PV.

The initial oil production increase is steep, leading to a peak just above 1.2 on the normalized scale, slightly lower than in Zones 6 and 7. After reaching the peak, a sharp production drop occurs, much steeper than in Zone 7 but like Zone 6. The effect of the water slug (0.05 PV) causes a sudden drop in production ("Slope Decline Water Injection"), indicating that the polymer front lost effectiveness faster than in previous cases because water re-opened pathways, causing oil bypassing. The WAP transition slows down production decline, but the post-WAP peaks are lower compared to Zones 6 and 7. The WAP helped moderate decline, but not as effectively as in Zones 6 and 7. The decline after WAP is significant and steeper than in Zone 7, like Zone 6, suggesting that water breakthrough was more dominant in this case, and effectiveness decreased faster because of the earlier water injection sequence.

Table 9 summarizes and contrasts the production response across three zones—Zone 8–9, Zone 7, and Zone 6—each having undergone polymer injection followed by Water-Alternating-Polymer (WAP) processes. The analysis includes five key production features.

3.4.3 Zone with tertiary polymer injection initiating below 0.5 PV water injected, then continuous polymer injection for less than 0.2 PV, and then Water-Alternating-Polymer for 0.3 PV

Figure 27 represents the field production performance of Zone 11, where tertiary polymer injection is initiated below 0.4 PV of water injected, less than in previous zones. The steep production increase leads to a peak above 1.4, like Zone 6, but slightly higher than Zones 8–9. This peak is among the highest observed, suggesting that polymer injection had strong early performance because of the lack of preexisting water saturation, but it is not conclusive because Zone 6 and Zones 8–9 had a greater period of water injection.

Feature	Zones 8–9 (Figure 26)	Zone 7 (Figure 24)	Zone 6 (Figure 25)
First production peak	Slightly lower (1.2) because of early water injection effects	Higher (1.3) with more stable polymer performance	Higher (1.4) but followed by a fast decline
Initial decline after polymer injection	Steep decline because of polymer dilution	More gradual decline with better polymer effectiveness	Rapid decline because of polymer retention issues
Water slug impact	Caused a sharp drop before WAP	No water slug, smoother transition to WAP	No water slug, but WAP had mixed results
WAP transition effect	Limited production recovery, weaker than Zone 7	Moderate stabilization, prolonged efficiency	Some recovery, but rapid decline afterward
Final decline trend	Steeper than Zone 7, similar to Zone 6	More gradual decline	Fast decline, similar to Zone 8–9

Table 9.
Comparative evaluation of polymer and WAP injection performance in zones 8–9, 7, and 6. The table summarizes key production behavior metrics such as initial peak, decline phases, water slug impact, and the final performance trend. Zone 7 displayed the most stable performance with a strong polymer response and effective WAP transition. In contrast, Zones 8–9 and 6 showed steeper declines, affected by early water injection or retention issues, with Zones 8–9 particularly impacted by a water slug prior to WAP.

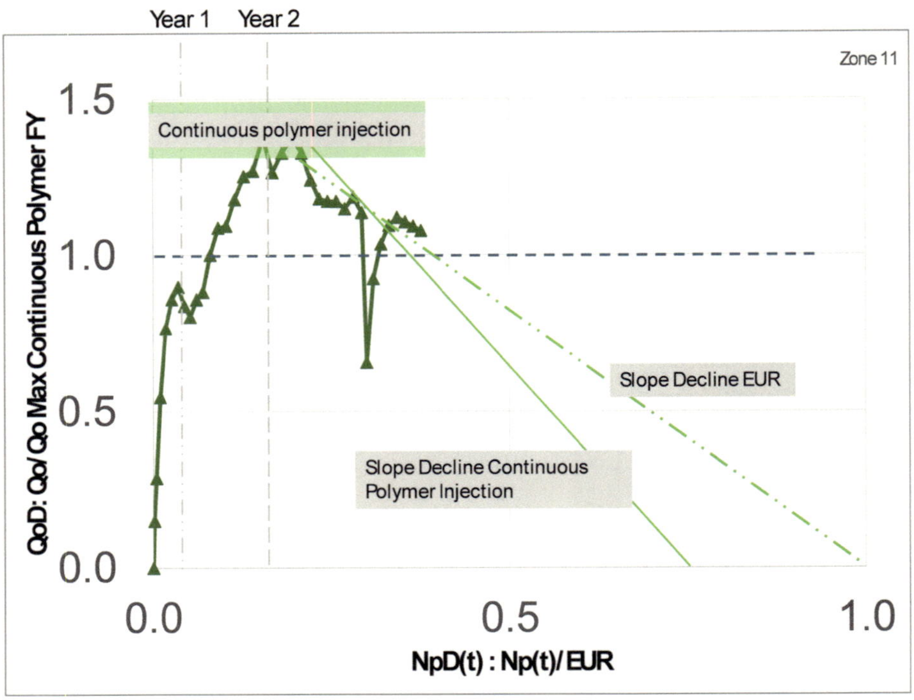

Figure 27.
Normalized production response during continuous polymer injection in Zone 11. The graph shows the production evolution relative to the maximum output observed during polymer flooding. After an initial peak, production follows a decline with a defined slope. Two trend lines highlight (1) the decline rate during continued polymer injection and (2) the extrapolated slope toward estimated ultimate recovery (EUR). The chart helps evaluate the sustainability and long-term efficiency of the polymer flooding strategy after [30].

Feature	Zone 11 (Figure 27)	Zone 7 (Best performer (Figure 24))	Zone 6 (Similar fast decline (Figure 25))	Zones 8–9 (Intermediate case (Figure 26))
First production peak	Highest (1.4)	High (1.3)	High (1.4)	Lower (1.2, because of early water injection)
Polymer production retardation	Weak, decline is steep	Strong, slower decline	Moderate	Present but affected by water slug
Water slug impact	Not applicable (no water slug used)	Not applicable	Not applicable	Caused additional fluctuations
WAP impact	No WAP, leading to faster decline	Strongest recovery, best stabilization	Some recovery, but not sustained	Moderate impact, but less stable

Table 10.
Comparative summary of production behavior across four zones under polymer and WAP injection strategies. This table highlights differences in initial production peaks, polymer flood performance, and the effectiveness of WAP implementation. Zone 7 demonstrates the most robust performance, marked by slower decline and strong WAP stabilization. In contrast, Zone 11, despite having the highest peak, suffered a steep decline without WAP support. Zone 6 showed only a moderate polymer response, and Zone 8–9 was adversely impacted by an early water slug and less stable WAP response.

The first decline follows a steep straight-line trajectory, like Zone 6 and Zones 8–9. The absence of WAP or water slugs causes a more pronounced and immediate drop, suggesting that polymer depletion is faster without additional support.

To verify whether the decline slope matches previous cases, we analyzed the decline straight-line slopes numerically by comparing them with Zones 6, 7, and 8–9.

Table 10 provides a comparative performance assessment of polymer injection and Water-Alternating-Polymer operations across four reservoir zones: Zone 11 (current), Zone 7 (best performer), Zone 6 (fast decline case), and Zones 8–9 (intermediate case). Each zone is evaluated across four key production features: initial production peak, polymer-induced production retardation, water slug impact, and the impact of WAP. The decline in Zone 11 is the steepest, meaning polymer flooding alone is not enough to sustain production. Zones with WAP had better performance, suggesting that introducing WAP could significantly improve recovery in Zone 11.

The decline slope for Zone 11 is identical to Zones 6 and 8–9 (−3.0), proving that continuous polymer injection leads to the same post-peak production losses across different conditions. Only Zone 7 (with WAP) exhibits a slower decline (−2.0), reinforcing the benefit of hybrid injection strategies.

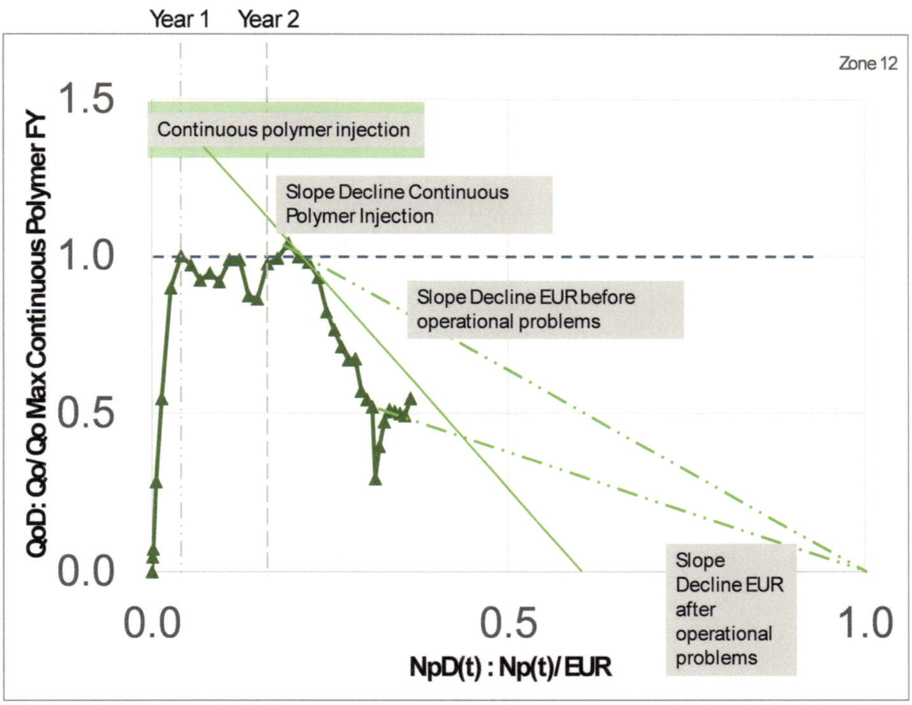

Figure 28.
Production performance of Zone 12 under continuous polymer injection. The graph illustrates a moderate peak followed by a sharp decline in normalized oil production. Slope analysis reveals a steep decline during polymer flooding (−2.67) and two projections of ultimate recovery (EUR): one before operational disruptions (−1.375) and a significantly flatter slope after (−0.6), indicating possible pressure stabilization or improved flow redistribution. This dual-slope analysis supports real-time reassessment of EUR and project efficiency following operational deviations after [30].

Feature	Zone 12 (Figure 28)	Zone 7 (Best performer (Figure 24))	Zone 6 (Fast decline (Figure 25))	Zones 8–9 (Intermediate case (Figure 26))	Zone 11 (Similar case (Figure 27))
First production peak	1.1	1.3 (Higher, better polymer efficiency)	1.4 (Higher but short-lived)	1.2 (Early water injection)	1.4 (Similar)
Polymer decline slope	−2.67 (Slightly less steep)	−2.0 (Best case, WAP helped)	−3.0 (Steepest decline)	−3.0	−3.0
EUR decline before issues	−1.375 (Moderate decline)	−2.0	N/A	N/A	N/A
EUR decline after issues	−0.6 (Flattest decline, suggests pressure stabilization)	N/A	N/A	N/A	N/A

Table 11.
Comparative evaluation of production performance in five zones under polymer injection. The table highlights variations in peak response, decline rates during polymer flooding, and projected EUR decline before and after operational issues. Zone 7 demonstrated the best overall stability, while Zones 6, 8–9, and 11 exhibited steeper declines. Zone 12, despite a modest initial peak, showed favorable EUR flattening after issues, suggesting system resilience or operational recovery.

3.4.4 Zone with tertiary polymer injection initiating below 0.4 PV water injected, then continuous polymer injection for less than 0.3 PV, and continue injecting

Figure 28 represents Zone 12, where tertiary polymer injection is initiated below 0.4 PV of water injected. The steep production increase leads to a peak around 1.1 on the normalized scale, which is slightly lower than in Zone 11. The decline rate of continuous polymer injection is −2.67, which is slightly less pronounced than Zones 6, 8–9, and 11 (−3.0) but more pronounced than Zone 7 (−2.0), which had WAP to slow decline. This might indicate that polymer alone is not sufficient for long-term stabilization. The slope before operational issues is −1.375, which is more gradual than the polymer decline (−2.67), indicating some production stability. Still steeper than Zone 7 (−2.0), meaning polymer depletion was already reducing effectiveness before operational issues. The final slope, after operational problems, is −0.6, indicating a significantly slower decline.

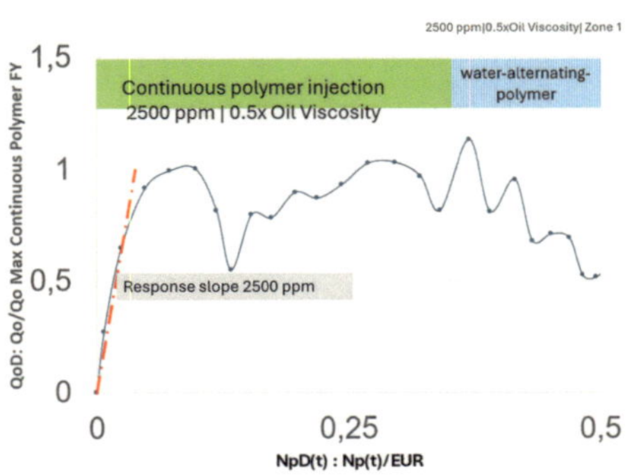

Figure 29.
Zone with 2500 ppm polymer injection at 0.5x oil viscosity. Early production peak followed by erratic response and steep decline after [30].

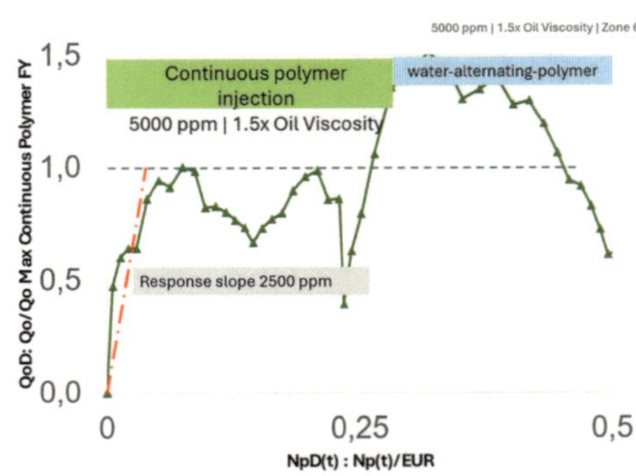

Figure 30.
Zone with 5000 ppm polymer at 1.5x oil viscosity. A clear peak and stable slope reflect improved conformance and effectiveness of burst injection after [30].

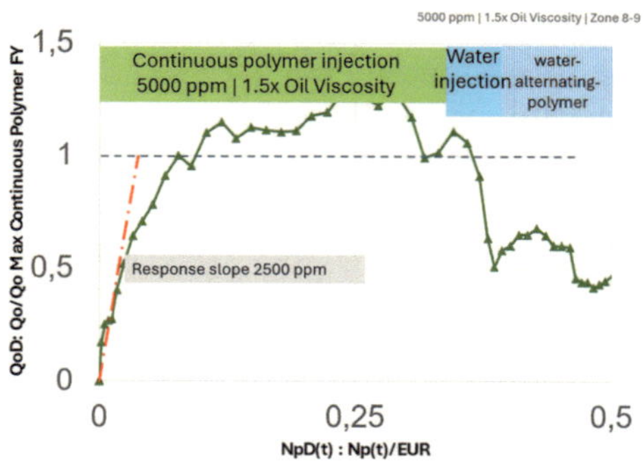

Figure 31.
Burst polymer injection followed by water and WAP. Zone 8–9 shows smoother slope transitions and higher cumulative production than the 2500 ppm case after [30].

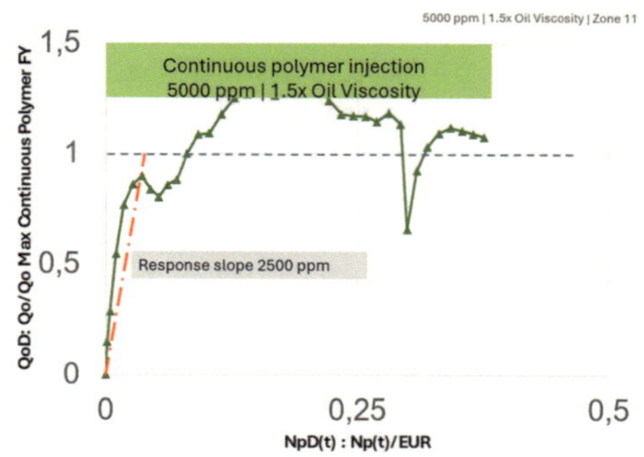

Figure 32.
Zone 11 with 5000 ppm polymer at 1.5x oil viscosity. A stable peak and sustained response validate the benefit of burst polymer design after [30].

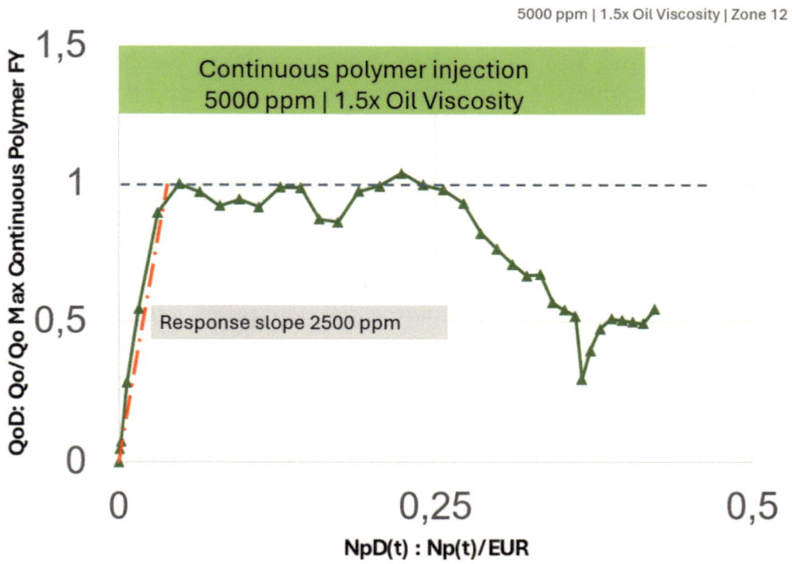

Figure 33.
Zone 12 with 5000 ppm polymer injected at 1.5x oil viscosity. The response shows a broad production plateau with delayed decline, confirming the effect of burst injection even without a WAP cycle. The red line represents a 2500 ppm polymer flood, where polymer viscosity is only half that of the oil after [30].

Table 11 presents a comparative analysis of production performance across five reservoir zones—Zone 12 (current graph), Zone 7 (best performer), Zone 6 (fast decline), Zone 8–9 (intermediate case), and Zone 11 (similar to Zone 6)—based on key production metrics from polymer injection projects.

3.4.5 Field response comparison for burst polymer injection: Zone with 2500 ppm polymer and concentration 0.5 times oil viscosity versus zones with 5000 ppm polymer injection and 1.5 times oil viscosity

This section compares the field performance of burst polymer injection under different design conditions [6]. **Figure 29** represents a baseline case with 2500 ppm polymer at 0.5 times oil viscosity, while **Figures 30–33** correspond to 5000 ppm polymer at 1.5 times oil viscosity, reflecting the burst polymer injection regime. The curves in **Figures 31–34** started above the red reference, indicating a faster initial production buildup than the 2500 ppm baseline. This is especially notable for polymer concentrations of 5000 ppm (1.5× oil viscosity), referred to in this work as burst polymer injection. A higher initial slope translates to more immediate, robust, and quicker oil rate ramp-up. Some cases surpass the normalized production factor of 1.0 on the y-axis. This indicates that these cases have improved on the early buildup but also maintained rates surpassing the initial 12-month response. Burst polymer injections (polymer viscosity > oil viscosity) consistently show higher initial slopes relative to the 2500 ppm reference. By establishing a more stable displacement front, they reduce early water breakthrough and re-invade previously bypassed zones faster. The fact that certain curves exceed 1 underscores that a well-chosen high-viscosity polymer can outperform the baseline not just in ultimate recovery but even in initial production peaks.

The high concentration and burst viscosity produce a stronger, more stable front, followed by WAP cycles that delay decline and improve conformance. The recovery is superior to **Figure 29**, both because the curve lies above the response slope of 2500 ppm and because production is sustained.

Figure 33 (5000 ppm, 1.5x oil viscosity, Zone 12) in this case shows the flattest and most sustained production plateau among all zones. It demonstrates that burst injection alone, if optimized and uninterrupted, can yield strong and prolonged responses.

3.5 Field-scale secondary polymer injection pilot versus tertiary polymer injection pilot

This section contrasts the performance of two field pilots—one implementing secondary polymer injection and the other using tertiary polymer injection [28, 30]. **Figure 34** plots recovery factor (fraction of original oil in place produced) against

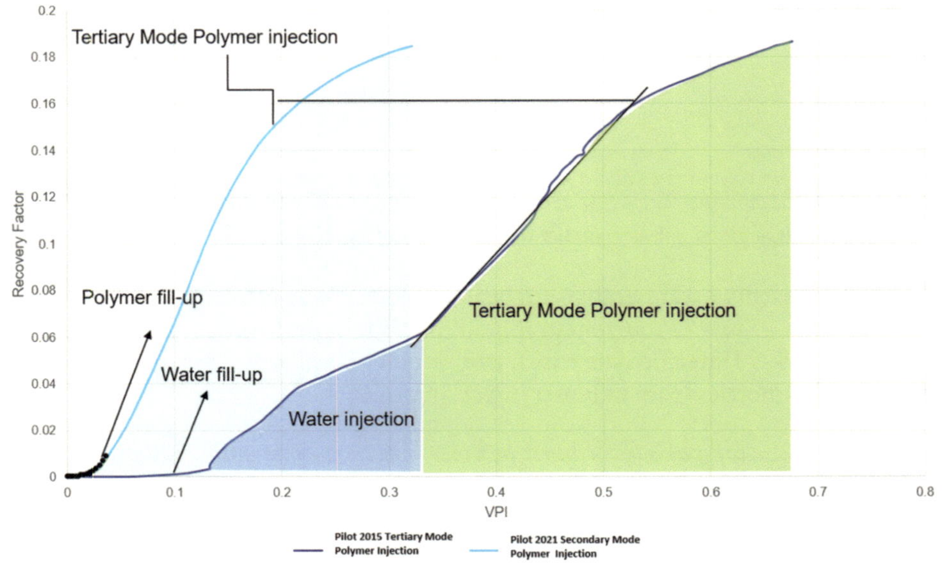

Figure 34.
X-axis (VPI): Volume of fluid injected as a fraction of pore volume. Y-axis (Recovery Factor): Cumulative oil recovered as a fraction of original oil in place (OOIP).

Phase	Slope description	Efficiency interpretation
Water (0–0.13 VPI)	Very low slope	Poor displacement, channeling, or fingering
Polymer fill-up (0.13–0.33)	Gradual slope increase	Improving mobility ratio, but still inefficient overall
Tertiary polymer (>0.33 VPI)	Steep slope	Effective displacement after mobility control kicks in
Secondary polymer (2021)	Steep from start, sustained later	Superior sweep and early oil mobilization

Table 12.
Secondary polymer injection versus tertiary polymer injection.

volume of fluid injected (VPI) for two polymer injection pilots (2015 and 2021) under different injection modes (tertiary and secondary).

The light-blue-shaded area corresponds to the secondary polymer injection pilot. By contrast, the green-shaded area illustrates tertiary polymer injection, introduced after a more extensive waterflood. Observing the steepness and timing of each recovery curve offers clear insight into the advantages of earlier polymer injection versus delaying polymer until after significant waterflooding.

Table 12 summarizes the tertiary mode polymer injection versus the secondary mode polymer injection. Secondary polymer injection is more efficient and delivers higher recovery earlier.

Core D (polymer-only injection from the beginning) mirrors the secondary polymer injection in the 2021 pilot; in both cases, it achieved faster and higher initial recovery compared to tertiary polymer cases. Core D achieved 0.844 RF with only 2 PV injected, while Core B needed 13.3 PV to reach 0.87 RF, emphasizing the efficiency of early polymer injection. The secondary polymer pilot curve shows a much steeper early slope, just like Core D had the highest RF per PV injected.

4. Conclusions

These concepts transformed polymer flooding into a flexible, factory-like process, combining hybrid injection design with modular and mobile plants. This work demonstrates how the concept is scaled, turning pilots into a robust operating model with a set of receipts that can be repeated with consistent results. The proposed hybrid polymer injection strategy unites two approaches that are typically considered mutually exclusive: Burst polymer with water injection cycles, and the Polymer-On/Polymer-Off alternation (with Polymer-On corresponding to the burst phase). This approach offers several advantages: reduced facility demands, optimized polymer usage, and the use of mobile, modular units that can be relocated across long distances. It also provides an exit strategy if a zone fails to meet production targets.

Field experience and pore-scale modeling indicate the following:

Polymer dilution after water injection: Injecting polymer after water causes 20–40% dilution due to mixing with existing water. This highlights the need for burst polymer injection to maintain effectiveness.

Polymer viscosity requirement: Downhole polymer viscosity should match or exceed oil viscosity. Field results show that a viscosity around 1.5 times the oil's produces excellent outcomes, though optimal ratios vary by project.

Sweep efficiency and pressure support: High-viscosity polymer maintains pressure behind the displacement front, supports viscous crossflow, and preserves front stability—key for improving sweep efficiency.

Molecular entanglement: Concentrated polymer injections promote entanglement within pores, boosting sweep efficiency while reducing degradation and adsorption. These cooperative effects outperform other fluids of similar viscosity.

Water injection limitations: Both pore-scale simulations and field data show that water injection alone leads to severe viscous fingering and early water breakthrough, causing rapid oil production decline.

Polymer injection benefits: Polymer stabilizes the displacement front, enabling uniform oil displacement, better mobility control, and delayed water breakthrough.

Water-Alternating-Polymer (WAP) advantages: WAP improves oil recovery by reactivating unswept zones and remobilizing bypassed oil. Field-scale results show reduced decline rates after polymer flooding when followed by WAP. This improvement stems from saturation redistribution: high-viscosity polymer pushes oil into uncontacted zones, while follow-up water sweeps surrounding areas. Alternating injections create new flow paths and break up stagnant oil pockets more effectively than continuous injection. Notably, WAP induces complex flow behind the front, resulting in higher recovery than predicted by numerical simulations. The global mobility minimum that develops behind the saturation front induced by alternation creates a favorable displacement zone, enhancing sweep and thereby improving oil recovery.

Cost and operational efficiency: Intermittent high-concentration polymer injections (e.g., >5000 ppm) lower the cost per barrel recovered compared to continuous injection, maximizing impact by targeting the most responsive zones. It is important to note that the mobile plant strategy, with short injection cycles, is naturally aligned with an exit strategy. Because these units are modular and movable, they can deliver early production gains and then relocate or withdraw them without leaving behind stranded capital. In practical terms, it means we can capture value quickly, and when the asset is no longer strategic, we can step out cleanly and with minimal sunk cost. Mobile plants + short cycles = fast value, clean exit, no stranded capital.

4.1 Summary of injection modes

Waterflooding: Fast breakthrough, severe fingering, rapid decline.

Burst polymer injection: Uniform displacement, improved sweep, higher recovery, slower decline.

WAP: Re-engages unswept zones, adds incremental recovery, sustains long-term production, and flattens decline curves.

Visual evidence provided in later sections connects pore-scale displacement efficiency with field-scale recovery (EUR). These findings demonstrate that improvements at the pore level directly translate to higher field recovery, validating the burst-WAP approach. By adopting hysteresis principles and this size of modular scale, innovative injection strategies, and the standardization provided by modular plants, the oil industry can overcome previous limitations and significantly expand the global implementation of polymer flooding at low-risk.

We hope these insights encourage practitioners to adopt intermittent injection strategies that improve oil recovery while reducing energy use and infrastructure requirements.

Acknowledgements

I want to thank Viviana Serrano, Ana Maria Ruiz, Paula Guillen, Damian Llanes, Alejandro Tosi, Cristian Allende, Marcio Armisen, Patricia Pagliero, Gisela Villarroel, Jazmin Propato, Barbara Mrla and Luciano Minor who in different ways have contributed to implementing these ideas and these projects in the field.

Author details

Juan Ernesto Juri
UNC-IFP-School, Heriot-Watt University, Alumnus

*Address all correspondence to: juan.ernesto.juri@gmail.com

References

[1] Carpenter C. Four-injector polymer pilot expands to 80 new injectors using modular concept. Journal of Petroleum Technology. 2020;**72**(06):76-77

[2] Juri JE, Schein F, Pedersen G, Ruiz A, Serrano V, Vazquez P, et al. Evidence that high polymer viscosity accelerates and increases oil response in Grimbeek Manantiales Behr. In: IOR 2021-20th European Symposium on Improved Oil Recovery. Vol. 2021. No. 1. European Association of Geoscientists & Engineers; 2021. pp. 1-14

[3] Llanes JD, Viñales A, Juri JE. Assisted 3D model construction and facies propagation in Golfo San Jorge Basin reservoirs for modelling EOR. In: SPE Improved Oil Recovery Conference. SPE; 2022:D031S027R001

[4] Teraoka I. Polymer Solutions: An Introduction to Physical Properties. Hoboken (NJ): John Wiley & Sons; 2002

[5] Grosberg AY, Khokhlov AR. Introduction to Polymer Physics. New York: Oxford University Press; 1994

[6] Rubinstein M, Colby RH. Polymer Physics. Oxford: Oxford University Press; 2003

[7] de Gennes PG. Scaling Concepts in Polymer Physics. Ithaca (NY): Cornell University Press; 1979

[8] Flory PJ. Principles of Polymer Chemistry. Ithaca (NY): Cornell University Press; 1953

[9] Juri JE, Ruiz A, Pedersen G, Pagliero P, Limeres A, Bernhardt C, et al. Grimbeek—120 cP oil in a multilayer heterogeneous fluvial reservoir. First successful application of polymer flooding at YPF. In: IOR 2017-19th European Symposium on Improved Oil Recovery. Vol. 2017. No. 1. European Association of Geoscientists & Engineers; 2017. pp. 1-20

[10] Serrano V, Campos H, Allende C, Juri JE. Sustainable on-off polymer injection reduces more than 50% CO_2 emissions, materials demand, and water by energy storage in formation compressibility achieving equal recovery factor as continuous polymer injection. In: SPE Improved Oil Recovery Conference. SPE; 2024:D031S027R005

[11] Corbo DN, Lathion R, Games F, Martinuzzi V. High-resolution discrete fracture network application for polymer flooding design in a naturally fractured carbonate. In: SPE Gas & Oil Technology Showcase and Conference. SPE; 2023:D021S029R003

[12] van Dijke MIJ, Juri JE, Sorbie KS. Shortest path algorithm for pore-scale simulation of water-alternating-gas injection. In: ECMOR XI-11th European Conference on the Mathematics of Oil Recovery. European Association of Geoscientists & Engineers; 2008. p. cp-62

[13] Juri JE, van Dijke MIJ, Sorbie KS. Inversion of the lattice network wettability subjected to the capillary pressure of the entire flooding cycle: Hamiltonian Monte Carlo posterior sampling and prediction of the relative permeability. Journal of Petroleum Science and Engineering. 2016;**146**:1037-1062

[14] Juri JE, van Dijke MIJ, Sorbie KS, Masalmeh SK. Pore-network model predictions of multi-phase flow functions for carbonate reservoirs based

on accurate matching of the pore size distribution. In: Computational Methods in Water Resources Conference. 2010

[15] Juri JE, Schein F, Ruiz A, Serrano V, Guillen P, De Miranda V, et al. A faster and higher oil response in Grimbeek led the transition to tertiary recovery in Manantiales Behr. In: First EAGE Online Workshop on EOR in Latin America: Research, Planning, Implementation and Surveillance. Vol. 2020. No. 1. European Association of Geoscientists & Engineers; 2020. pp. 1-6

[16] Juri JE, Schein F, Ruiz A, Serrano V, Guillen P, De Miranda V, et al. A faster and higher oil response in Grimbeek lead the transition to tertiary recovery in Manantiales Behr. In: First EAGE Online Workshop on EOR in Latin America: Research, Planning, Implementation and Surveillance. Vol. 2020. No. 1. European Association of Geoscientists & Engineers; 2020. pp. 1-6

[17] Juri JE, van Dijke MIJ, Sorbie KS, Masalmeh SK. Classic and hybrid MCMC methods to approximate pore size distributions of carbonate reservoirs using pore-network models. In: ECMOR XII-12th European Conference on the Mathematics of Oil Recovery. European Association of Geoscientists & Engineers; 2010. p. cp-163

[18] Juri JE, van Dijke MI, Sorbie KS. Inversion of lattice network structure subjected to carbonate mercury intrusion capillary pressure: Hamiltonian Monte Carlo posterior sampling. Transport in Porous Media. 2015;**106**:73-106

[19] Witten T Jr, Sander LM. Diffusion-limited aggregation, a kinetic critical phenomenon. Physical Review Letters. 1981;**47**:1400

[20] Juri JE, Ruiz A, Serrano V, Guillen P, Thill M, Kichick L, et al. A successful 18% STOOIP 4-injector polymer pilot expands to 80 new injectors in 6 years adopting a modular concept in Grimbeek fluvial reservoirs. In: International Petroleum Technology Conference. Dhahran, Kingdom of Saudi Arabia; 2020

[21] Juri JE, Ruiz A, Schein F, Serrano V, Thill M, Guillen P, et al. Grimbeek successful polymer pilot extends to 80 injectors in factory-mode development at CGSJ basin. In: IOR 2019-20th European Symposium on Improved Oil Recovery. Vol. 2019. No. 1. European Association of Geoscientists & Engineers; 2019. pp. 1-14

[22] Juri JE, Ruiz A, Pedersen G, Pagliero P, Blanco H, Eguia V, et al. Grimbeek2: First successful application polymer flooding in multilayer reservoir at YPF. Interpretation of polymer flooding response. In: SPE Latin America and Caribbean Petroleum Engineering Conference. SPE; 2017:D011S002R006

[23] Juri JE, Ruiz AM, Hernandez MI, Kaminszczik S. Estimation of polymer retention from extended injectivity test. In: SPE Asia Pacific Enhanced Oil Recovery Conference. SPE; 2015:D021S009R001

[24] Juri JE, Schein F, Pedersen G, Ruiz A, Serrano V, Vazquez P, et al. Evidence that high polymer viscosity accelerates and increases oil response in Grimbeek Manantiales Behr. In: IOR 2024. Vol. 2024. No. 1. European Association of Geoscientists & Engineers; 2024. pp. 1-14

[25] Villarroel G, Propato J, Vocaturo G, Mortaloni M, Juri JE, Fernández E, et al. El Alba Valle 200 cp, polymer units' rotation outside Grimbeek in a new heterogeneous, fluvial reservoir. In: Third EAGE Workshop on EOR. Vol. 2024. No. 1. European Association of Geoscientists & Engineers; 2024. pp. 1-7

[26] Serrano V, Campos H, Juri JE, Allende C. Sustainable on-off polymer injection reduces more than 50% CO_2 emissions, materials demand, and water use. In: Third EAGE Workshop on EOR. Vol. 2024. No. 1. European Association of Geoscientists & Engineers; 2024. pp. 1-5

[27] Serrano V, Ojeda N, Tosi A, Guillen P, Hernandez L, Zurletti A, et al. Surfactant-polymer flooding—Coupling the development to a previous mobile-modular polymer development. In: SPE Improved Oil Recovery Conference. SPE; 2024:D041S035R002

[28] Juri JE, Serrano V, Guillen P, Campos H, Zurletti A, Martinez D, et al. Enhanced fill-up response and recovery with secondary polymer injection vs. tertiary polymer injection: A field-scale pilot. In: SPE Improved Oil Recovery Conference. SPE; 2024:D031S023R001

[29] Llanes J, Gonzalez G, Campos H, Minor L, Juri JE. Modular polymer expansion strategy: Full-field 3D reservoir modeling and numerical simulation. In: SPE Argentina Region Young Professionals Symposium. SPE; 2024:D011S002R004

[30] Argentina.gob.ar. Información Estadística: Exploración y Producción de Hidrocarburos [Internet]. Buenos Aires: Ministerio de Economía, Secretaría de Energía; [cited 2025 Oct 6]. Available from: https://www.argentina.gob.ar/economia/energia/planeamiento-energetico/sistema-unificado-de-informacion-energetica/exploracion-y-0

Chapter 9

Perspective Chapter: Repurposing Wells for Geothermal Use – The Role of Geological Conditions on Wellbore Heat Exchanger Performance

Martina Gizzi, Fatemeh Hosseinpour, Federico Vagnon, Glenda Taddia and Stefano Lo Russo

Abstract

Repurposing wells in Italian oilfields presents a significant opportunity for renewable energy production. This chapter investigates the thermal performance of coaxial and U-shaped wellbore heat exchangers (WBHEs) for geothermal energy recovery in decommissioned wells. Three sites with distinct geological characteristics—Villafortuna 1, Tempa Rossa, and Gela 38—were selected to assess the influence of site-specific factors, well depth, and exchanger design on the heat transfer process. Semi-analytical models are implemented in Python codes to evaluate the thermal profiles of the working fluid in both configurations. Results indicate that the geothermal gradient and well depth are key parameters affecting heat exchange performance, with Villafortuna 1 exhibiting the highest outlet temperature due to its elevated geothermal gradient. For specific working fluid's flowrate values, the U-shaped WBHE consistently outperformed the coaxial system, achieving higher outlet temperatures due to an increased heat exchange surface and the inclusion of a horizontal section at maximum depth. While fluid properties and thermal conductivity were confirmed as critical parameters, pressure drop effects were not explicitly considered and warrant further investigation. The approach proposed serves as a useful methodological tool for the preliminary assessment of the feasibility of converting selected hydrocarbon wells; the described insights offer valuable guidance for improving WBHEs design and advancing geothermal energy recovery technologies, supporting the transition toward sustainable energy solutions.

Keywords: renewable energy, geothermal energy, abandoned hydrocarbon well, closed-loop geothermal system, wellbore heat exchanger

1. Introduction

With the cessation of extraction activities, a significant number of hydrocarbon wells are decommissioned and abandoned, often resulting in severe environmental and economic consequences. Precise estimates of the number of decommissioned wells each year are not readily accessible in the literature. Where available, the figures reported by different authors in various sources seem inconsistent. Following what was reported by Zhu et al. [1], about 20–30 million abandoned oil wells exist worldwide. Reference [2] reported that the total number of documented orphaned wells in the United States exceeded 80,000 from June to September 2021 and 122,000 from January to April 2022, representing 2 and 3% of all the estimated abandoned wells in the United States. In Gianoutsos et al. [3], 117,672 orphaned wells in 27 States were documented by the US Geological Survey (USGS) [4, 5]. Basic data on onshore wells in European countries can be obtained from the primary regulatory authority's data centre or website, such as the UK Oil and Gas Authority or the Italian Ministry of Environment and Energy Security. This data includes key essential attributes associated with each well while avoiding the provision of information regarding the status of the infrastructure following abandonment or decommissioning [6, 7].

Addressing the legacy of decommissioned sites is an environmental imperative and an opportunity for innovation and progress. Effective management of these sites can mitigate environmental impacts while contributing to broader sustainability goals. For instance, repurposing abandoned infrastructure can support renewable energy initiatives, including geothermal energy production and carbon capture and storage projects. By integrating decommissioning efforts with the transition to cleaner energy, the industry can demonstrate its commitment to responsible resource management, transforming a historical challenge into a pathway for sustainable development. Geoscience disciplines possess significant potential to contribute to this effort, mainly through their expertise in subsoil characterisation. Rather than being abandoned, unused wells can be repurposed for geothermal energy production and heat storage. This provides a sustainable energy source to support local communities and industries during the energy transition.

Geothermal energy is the heat stored beneath the Earth's surface, while the subsoil's temperature difference is termed a 'Geothermal Gradient'. The potential for its economic use depends on various factors, including the geological and thermophysical conditions, such as the subsoil's temperature, the flow rate, and the degree of mineralisation of geothermal waters in the analysed area. These conditions influence the viability of utilising thermal resources in different geological contexts, offering multiple possibilities for harnessing this renewable energy source. Geothermal energy, electrical, and direct-use technologies are growing worldwide in application, including district heating and/or space heating using heat pumps, commercial treatments, agricultural purposes (e.g., greenhouses), fish farming, and other purposes [8]. The direct use of geothermal energy is becoming increasingly accessible as it relies on existing technologies and straightforward engineering solutions. One of the most significant costs in non-electrical geothermal projects is the drilling and casing of low-temperature wells. However, the abundance of existing wells, infrastructure, and geological data from oilfields presents a unique opportunity for geothermal development. By leveraging these resources, new geothermal projects can reduce capital expenditures, minimise risks, and avoid many challenges associated with starting a new project. This approach not only makes geothermal energy more affordable but also enhances the sustainability and efficiency of these projects [9–11]. Oil companies have the option to convert abandoned wells into geothermal wells by sealing the bottom with insulation and installing a closed-loop pipe system, that is, wellbore heat exchanger (WBHE), applicable to both horizontal and vertical wells [12, 13]. Unlike traditional open-loop geothermal systems, closed-loop systems circulate heat carrier fluids within a WBHE without extracting ground fluids from the surrounding geological formations. This configuration also ensures that working fluids are kept separate from the surrounding geological formations, reducing the potential for corrosion and scaling problems. Given their demonstrated advantages, more studies on WBHE technologies continue to be published in the literature. Pioneering authors like [14–17] have started to explore the potential of coaxial WBHE for geothermal energy extraction. Drawing on the studies published over the last 20 years, the key findings were summarised by Harris et al. [18], along with several other researchers who presented a review of previous scientific work on the application of WBHE in abandoned wells [19–21]. Despite promising results, as shown by Alimonti et al. [22], challenges such as low heat recovery efficiency persist, requiring further research and improvement. Besides, the development of the Eavor Loop design has introduced new challenges in estimating potential outlet temperatures, particularly considering horizontal wells drilled in thermally favourable geological layers. Introduced in 2019, such innovative closed-loop geothermal technology is designed to extract geothermal energy more efficiently than traditional systems. It utilises an underground network of pipes through which a working fluid circulates in a closed circuit, eliminating the need for hydraulic fracturing fluids. Similar to other single-well WBHEs, the primary limitation is that heat is extracted solely through conduction, which limits the heat recovery rate due to the absence of convection. Besides, repurposing long horizontal wells and implementing a U-tube-shaped closed-loop system (U-shaped WBHE) could increase the heat exchange surface area and improve heat recovery [23, 24].

Despite some recent successful theoretical geothermal experiments in oilfields worldwide, specific challenges remain in the large-scale utilisation of geothermal resources in oilfields. The primary factors hindering geothermal energy exploitation in oilfields are lower heat recovery efficiency than conventional geothermal plants, inadequate assessment and planning, and a lack of knowledge about wells' integrity and accessible underground data.

Converting an existing facility into a geothermal plant is challenging, particularly in regions close to urban areas [25]. To perform a preliminary evaluation of the potential for converting a hydrocarbon well to a geothermal closed-loop system located in a specific geological context, it is essential to use analytical and/or semi-analytical tools for pre-assessing productivity, such as determining the maximum working fluid outlet temperature. This chapter builds upon the research conducted over time on coaxial and U-shaped WBHE. After discussing methods that can be used to describe the heat exchange processes, it presents a simplified approach for assessing the feasibility of a selected well to be converted for new geothermal purposes. In detail, it introduces the functions of developed open-access Python codes for the analysis of coaxial and U-shaped WBHE systems. Data from three decommissioned hydrocarbon wells, that is, Villafortuna 1, Tempa Rossa, and Gela 38, in Italy, were used to test such tools. The results, focusing on the output temperature of the selected working fluid, emphasise the significant influence that geological context and technical input parameters concerning the fluid and the geometric setup can have on heat exchange processes.

2. Heat transfer processes in deep reservoirs

In deep hydrocarbon reservoirs, temperature differences between materials result in the movement of thermal energy, known as heat transfer [26]. Whenever a temperature difference exists between wellbores and the surrounding geological formations, energy exchange occurs. The processes of conduction and convection drive energy transfer within wellbores during heat extraction. Conduction refers to heat transfer through direct contact between substances, such as solids or stationary fluids, where a temperature gradient is present. The rate of heat transfer per unit area, q_x, by conduction in the x-dimension (one-dimensional perspective) through a plane wall is described by Eq. (1) [27]:

$$q_x = -\lambda_s A \frac{dT}{dx} \tag{1}$$

The term dT/dx represents the temperature gradient in the x-direction, which indicates heat flow, while the positive constant λ_s denotes the material's thermal conductivity. This relationship is known as Fourier's law of heat conduction, named after the French mathematical physicist Joseph Fourier, who significantly advanced the analysis of conductive heat transfer. The inclusion of a negative sign ensures compliance with the second law of thermodynamics, indicating that heat flows from warmer to cooler areas along the temperature gradient.

Convection refers to heat transfer between two surfaces, mediated by a moving fluid and achieved through molecular interactions. The primary mechanisms for this heat transfer are diffusion and advection, with advection representing the energy transfer resulting from bulk fluid movement in the presence of a temperature gradient. The overall impact of convection is expressed using Newton's law of cooling, described by Eq. (2) [27]:

$$q = hA(T_w - T_\infty) \tag{2}$$

In this context, T_∞ refers to the temperature of the free stream outside the velocity boundary layer, while T_w signifies the surface temperature where convection occurs. The convective heat transfer coefficient, represented by h, is influenced by the geometry of the system, the fluid's thermodynamic properties, the solid medium's thermal characteristics, and the system's boundary conditions. Standard tables and empirical values of h values exist for fluids and flow situations that are commonly encountered. Additionally, for circular tubes, h can be determined through the Nusselt number Nu and correlation functions found in the literature (Eq. (3)):

$$h_f = \frac{Nu\lambda_f}{2r} \tag{3}$$

where $2r$ is the equivalent diameter of the circular tube.

The relationship between conduction and convection in fluid flow through circular tubes, characterised by a temperature difference between the fluid and tube walls, can be approximated using experimentally derived correlations that vary by flow conditions and geometry. The most prevalent correlations for forced convection in circular tubes are designed for turbulent and laminar flow regimes [28].

Understanding how heat transfer modes affect wellbores in oilfields is essential for estimating the exchange rate under specific geological conditions while planning a new geothermal energy project. Additionally, accurately interpreting the physical behaviour of heat transfer in various WBHE configurations is vital for assessing the viability of utilising abandoned wells to exploit geothermal resources in sedimentary basins. The subsequent chapter outlines the coaxial or double-pipe and U-shaped configurations of vertical wellbore heat exchangers utilised for harnessing geothermal energy from decommissioned oilfield boreholes. Following this, pertinent references were presented to the principal models employed to simulate the heat transfer between WBHE pipes and the adjacent subsurface.

3. Closed-loop geothermal energy systems: Coaxial and U-shaped WBHEs

Researchers have recently concentrated on coaxial or double-pipe and U-shaped configurations of vertical WBHE to tap into geothermal energy from abandoned oilfield boreholes. These systems efficiently extract heat from the ground without needing to extract or re-inject geothermal fluids [29]. In coaxial WBHEs, the working fluid is introduced into the outer pipe (the injection pipe), flowing downward to the bottom of the exchanger, where the surrounding rocks heat it. Upon reaching the bottom of the well, the heated fluid ascends through a narrower inner (extraction) pipe. The annular space between the inner pipes is filled with grout, commonly bentonite, and the well's bottom is sealed. Heat transfer occurs between the geological formation and the fluid in the injection pipe and between the fluids in both pipes. Conversely, U-shaped systems function differently. In these systems, the working fluid is pumped down one tube and returned through another. This U-shaped design features two vertical sealed sections and a horizontal wellbore that reaches the target thermal reservoir. The working fluid circulates in the sealed wellbore, moving from the vertical inlet through the horizontal section and into the production well. As it circulates, the working fluid absorbs heat from the thermal reservoir before exiting to supply a power generator or heat exchanger (**Figure 1**).

Research on actual WBHE systems, particularly experimental and pilot studies, remains limited, with most investigations focusing on theoretical modelling. The study conducted by Kolo et al. [21] significantly advances the understanding of the various methods used to model heat flux between WBHE pipes and the nearby subsurface. The chapter reviews analytical tools for heat transfer analysis in WBHEs, along with numerical discretisation techniques that effectively model the heat transport processes in coaxial WBHEs and their surrounding formations. The earliest advancements in integrating horizontal sections with vertical heat exchangers emerged around the middle of the last decade. In Feng et al. [30], researchers proposed using coaxial heat exchanger designs within directionally drilled wells featuring a vertical well paired with a horizontal branch. Specifically, they recommended placing the heat exchangers within a relatively short section of the horizontal well section (ranging from 152.5 to 610 m) to harness the highest temperatures in a geothermal reservoir. In 2020, Toews et al. [24] shared findings from a full-scale demonstration project involving a multilateral closed-loop geothermal system. Situated near Sylvan Lake, Alberta, Canada, the project includes two 2.5 km long multilateral horizontal wellbores linked to two 2.4 km deep vertical wellbores, forming a U-shaped closed-loop geothermal system. The heat exchangers employed in the wellbores depend solely on conductive heat transfer, and a straightforward analytical equation was utilised to estimate the outlet temperature from the horizontal lateral well. This equation follows the method established by Ramey [31] for vertical wellbores, based on the critical assumption that heat transfer is due to quasi-steady-state radial heat conduction from the rock into the fluid moving through the wellbore. Therefore, assumptions and simplifications must be made to avoid complex derivations and calculations when developing such a mathematical model of the WBHE that can be

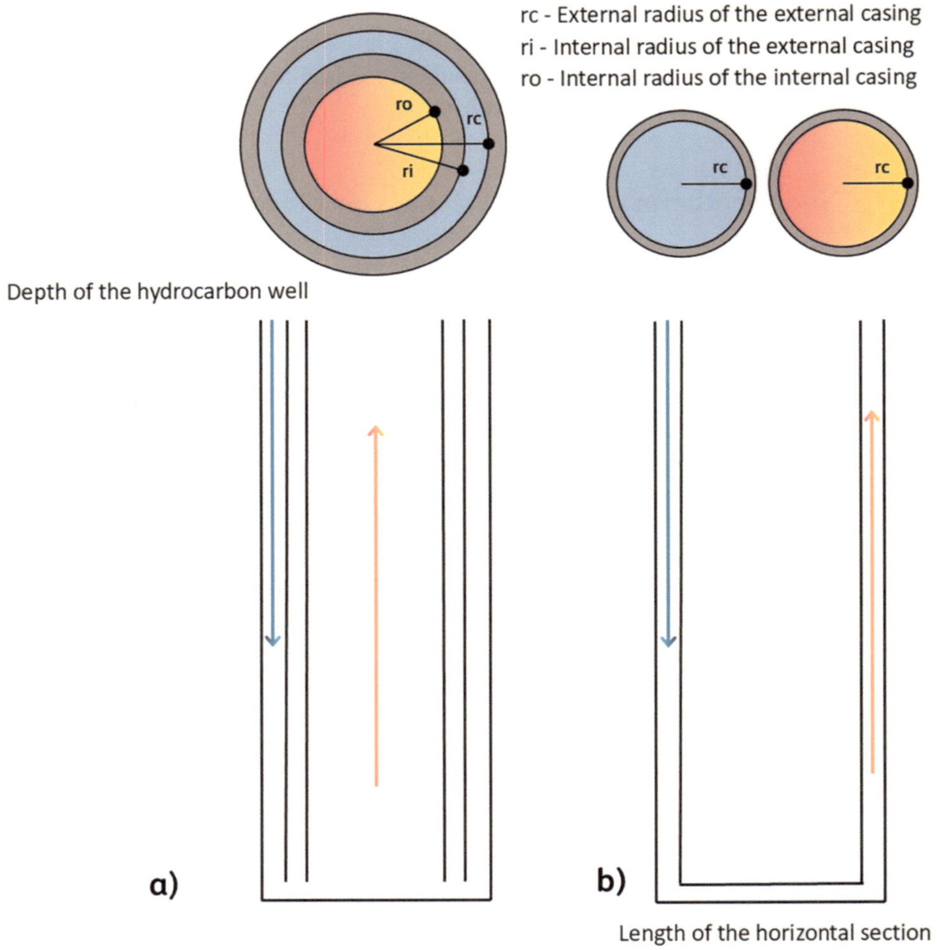

Figure 1.
Simplified configuration of a coaxial (a) and U-shaped WBHE (b).

solved analytically. Despite the limitations, the authors found that it still closely matches the Eavor Lite field data and the more complex numerical models described in References [23, 24]. The main limitation of the proposed analytical model is that it ignores heat transfer in the vertical wellbore segments. Reference [18] proposed an approach for extracting geothermal energy from abandoned oil and gas wells using directionally drilled wells, thereby improving the existing designs by eliminating internal piping and maximising contact with high ground temperatures. The proposed geometry was studied numerically to determine the system's outlet temperatures and heat extraction rates for predicting thermal energy extraction and electrical power production. Reference [32], working on multiple laterals, demonstrated that closed-loop geothermal systems can provide reasonable temperature and heat duty for over 30 years when installed in a suitable geological setting. Using two analytical methods, results indicate that the closed-loop geothermal system is sensitive to reservoir thermal conductivity, which controls the level of outlet temperature and interference between wells over time. In Ref. [33], the two configurations of coaxial and U-shaped exchangers were set up to analyse the performance and evaluate the best operational configuration. As already proposed in Ref. [34], the modelling of the two systems for capturing heat from the subsoil is based on a one-dimensional, quasi-stationary model, which assumes that the heat exchange with the rock system takes place in a radial direction by conduction and varies in time. The heat transfer between the ground and the WBHE is evaluated using a semi-analytical approach based on the thermal resistances of the components of the WBHE described in Ref. [35] and also applied in Ref. [7].

4. Heat transfer in coaxial and U-shaped WBHEs

As reported by Nian and Cheng [36] and Blank et al. [37], the general energy balance of the fluid in the injection pipe of a coaxial WBHE can be expressed in Eq. (4):

$$\frac{\partial\left((\rho c)_f A_0 T_{fo}\right)}{\partial\tau} + \frac{\partial\left((\rho c)_f A_0 v_f T_{fo}\right)}{\partial z} = -\frac{dQ}{dz} + \frac{dQ_{i0}}{dz} \tag{4}$$

where T_{fo} refers to the fluid temperature in the outer pipe, A_0 denotes the outer pipe area, and v_f indicates the fluid velocity. The term *dQ/dz* is the heat extraction from the formation per unit well depth (W/m). Furthermore, *dQio/dz* defines the heat flux from the inner pipe to the outer pipe; despite insulation being employed to minimise heat loss from the inner pipe fluid, some heat can transfers between the two pipes. Consequently, the energy equation for the inner tube can be represented by the following Eq. (5):

$$\frac{\partial\left((\rho c)_f A_i T_{fi}\right)}{\partial\tau} + \frac{\partial\left((\rho c)_f A_i v_f T_{fi}\right)}{\partial z} = -\frac{dQ_{i0}}{dz} \tag{5}$$

By assuming steady heat transfer and constant heat flux in wellbore components (insulation, casing, grout), heat extraction from the formation is assumed to equal the heat flux at the wellbore's outer surface (the boundary between the wellbore and rock formation) to the injected fluid (Eq. (6)):

$$\frac{dQ}{dz} = 2\pi r_w k_w \left(T_{f0} - T_w\right) = \left(T_{f0} - T_w\right)/R_w \tag{6}$$

where T_w is the temperature at the interface of the wellbore/formation, r_w is the radius of wellbore outside, k_w is the heat transfer coefficient between outer pipe fluid and wellbore exterior, R_w is the thermal resistance between the outer pipe and surrounding rocks.

At the bottom of the well, the heated fluid enters and moves through the internal pipe of the coaxial WBHE. As it ascends to the wellhead, heat is only transferred through the wall of the inner pipe. Therefore, *dQio/dz* is calculated by factoring in the temperature difference between the fluids in the outer and inner pipes and the estimated thermal resistance of the insulation (Eq. (7)):

$$\frac{dQ_{i0}}{dz} = 2\pi \mathrm{r}_0 k_{i0}\left(T_{fi} - T_{f0}\right) = \left(T_{fi} - T_{f0}\right)/R_{i0} \tag{7}$$

where T_{fi} is the fluid temperature in the inner pipe, k_{io} is the heat transfer coefficient between the outer pipe and inner pipe, and R_{io} is the thermal resistance between the outer pipe and inner pipe.

4.1 Coaxial WBHE: Coefficient of heat exchange between the outer-pipe fluid and the wellbore exterior

An accurate estimation of the parameter k_w is crucial for assessing heat exchange between the fluid in the outer pipe and the geological formations. In the case of coaxial WBHE, the heat exchange coefficient of the injection pipe can be represented as the sum of heat transfer components based on thermal resistance values *(R_w)* (Eq. (7)) [35, 36]:

$$R_w = R_s + R_a + R_c \tag{8}$$

R_s, a time-dependent function, reflects the thermal resistance from conductive heat transfer in the rock, while R_a indicates the thermal resistance caused by convection into the pipe. R_c represents the thermal resistance from conductive heat transfer through the well's casing. The conductive term prevails in evaluating the total thermal resistance; consequently, the thermal exchange is directly proportional to the convective transfer coefficient.

Conductive thermal resistance *(R_s)* can be expressed as follows (Eq. (9)):

$$R_s = \frac{1}{2\lambda_s} In \frac{2\sqrt{\alpha_s t}}{r_w} \tag{9}$$

where λ_s (W/mK) is the thermal conductivity of the rock and, α_s (m/s) is the thermal diffusivity of the rock, r_w is the radius of the outside wellbore (**Table 1**). In Eq. (9), the numerator of the argument of the natural logarithm represents the time-dependent radius of the thermal influence of the well (r_s). This parameter considers the change, over time, of the heat flux into WBHE surrounding geological formations.

Convective thermal resistance *(R_a)* can be determined by the following equation (Eq. (10)):

Coaxial wellbore heat exchanger-geometric parameters	Symbol	Unit of measure
Outer pipe area	A_O	[m^2]
Inner pipe area	A_i	[m^2]
Radius of outside wellbore	r_w	[mm]
The external radius of the external casing	r_c	[mm]
The internal radius of the external casing	r_i	[mm]
The internal radius of the internal casing	r_O	[mm]
Thicknesses of the pipe exchanger	d	[mm]
Depth	z	[m]

Table 1.
Coaxial WBHE–geometric parameters.

$$R_a = \frac{1}{2r_c h_f} \tag{10}$$

where r_c is the external radius of the external casing, h_f is the convective heat transfer coefficient, calculated using Eq. (3) and by using the Nusselt number (Nu) and the form of the Dittus-Boelter equation assuming turbulent flow inside the tubes (Reynolds number ≥ 104) [28]:

$$Nu = 0.023\, Re^{\,0.8} Pr^{0.4} \tag{11}$$

where $Pr = \frac{\rho c_f \mu}{\lambda_f}$ and $Re = \frac{\rho v_f 2r_c}{\mu}$.

Finally, thermal resistance to heat conduction through the casings of the well (R_c) can be determined as follows:

$$R_c = \sum_{i=1}^{n} R_{\lambda i} = \frac{1}{2} \sum_{i=1}^{n} \frac{1}{\lambda_i} In \frac{r_{c,i+1}}{r_{c,i}} \tag{12}$$

where λ_i is the thermal conductivity of the rock in correspondence with the different casings of the well. Generally, due to the high thermal conductivity of the steel piping, the total thermal resistance of the casing is negligible compared with the rock's thermal resistance.

As a result, the heat exchange coefficient k_w can be determined as follows [38, 39]:

$$\frac{1}{k_w} = \frac{2r_c}{2\lambda_s} In \frac{4\sqrt{a_s t}}{2r_w} + \frac{1}{h_f} \tag{13}$$

where $r_c = r_w$ as the thickness of the external tube is negligible.

4.2 Coaxial WBHE: Coefficient of the heat exchange between the outer-pipe fluid and the inner pipe

Unlike in the injection pipe, the total heat flux in the upward pipe (extraction pipe) is determined by a conductive component of the composite pipe and by two convective components: one on the internal wall and one on the external wall of the WBHE. Consequently, the total heat exchange coefficient k_{iO} for the extraction pipe can be calculated as follows (Eq. (14)):

$$\frac{1}{k_{i0}} = \frac{r_0}{r_{0+d}} \frac{1}{h_i} + r_0 \sum_{i=1}^{n} \frac{1}{\lambda_i} In \left(\frac{r_{i+1}}{r_i} \right) + \frac{1}{h_0} \tag{14}$$

where r_O is the radius of the internal casing, r_i the internal radius of the external casing, d is the thicknesses of the pipe exchanger, h_O and h_i are the coefficients of convective heat transfer to the inner and outer wall, respectively, and λ_i is the thermal conductivity of the pipe material (air and steel) (**Table 1**).

4.3 U-shaped WBHE: Coefficient of heat exchange between pipe fluid and the wellbore exterior

As outlined in Chapter 3, a U-shaped WBHE involves pumping fluid through one tube while it exits *via* another. The U-shaped configuration includes two vertical sealed sections and a horizontal wellbore that accesses the targeted thermal reservoir at the maximum well's depth. The above-mentioned vertical sections can be depicted as two hydrocarbon wells spaced apart, as detailed in Ref. [20]. The working fluid flows within the sealed wellbore, travelling from the vertical inlet through the horizontal segment to the production well. During this flow, it gathers and loses heat from the thermal reservoir along its path, according to the subsoil's temperature conditions at different depths. Consequently, in describing the heat extraction from the formation, it is possible to assume that it is equal to the heat flux at the wellbore's outer surface (the boundary between the wellbore and rock formation) to the injected fluid for both the vertical and horizontal sections. Furthermore, the heat flux from the inner pipe to the outer pipe, as described for coaxial WHBE in Eq. (4), has not been defined. As a result, the heat exchange coefficient k_w defined in Eq. (13) can be applied to assess heat exchange between the fluid in the pipes and the geological formations for a U-shaped WBHE.

4.4 Python code implementation and input parameters definition

The previously described methods can be used to determine the preliminary temperature profiles of a selected working fluid in a specific wellbore heat exchanger. This calculation process can be automated with computational tools built using C or Python programming languages. Python, developed by Guido van Rossum in 1991 [40], is an object-oriented, high-level language which was utilised to model the analysed systems of this study. The available code versions allow for the replication and application of these analyses in new and varied research contexts. Once the input parameters are established, the temperature profile of the carrier fluid can be obtained. Below, in **Table 2** and **Figure 2**, a list of the necessary input parameters and a graphical representation of the Python code for the discussed WBHEs are proposed. In selecting input parameters and applying the above-mentioned codes, primary considerations and assumptions must be considered when modelling the coaxial and U-shaped profiles using the selected models:

- Geological Model (GM): The geological model must be defined. Since a stratigraphic variation with depth is observed, it is necessary to proceed with characterising the thermal properties of the specific geological formations involved. The horizontal section of the U-shaped WHBE will affect the geological formation it crosses. The thermal properties defined for such formations will influence the heat exchange processes.

- Geothermal Gradient (GG): Since a geothermal gradient condition characterises each specific geological context, it must be appropriately defined for the ongoing analysis. The temperature at depth Z was estimated based on the local geothermal gradient, using the following Eq. (15), where AT and B factors are defined in **Table 2**:

$$T_a(Z) = B + AT \cdot Z \tag{15}$$

- One of the sets of configurations proposed by (35) was adopted for the definition of geometric parameters related to the coaxial heat exchanger. Further studies and configuration proposals are available in References [41, 42]. The diameter size significantly influences fluid velocity, thus allowing for additional analyses that consider different configurations that can be carried out.

Input parameters	Symbol	Unit of measure	Value
Wellbore heat exchanger			
Depth	z	m	6202
Length of the horizontal section	l	m	1000
The external radius of the external casing	r_c	mm	75.0
The internal radius of the external casing	r_i	mm	69.5
The internal radius of the internal casing	r_0	mm	38.5
Thermal conductivity of the pipe material	λ_m	$W\ m^{-1}\ K^{-1}$	15
Thermal conductivity of the insulation	λ_i	$W\ m^{-1}\ K^{-1}$	0.025
Working fluid – water			
Flow rate	q	$Kg\ s^{-1}$	3–1.5
Inlet temperature	T	°C	15
Volumetric heat capacity of the fluid	ρc_f	$J\ kg^{-1}\ K^{-1}$	4215.45
Thermal conductivity of the fluid	λ_f	$W\ m^{-1}\ K^{-1}$	0.68
Density	ρ_f	$Kg\ m^{-3}$	958.40
Viscosity	μ	Pas	2.82e−4
Simulation time			
Time stop	t	Year	3
Geological formations			
Thermal conductivity of the rock	λ_s	$W\ m^{-1}\ K^{-1}$	GM
Volumetric heat capacity of the rock	ρc_s	$J\ kg^{-1}\ K^{-1}$	GM
Density	ρ_s	$Kg\ m^{-3}$	GM
Geothermal gradient			
Soil temperature profile slope	AT		GG
Soil temperature profile intercept	B	°C	CG

Table 2.
WBHEs–input parameters.

- Heat transfer mode: Heat moves through conduction in the reservoir, while conduction and convection occur within the wellbore tubes. The model presumes steady-state conditions and initial thermal equilibrium, ignoring time-dependent temperature variations; each point within the tubes consistently maintains the same temperature throughout the system's operation.

- Temperature distribution: The temperature profile is assumed to remain uniform in the radial direction. Turbulent flow within the tubes, described using (11), promotes enhanced mixing, which reduces the radial gradient. Therefore, temperature changes are only observed in the vertical direction of the upward and downward tubes, leading to a unidirectional temperature profile.

- Carrier fluid: The properties of the chosen carrier fluid, for example, water, are assumed to be constant. No variations arise from pressure or temperature gradients throughout the process.

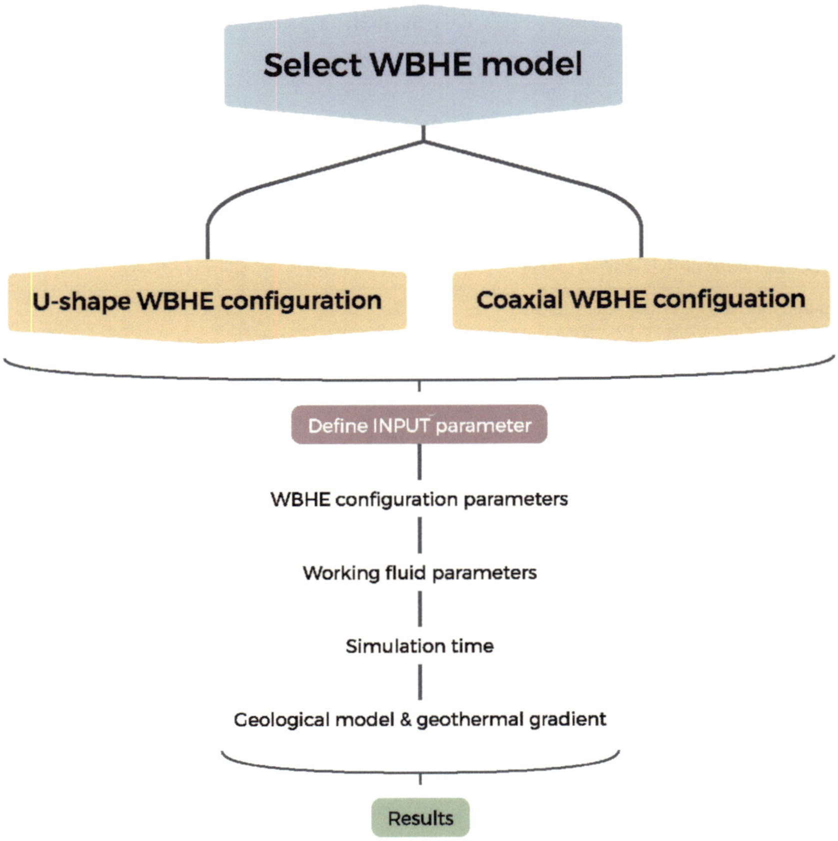

Figure 2.
Simplified structure of the developed Python code.

- Thermal Resistance: The model considers the thermal resistance from pipe thickness negligible, given that the tube material is assumed to have a high conductivity value. Therefore, its resistance is considered minor compared to other resistances in the system.

5. Case studies and applications

As mentioned in the introduction chapter, data on onshore wells in European countries can be accessed through the primary regulatory authority's data centre or website. The National Mining Office of the Italian Ministry for Economic Development (MISE) and the VIGOUR project's website provide technical information regarding productive and decommissioned hydrocarbon wells in Italy. By analysing the data available at the end of 2024 on the official website of the Italian Ministry for Economic Development (MISE), one can obtain information about 1554 wells; 43.8% of these are producing, while the remaining 54.8% are non-producing wells (**Figure 3**) [43].

Furthermore, lithostratigraphic units and temperature data visualisation can be accessed through the Italian National Geothermal Database (BDNG), established in the 1980s and managed by the Institute of Geosciences and Earth Resources (IGG) of

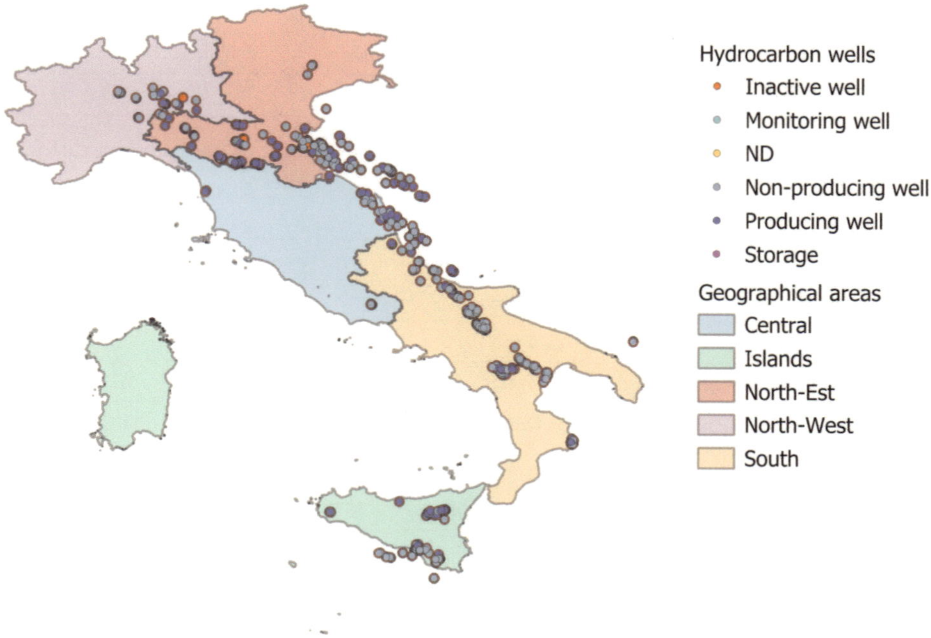

Figure 3.
Hydrocarbon wells distribution in Italy at the end of 2024.

the National Research Council (CNR) of Italy. In addition, stratigraphic profiles sourced from GEOTHOPICA, the Italian geothermal data infrastructure portal launched in 2007 by CNR-IGG, can provide extensive lithological information from well logs [44]. Beginning with temperature and stratigraphic profiles, the thermophysical properties of geological formations can be characterised, thereby defining the geological model (GM) and geothermal gradient (GG) for each selected case study.

Considering the specific geographical and geological context, it is necessary to obtain measurements and/or bibliographically available information for the characterisation of the thermal parameters of the geological formations involved. To determine heat conductivity values at the regional and local scales of Italian sedimentary basins' geological formations, Di Sipio et al. [45] investigated the thermal conductivity of 200 rock samples collected from four different regions of Southern Italy (Calabria, Campania, Apulia, and Sicily), measuring in both dry and wet conditions. Moreover, Pasquale et al. [46] utilised the framework of the MIUR–2008 project 'Geothermal resources of the Mesozoic basement of the Po Basin: groundwater flow and heat transport' to accurately estimate the thermophysical properties of a wide variety of sedimentary and volcanic rocks from the Po Basin in the Northern part of Italy, through laboratory measurements of density and porosity. In 2020, Dalla Santa et al. [47] included a new database of thermal properties for rocks and unconsolidated sediments in their work. As already reported in Reference [7], the values defined by the above-mentioned authors were used to determine the thermophysical parameters of the geological formations and the geological model associated with each of the selected case studies (**Tables 3–5**). Furthermore, the temperature profiles available for download from the Geothopica geoportal were used to establish local geothermal gradient conditions.

State	Productive, not supplying well			
Location	Piemonte Region			
Field	VILLAFORTUNA			
Depth	Lithostratigraphic formation			
		λ_s	ρc_s	ρ
m		W/mK	J/kg/K	kg/m^3
609	Terrigenous sedimentary deposits	0.30	800	1700
1258	Sand (dry)	0.30	800	1700
1405	Clay Sand	1.61	1696	1890
5493	Clastic sedimentary rocks (Sandstone, Conglomerates and Silty Marl)	3.16	821.11	2359
6202	Carbonate rocks -Calcarenite/Dolostone	3.50	810.48	2480

Table 3.
Villafortuna 1 hydrocarbon well-lithostratigraphic profile [7].

State	Productive, not supplying well			
Location	Basilicata Region			
Field	GORGOGLIONE			
Depth	Lithostratigraphic formation			
		λ_s	ρc_s	ρ
m		W/mK	J/kg/K	kg/m^3
23	Superficial sedimentary deposits	0.30	800	1700
2912	Sandstones interspersed with shale and clays	3.00	808.6	2330
5042	Clays, argillites and calcarenites	2.34	829.4	1917

Table 4.
Tempa Rossa 1D hydrocarbon well-lithostratigraphic profile [7].

State	Productive well			
Location	Sicilia Region			
Field	GELA TERRA			
Depth	Lithostratigraphic formation			
		λ_s	ρc_s	ρ
m		W/mK	J/kg/K	kg/m^3
1632	Terrigenous sedimentary deposits—Sandy Clay	1.61	1696	1890
5451	Calcareous Marl	2.17	830	1801
6189	Carbonate rocks–calcarenite/dolostone	3.50	810.48	2480
6282	Clastic sedimentary rocks—argillaceous sandstone	3.00	821.11	2330

Table 5.
Gela 38 hydrocarbon well-lithostratigraphic profile [7].

The Villafortuna-Trecate field is situated in the Piemonte Region (Novara province, North-West geographic area). This hydrocarbon system represents one of the most significant oil accumulations within the Italian Middle Triassic carbonate petroleum system. The petroleum system has developed within the Triassic geological succession. Generally, the primary reservoir linked to the Villafortuna-Trecate field is identified to be at a depth ranging from 5800 to 6100 m, with a temperature of approximately 160–170°C [48]. As evidenced by information regarding lithostratigraphic units presented in **Table 3**, the stratigraphic sequence associated with Villafortuna 1 well consists predominantly of clastic sedimentary and carbonate rocks. The well under examination reaches a maximum depth of 6202 m (**Table 3**).

The Tempa Rossa 1D hydrocarbon well is situated within the Tempa Rossa oilfield in the Basilicata Region, in the southern area. This system lies within the Mesozoic carbonate substratum of the foredeep/foreland area and the external thrust belt of the southern Apennines. It hosts Italy's most significant oil and gas accumulations, namely the Val d'Agri and Tempa Rossa oilfields. The reservoirs are characterised by fractured limestones of the buried Apulia Platform, which extend in time from the Cretaceous to the Miocene [49]. Unlike Villafortuna 1, lithostratigraphic units are mainly composed of sandstone with associated shales (**Table 4**). The maximum depth reached by the analysed well is 5042 m.

The examined case study of the Gela 38 hydrocarbon well (Gela oilfield, Sicily region, Islands) pertains to the Gela field, a Late Triassic–Early Jurassic petroleum system linked to the main phase of Tethyan rifting. The reservoir consists of fractured, massive dolomites from the Upper Triassic Gela Formation [50]. Based on the available information regarding the lithological and temperature data, the stratigraphic succession in the area consists of marl, calcareous marl, and clays. The maximum depth reached by the analysed well is 3446 m (**Table 5**).

6. Results

6.1 Coaxial and U-shaped WBHEs: Output temperature analysis

The temperature profiles for the WBHEs were derived from specific geological properties obtained from selected case studies' geological models. Thermal properties for various rock formations were adopted based on those provided in **Tables 3–5**. According to the information reported in **Table 2**, among the input parameters, the inlet flow rate of the working fluid (water) and its temperature were assumed to be 3.0 kg/s and 15°C, respectively. The analysis of the coaxial WBHE performance across the three hydrocarbon wells reveals significant differences in performance, primarily driven by the geological characteristics of each site. The efficiency of heat exchange in this configuration relies on the geothermal gradient, and the thermal conductivity of the surrounding rock formations. The outlet temperatures at the surface for the three sites are as follows (**Figure 4**).

Among the three sites, Villafortuna 1 exhibits the highest thermal performance, with an outlet temperature of 89.93°C. This result suggests that the site benefits from a higher well's depth and geothermal gradient, enabling the fluid to absorb substantial heat as it descends through the wellbore. The steep temperature increase indicates effective heat transfer from the ground, making this location suitable for geothermal energy extraction.

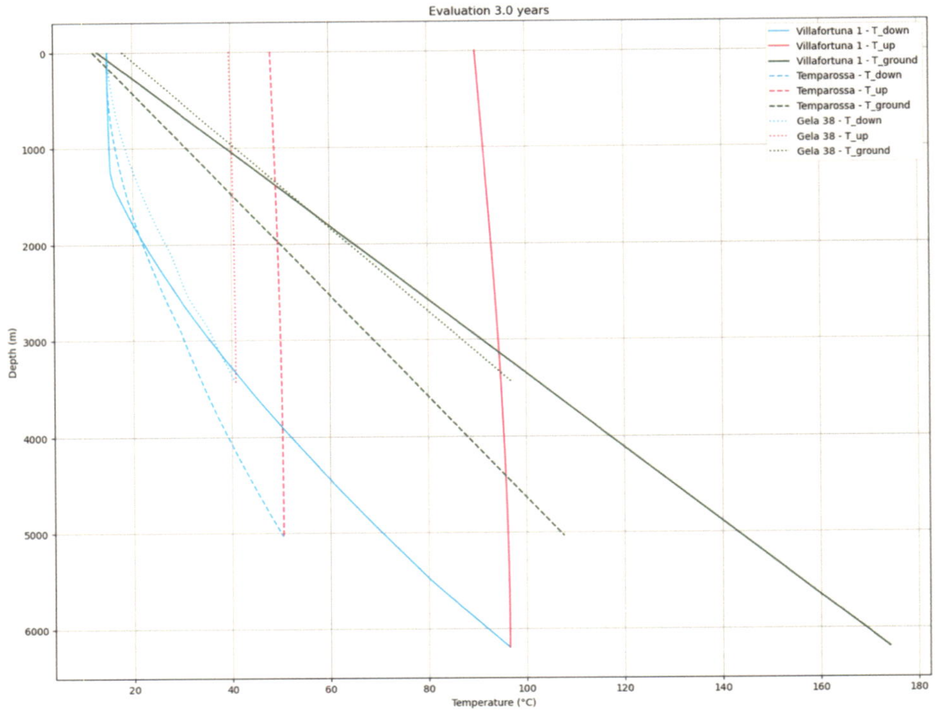

Figure 4.
Temperature distribution of the working fluid in the coaxial WBHE (flow rate: 3 kg/s).

Tempa Rossa presents intermediate thermal performance, with an outlet temperature of 48.20°C. The lower-temperature rise compared to Villafortuna 1 suggests this site has a less favourable geothermal gradient or lower thermal conductivity in the surrounding geological formations. Consequently, the working fluid absorbs less heat from the ground, resulting in a more moderate increase in temperature upon returning to the surface.

Gela 38 exhibits the least thermal enhancement, with an outlet temperature of 39.79°C. Although the initial ground temperature is higher than that at the other two sites (18°C), the overall temperature gain during fluid circulation remains limited. As a result, this well has the least potential for geothermal energy recovery among the three.

The analysis of the U-shaped WBHE across the three hydrocarbon wells highlights greater efficiency in heat exchange compared to the coaxial WBHE. This improvement is likely attributed to the increased contact area between the fluid and the surrounding rock formations, along with the enhanced heat transfer mechanisms inherent in the U-shaped configuration. The outlet temperatures at the surface for the three sites are as follows (**Figure 5**).

Villafortuna 1 exhibits the highest thermal performance, with an outlet temperature of 99.75°C. This result is significantly higher than that of the coaxial WBHE (89.93°C) and suggests that the U-shaped configuration optimises heat absorption from the ground. The improved heat transfer efficiency of this design enables the working fluid to attain higher temperatures, further enhancing its potential for geothermal energy applications.

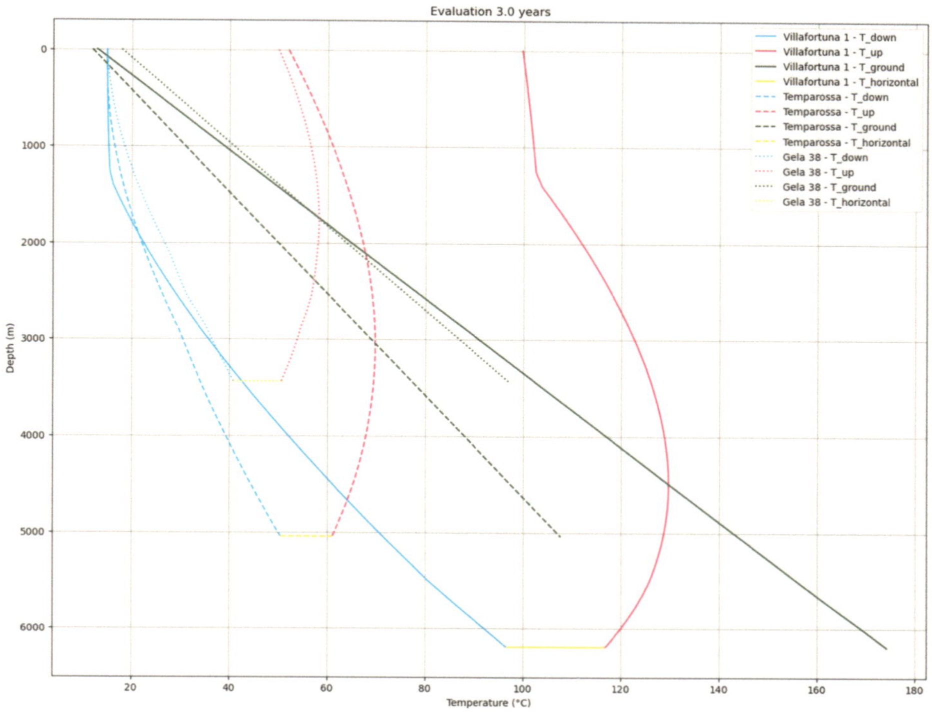

Figure 5.
Temperature distribution of the working fluid in the U-shaped WBHE (flow rate: 3 kg/s).

Tempa Rossa demonstrates an intermediate thermal performance, with an outlet temperature of 51.99°C. In comparison to the coaxial WBHE (48.20°C), the U-shaped design provides a moderate increase in thermal efficiency. The enhancement in heat exchange is likely due to improved thermal interaction between the fluid and the ground. However, the geothermal gradient at this site remains lower than that at Villafortuna 1, thus limiting the maximum achievable temperature.

Gela 38 shows the most significant relative improvement in heat exchange efficiency, with an outlet temperature of 49.99°C compared to 39.79°C in the coaxial WBHE.

The inclusion of a horizontal section at maximum depth in the U-shaped WBHE configuration plays a role in enhancing heat exchange efficiency. This section allows the working fluid to remain in contact with the hottest subsurface layers for a more extended period, increasing its thermal absorption before it begins its ascent to the surface. The thermal contribution of the horizontal section is evident when comparing the outlet temperatures of the U-shaped and coaxial WBHE systems (**Table 6**). The most significant improvements are observed in Villafortuna 1 and Gela 38, where the temperature at the outlet increased by 9.82°C and 10.20°C, respectively, compared to the coaxial system. This indicates that in sites with a high geothermal gradient at depth, the horizontal section enhances the heat transfer process. In contrast, Tempa Rossa exhibits a more modest improvement of 3.79°C, suggesting that when the geothermal gradient is weaker, the additional horizontal section has a less pronounced effect on heat absorption. Furthermore, the effectiveness of this configuration is highly dependent on the thermal conductivity of the subsurface materials; if the rock formations have poor heat transfer properties, the additional section may not yield significant advantages. Another consideration is the increase in system complexity

Site	Horizontal temperature	Temperature difference (U-shaped—coaxial)
Villafortuna 1	116.73°C	*+9.82°C* (99.75–89.93°C)
Tempa Rossa	60.99°C	*+3.79°C* (51.99–48.20°C)
Gela 38	50.56°C	*+10.20°C* (49.99–39.79°C)

Table 6.
Thermal comparison of U-shaped and coaxial configurations.

and potential installation costs, as incorporating a horizontal section necessitates more space and a more intricate design compared to simpler wellbore heat exchanger configurations.

6.2 Influence of flow rate reduction on geothermal heat recovery

Operational parameters play a significant role in determining the efficiency of the heat exchange process. Among these parameters, the inlet flow rate stands out as one of the most influential factors. Thermal profiles were recalculated by reducing the inlet flow rate from 3 kg/s to 1.5 kg/s. This reduction in flow rate was intended to increase the fluid's residence time in the system, thereby enhancing heat absorption from the surrounding rock formations.

The results varied according to the geological characteristics of each site. Among the coaxial WBHE configurations, Villafortuna 1 exhibited the most remarkable improvement, with its outlet temperature rising to 122.56°C from the original

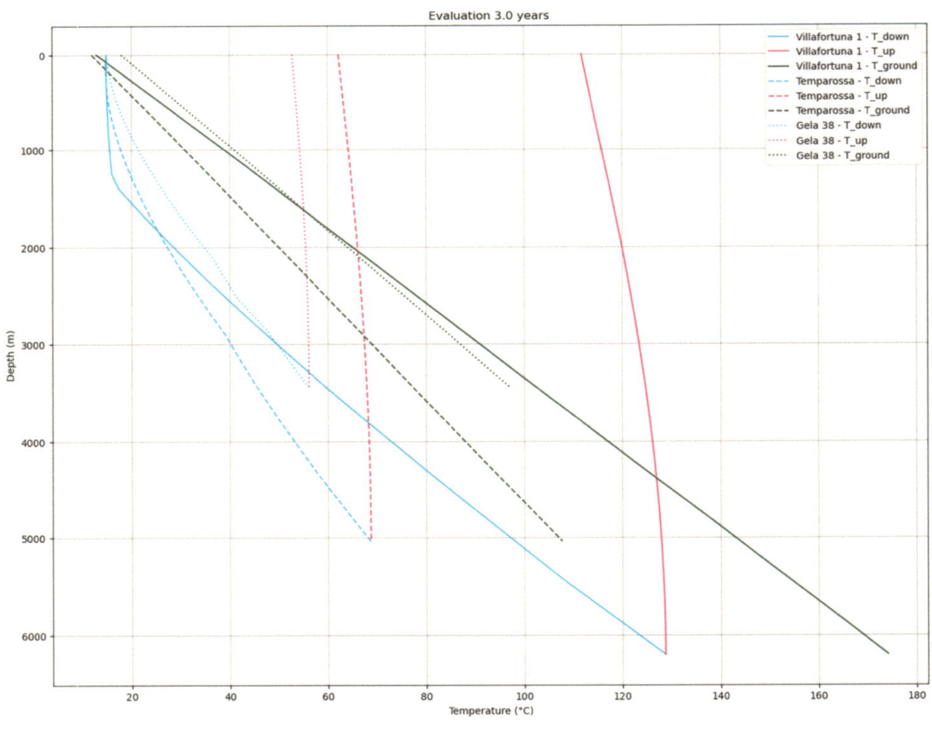

Figure 6.
Working fluid temperature distribution in the coaxial WBHE (flowrate: 1.5 kg/s).

89.93°C (**Figure 6**). The extended residence time of the fluid, resulting from the reduced flow rate, allowed for enhanced heat absorption from the surrounding rock, maximising the system's thermal efficiency. Tempa Rossa also experienced a significant boost in performance, with the outlet temperature increasing from 48.20 to 66.43°C. Although this improvement is considerable, the site's lower thermal conductivity and moderate geothermal gradient limited the overall gain compared to Villafortuna 1. Gela 38 showed the least improvement, with its outlet temperature rising from 39.79 to 54.99°C. While the reduced flow rate improved heat recovery, further improvements—such as enhanced insulation or extended well depth—may be necessary to achieve greater efficiency at this location.

In the U-shaped WBHE configuration, notable differences compared to the coaxial WBHE were observed (**Figure 7**). Villafortuna 1 previously demonstrated an outlet temperature of 99.75°C, outperforming its coaxial counterpart (89.930°C). Following the modification of the inlet flow rate, the outlet temperature in the U-shaped configuration now reaches 85.51°C. The reduction may reflect the trade-off between increased heat absorption efficiency and potential cooling effects induced by reduced fluid velocity. Tempa Rossa previously recorded an outlet temperature of 51.99°C in the U-shaped configuration, slightly outperforming the coaxial WBHE result of 48.20°C. After adjusting the flow rate, the outlet temperature now stands at 44.14°C. Gela 38 previously demonstrated an outlet temperature of 49.99°C in the U-shaped configuration, significantly better than the coaxial WBHE's 39.79°C. After modifying the flow rate, the outlet temperature now reaches 49.13°C.

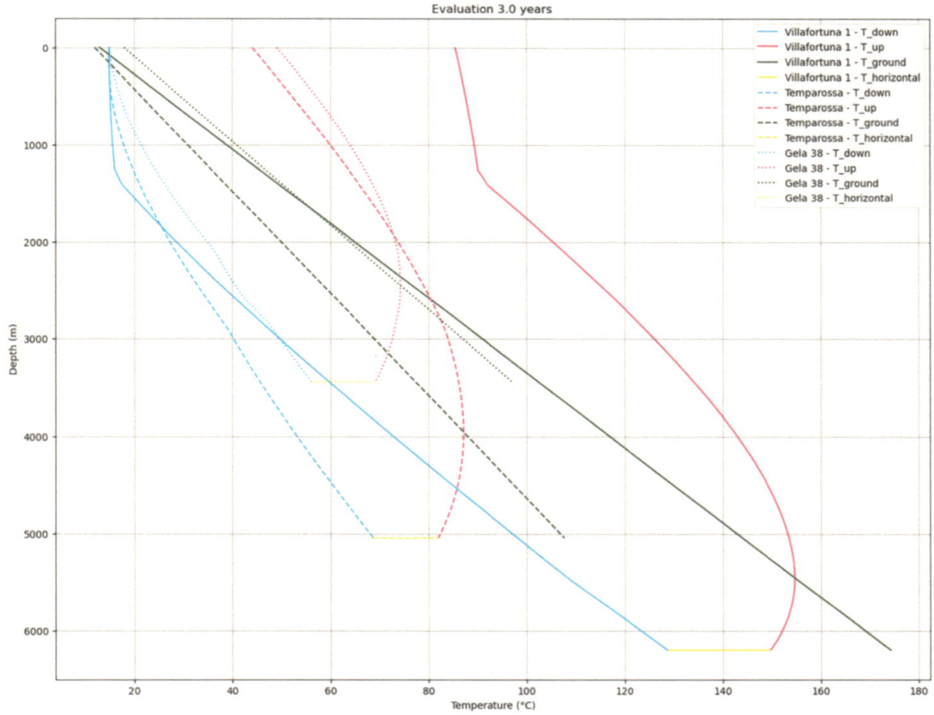

Figure 7.
Working fluid temperature distribution in the U-shaped WBHE flowrate: 1.5 kg/s).

7. Discussion

This study investigated the thermal performance of coaxial and U-shaped WBHEs in three hydrocarbon wells with distinct geological characteristics. The results demonstrated that site-specific geothermal properties, well depth, and exchanger design significantly influence heat exchange efficiency and the final outlet temperature of the working fluid. The geothermal gradient emerged as a primary factor affecting heat transfer performance. Villafortuna 1, characterised by the highest gradient among the three sites, exhibited the most substantial increase in outlet temperature in both configurations, confirming that greater geothermal energy availability enhances heat extraction. In contrast, Tempa Rossa showed more limited heat transfer efficiency, suggesting that the conductive properties of the surrounding formations play a crucial role in determining WBHE effectiveness. Gela 38, despite its relatively higher surface temperature, exhibited a lower increase in fluid temperature, indicating that the geothermal gradient at depth was insufficient to drive significant thermal gains. The comparison between coaxial and U-shaped WBHEs revealed that the U-shaped configuration consistently outperformed the coaxial system at all three sites. The key factors contributing to this enhanced performance include a larger heat exchange surface area, allowing for more significant thermal interaction between the fluid and the surrounding geological formations. The addition of a horizontal section at maximum depth, which extended the exposure time of the fluid to the highest temperature layers, increasing its thermal absorption capacity. The effectiveness of the horizontal section was particularly evident in Villafortuna 1 and Gela 38, where the increase in outlet temperature exceeded 9–10°C compared to the coaxial system. These results suggest that the horizontal section is most effective in sites with high geothermal energy availability, whereas its impact is reduced in lower-temperature environments. Additionally, the depth of the well was shown to influence overall heat exchange efficiency, as deeper wells generally resulted in higher outlet temperatures due to prolonged contact with heated geological layers. However, this effect was not linear, indicating that factors such as thermal conductivity, fluid properties, and flow dynamics contribute to overall system performance. Finally, the flow rate of the inlet working fluid played an important role in heat transfer efficiency in coaxial WBHEs. A critical factor to consider in the U-shaped WBHE configuration is the pronounced thermal dissipation effect along the fluid's upward path. This phenomenon becomes particularly relevant when fluid velocity is reduced, as the increased residence time in the riser pipe, which is exposed to the surrounding ground, can lead to significant heat loss. Such dissipation diminishes the system's overall efficiency and limits the benefits gained from reducing the inlet flow rate, especially in sites with moderate geothermal gradients or lower thermal conductivity in the surrounding formations. A fluid with a higher specific heat capacity absorbed and retained more thermal energy, while the viscosity of the fluid affected flow resistance and, consequently, overall thermal performance. Future research could further explore the optimisation of fluid properties to maximise energy extraction, particularly in low-enthalpy geothermal environments.

One crucial factor that was not explicitly addressed in this study is the impact of pressure drop on WBHE performance. Pressure losses can significantly influence fluid flow dynamics, heat exchange efficiency, and system energy consumption. This effect is particularly critical in U-shaped WBHE systems, where the extended horizontal section and longer fluid path increase the likelihood of pressure-related inefficiencies. Future research should integrate pressure drop analysis into WBHE performance models to better estimate real-world thermal outcomes and energy demands. Identifying optimal flow rates, enhancing insulation, and exploring alternative working fluids with lower viscosity may help mitigate pressure losses and improve overall system efficiency.

8. Conclusions

The findings of this study provide valuable insights for optimising WBHE systems, with practical implications for retrofitting hydrocarbon wells into geothermal energy recovery facilities. Besides, the simplified analytical approach and the associated model proposed serve as a useful methodological tool for the preliminary assessment of the feasibility of converting selected hydrocarbon wells into geothermal wells using WBHEs.

The results confirm that the geothermal gradient and well depth are the most critical factors influencing heat exchange performance, with higher gradients yielding more efficient heat extraction. The U-shaped WBHE configuration outperforms the coaxial system, demonstrating greater heat absorption due to an increased exchange surface and the introduction of a horizontal section at depth. The horizontal section at maximum depth significantly enhances heat transfer in high-gradient environments, whereas its contribution is less pronounced in low-temperature formations. However, factors such as pressure losses and working fluid properties require further investigation to ensure optimal performance under real-world operating conditions. Future studies should focus on exploring the role of different working fluids to maximise heat absorption and minimise losses due to viscosity, assessing the economic feasibility of retrofitting hydrocarbon wells, and considering cost-benefit analyses and energy recovery potential.

The results presented in this study are derived from the application of a mathematical model to case studies in which geological and thermal conditions were reconstructed. While these findings offer useful insights, their validation would benefit from future studies conducted on real sites, where the modelled heat exchanger geometries could be implemented and observed under operational conditions.

Author contributions

M.G., F.H., F.V., G.T., and S.L.R. formulated the research objectives and defined the methodological approach. M.G. and F.H. contributed to the identification of materials, the development of analytical tools, and the interpretation of results. The manuscript was written by M.G. All authors reviewed and approved the final version of the manuscript.

Conflict of interest

The authors declare no conflict of interest.

Notes

A version of the presented codes can be obtained upon request by contacting the corresponding author (martina.gizzi@polito.it).

Author details

Martina Gizzi*, Fatemeh Hosseinpour, Federico Vagnon, Glenda Taddia and Stefano Lo Russo
Department of Environment, Land and Infrastructure Engineering, Politecnico di Torino, Torino, Italy

*Address all correspondence to: martina.gizzi@polito.it

References

[1] Zhu Y, Li K, Liu C, Mgijimi MB. Geothermal power production from abandoned oil reservoirs using in situ combustion technology. Energies (Basel). 2019;**12**(23)

[2] Williams JH, Jones RA, Haley B, Kwok G, Hargreaves J, Farbes J, et al. Carbon-neutral pathways for the United States. AGU Advances. 2021;**2**(1): e2020AV000284. DOI: 10.1029/2020AV000284

[3] Gianoutsos NJ, Haase KB, Birdwell JE. Geologic sources and well integrity impact methane emissions from orphaned and abandoned oil and gas wells. Science of the Total Environment. 2024;**912**

[4] Jahan I, Mehana M, Kang M, Boutot J, Govert A, O'Malley D, et al. Data-driven estimates of undocumented orphan oil and gas wells with implications to the environment. Science of the Total Environment. 2025;**967**

[5] Merrill MD, Grove CA, Gianoutsos NJ, Freeman PA. Analysis of the United States documented unplugged orphaned oil and gas well dataset (ver. 1.1, April 2023): U.S. Geological Survey Data Report 1167. 2023. 10 p. Available from: https://pubs.usgs.gov/publication/dr1167/full. DOI: 10.3133/dr1167. ISSN: 2771-9448 (online)

[6] Cahill AG, Samano PSG. Prioritizing stewardship of decommissioned onshore oil and gas wells in the United Kingdom based on risk factors associated with potential long-term integrity. International Journal of Greenhouse Gas Control. 2022;**114**

[7] Gizzi M, Taddia G, Lo RS. Reuse of decommissioned hydrocarbon wells in Italian oilfields by means of a closed-loop geothermal system. Applied Sciences (Switzerland). 2021;**11**(5)

[8] Sharmin T, Khan NR, Akram MS, Ehsan MM. A state-of-the-art review on geothermal energy extraction, utilization, and improvement strategies: Conventional, hybridized, and enhanced geothermal systems. International Journal of Thermofluids. 2023;**18**

[9] Wang K, Yuan B, Ji G, Wu X. A comprehensive review of geothermal energy extraction and utilization in oilfields. Journal of Petroleum Science and Engineering. 2018;**168**:465-477. Available from: http://www.sciencedirect.com/science/article/pii/S0920410518304042

[10] Liu X, Falcone G, Alimonti C. A systematic study of harnessing low-temperature geothermal energy from oil and gas reservoirs. Energy. 2018;**142**: 346-355. DOI: 10.1016/j.energy.2017.10.058

[11] Gizzi M. Closed-loop systems for geothermal energy exploitation from hydrocarbon wells: An Italian case study. Applied Sciences (Switzerland). 2021;**11**(22)

[12] Jello J, Baser T. Utilization of existing hydrocarbon wells for geothermal system development: A review. Applied Energy. 2023:348

[13] Wei N, Guo B. Deliverable wellhead temperature—A feasibility study of converting abandoned oil/gas wells to geothermal energy wells. Sustainability (Switzerland). 2023;**15**(1)

[14] Kujawa T, Nowak W, Stachel AA. Analysis of the exploitation of existing deep production wells for acquiring geothermal energy. Journal of

Engineering Physics and Thermophysics. 2005;**78**(1):127-135

[15] Kujawa T, Nowak W, Stachel AA. Utilization of existing deep geological wells for acquisitions of geothermal energy. Energy. 2006;**31**(5):650-664

[16] Cheng WL, Li TT, Le NY, Xie K. An analysis of insulation of abandoned oil wells reused for geothermal power generation. Energy Procedia. 2014;**61**: 607-610. DOI: 10.1016/j.egypro.2014.11.1181

[17] Cheng WL, Li TT, Le NY, Wang CL. Studies on geothermal power generation using abandoned oil wells. Energy. 2013; **59**:248-254. DOI: 10.1016/j.energy.2013.07.008

[18] Harris BE, Lightstone MF, Reitsma S. A numerical investigation into the use of directionally drilled wells for the extraction of geothermal energy from abandoned oil and gas wells. Geothermics. 2021;**90**

[19] Lo Russo S, Gizzi M, Taddia G. Abandoned oil and gas wells exploitation by means of closed-loop geothermal systems: A review. Geoingegneria Ambientale e Mineraria. 2020;**160**:3-11

[20] Gola G, Di Sipio E, Facci M, Galgaro A, Manzella A. Geothermal deep closed-loop heat exchangers: A novel technical potential evaluation to answer the power and heat demands. Renewable Energy. 2022;**198**:1193-1209

[21] Kolo I, Brown CS, Nibbs W, Cai W, Falcone G, Nagel T, et al. A comprehensive review of deep borehole heat exchangers (DBHEs): Subsurface modelling studies and applications. Geothermal Energy. 2024;**12**

[22] Alimonti C, Soldo E, Bocchetti D, Berardi D. The wellbore heat exchangers: A technical review. Renewable Energy. 2018;**123**:353-381

[23] Toews M, Holmes M. Eavor-Lite performance update and extrapolation to commercial projects. GRC Transactions; **45**

[24] Toews M, Riddell D, Vany J, Schwarz B. Case study of a multilateral closed-loop geothermal system. In: Proceedings World Geothermal Congress 2020 Reykjavik, Iceland, April 26–May 2, 2020.

[25] Santos L, Dahi Taleghani A, Elsworth D. Repurposing abandoned wells for geothermal energy: Current status and future prospects. Renewable Energy. 2022;**194**:1288-1302

[26] Le NY, Cheng WL. Insights into heat transport for thermal oil recovery. Journal of Petroleum Science and Engineering. 2017;**151**:507-521

[27] Bergman TL, Lavine AS, Incropera FP, DeWitt DP. In: Dumas S, editor. Introduction to Heat Transfer. 6th ed. United State of America: John Wiley & Sons, Inc.; 2013

[28] Davis AP, Michaelides EE. Geothermal power production from abandoned oil wells. Energy. 2009;**34**(7): 866-872. DOI: 10.1016/j.energy.2009.03.017

[29] Xiao D, Xu J, Wang T, Cai C, Li L, Liu J. Research on design method for optimal flow rate of U-shaped geothermal well based on exergy analysis. Energy Exploration and Exploitation. 2022;**40**(2):656-681

[30] Feng G, Xu T, Gherardi F, Jiang Z, Bellani S. Geothermal assessment of the Pisa plain, Italy: Coupled thermal and hydraulic modeling. Renewable Energy.

2017;**111**:416-427. DOI: 10.1016/j.renene.2017.04.034

[31] Ramey HJ. Wellbore heat transmission. Journal of Petroleum Technology. 1962;**14**(04):427-435

[32] Yuan W, Chen Z, Grasby SE, Little E. Closed-loop geothermal energy recovery from deep high enthalpy systems. Renewable Energy. 2021;**177**: 976-991

[33] Alimonti C, Vitali F, Scrocca D. Reuse of oil wells in geothermal district heating networks: A sustainable opportunity for cities of the future. Energies (Basel). 2024;**17**(1)

[34] Alimonti C, Berardi D, Bocchetti D, Soldo E. Coupling of energy conversion systems and wellbore heat exchanger in a depleted oil well. Geothermal Energy. 2016;**4**(1)

[35] Alimonti C, Soldo E. Study of geothermal power generation from a very deep oil well with a wellbore heat exchanger. Renewable Energy. 2016;**86**: 292-301. DOI: 10.1016/j.renene.2015.08.031

[36] Le NY, Cheng WL. Evaluation of geothermal heating from abandoned oil wells. Energy. 2018;**142**:592-607

[37] Blank L, Meneses Rioseco E, Caiazzo A, Wilbrandt U. Modeling, simulation, and optimization of geothermal energy production from hot sedimentary aquifers. Computational Geosciences. 2021;**25**:67-104

[38] Charnyi I. Movement of the boundary of change in aggregate state with body cooling or heating, Izv. OTN AN SSSR. 1948;**2**

[39] Charnyi IA. Heating of a critical area of formation in pumping of hot water into a well. Neftyanoe Khozyaistvo (Petroleum Industry). 1953;(3)

[40] Van Rossum G. Python Tutorial, Technical Report CS-R9526. Centrum voor Wiskunde en Informatica (CWI); 1995

[41] Bram K, Draxler J, Hirschmann G, Zoth G, Hiron S, Kuehr M. The KTB Borehole-Germany's superdeep telescope into the earth crust. Netherlands: N. P.; 1995. Web

[42] Blanke T, Hagenkamp M, Döring B, Göttsche J, Reger V, Kuhnhenne M. Net-exergetic, hydraulic and thermal optimization of coaxial heat exchangers using fixed flow conditions instead of fixed flow rates. Geothermal Energy. 2021;**9**(1):19. DOI: 10.1186/s40517-021-00201-3

[43] Ministry of the Environment and Energy Security. Hydrocarbon Exploration and Production: List of Producing Wells in Italy. Available form: https://unmig.mase.gov.it/ricerca-e-coltivazione-di-idrocarburi/pozzi-produttivi/[Accessed: December 2024]

[44] Trumpy E, Manzella A. Geothopica and the interactive analysis and visualization of the updated Italian National Geothermal Database. International Journal of Applied Earth Observation and Geoinformation. 2017; **54**:28-37. DOI: 10.1016/j.jag.2016.09.004

[45] Di Sipio E, Galgaro A, Destro E, Giaretta A, Chiesa S, Manzella A. VIGOR team. Thermal conductivity of rocks and regional mapping. In: Proceedings of the European Geothermal Congress 2013, Pisa, Italy, 3–7 June 2013.

[46] Pasquale V, Gola G, Chiozzi P, Verdoya M. Thermophysical properties of the Po Basin rocks. Geophysical Journal International. 2011;**186**(1):69-81

[47] Dalla, Santa G, Galgaro A, Sassi R, Cultrera M, Scotton P, Mueller J, et al. An updated ground thermal properties database for GSHP applications. Geothermics. 2020;**85**

[48] Bello M, Fantoni R. Deep oil plays in Po Valley: Deformation and hydrocarbon generation in a deformed foreland. World Petroleum. 2002. p. 14-18

[49] Cazzini FF. The history of the upstream oil and gas industry in Italy. Geological Society - Special Publications. 2018;**465**(1):243-274

[50] Granath JW, Casero P. Tectonic Setting of the Petroleum Systems of Sicily. Deformation, Fluid Flow, and Reservoir Appraisal in Foreland Fold and Thrust Belts, Rudy Swennen, François Roure, James W. Granath. 2004